Introductory
Electronics

Introductory Electronics

ALDERT VAN DER ZIEL

Professor of Electrical Engineering
University of Minnesota
Graduate Research Professor
University of Florida

PRENTICE-HALL, INC.
ENGLEWOOD CLIFFS, NEW JERSEY

Library of Congress Cataloging in Publication Data

VAN DER ZIEL, ALDERT.
  Introductory electronics.

  (Prentice-Hall electrical engineering series)
  1. Electronics. I. Title.
TK7816.V35   621.381   73–3379
ISBN 0–13–501700–9

PRENTICE-HALL INTERNATIONAL, INC., *London*
PRENTICE-HALL OF AUSTRALIA, PTY LTD., *Sydney*
PRENTICE-HALL OF CANADA, LTD., *Toronto*
PRENTICE-HALL OF INDIA PRIVATE LIMITED, *New Delhi*
PRENTICE-HALL OF JAPAN, INC., *Tokyo*

Contents

PREFACE   ix

1. INTRODUCTION   1
   1.1. Properties of Semiconducting Materials   1
   1.2. Semiconductor Resistors   6
   1.3. Semiconductor Devices   8

2. DIODE CIRCUITS   15
   2.1. Actual Junction Diodes and Ideal Diodes   16
   2.2. Clipper Circuits   21
   2.3. Clamp Circuits   26
   2.4. Rectifiers   28
   2.5. Further Rectifier Circuits   34

3. FURTHER APPLICATIONS OF THE p-n DIODE   39
   3.1. Small-Signal Applications of the p-n Diode   39
   3.2. Pulse Response of the p-n Junction Diode   45

4. LINEAR AMPLIFIERS   **52**
    4.1. Linear Amplifier   *52*
    4.2. Power Considerations   *59*
    4.3. Frequency Response of Low-Frequency Amplifiers   *62*
    4.4. Miller Effect   *69*
    4.5. Amplifier with Tuned Interstage Network   *69*

5. SIMPLE FET CIRCUITS   **78**
    5.1. Junction FET and MOSFET   *78*
    5.2. Small-Signal Operation of the FET   *88*
    5.3. Biasing Circuits   *95*
    5.4. Power Considerations   *99*

6. MISCELLANEOUS FET CIRCUITS   **109**
    6.1. Constant Current Generator   *109*
    6.2. Source Follower Circuit   *110*
    6.3. High-Frequency Operation of FETs   *118*
    6.4. Differential Amplifier   *122*
    6.5. Integrated Circuit Involving Two Enhancement Mode
        MOSFETs   *124*
    6.6. FETs As *Off-On* Switches   *128*
    6.7. FET Cascaded Amplifiers   *130*

7. TRANSISTOR OPERATION AND SIMPLE CIRCUITS   **138**
    7.1. Current Flow in Silicon Transistors   *138*
    7.2. Small-Signal Operation of the Transistor   *148*
    7.3. Transistor Bias Circuits   *155*
    7.4. Power Dissipation in Transistors   *159*
    7.5. Small-Signal Device Parameters   *160*

8. VARIOUS TRANSISTOR CIRCUITS   **165**
    8.1. Various Integrated Transistor Circuits   *165*
    8.2. Large-Signal Transistor Circuit   *168*
    8.3. Emitter Follower   *171*
    8.4. Differential Amplifiers and Operational Amplifiers   *178*
    8.5. Transistor Power Amplifiers   *187*
    8.6. Transistor Chopper Circuits   *196*
    8.7. High-Frequency Circuits   *199*
    8.8. Transistor Cascaded Amplifiers   *204*

9. VACUUM TUBES   **212**
    9.1. Vacuum Diodes, Triodes, Tetrodes, and Pentodes   *212*
    9.2. Vacuum Tube Power Amplifiers   *218*

10. PULSE RESPONSE OF FETs AND TRANSISTORS   **226**
    10.1. Pulse Response in FETs   *226*
    10.2. Pulse Response of Transistors   *233*
    10.3. Transients in Basic Transistor Switching Circuits   *242*
    10.4. Totem-Pole Circuits   *248*
    10.5. Other Switching Devices   *251*

**11. BISTABLE, MONOSTABLE, AND ASTABLE
MULTIVIBRATORS**   **255**
    11.1. Bistable Multivibrator   *255*
    11.2. Monostable Multivibrator   *262*
    11.3. Astable Multivibrator   *266*

12. OSCILLATORS   **271**
    12.1. Oscillators and Feedback   *271*
    12.2. Admittance and Impedance Considerations in
        Oscillators   *277*

13. NOISE   **286**
    13.1. Fourier Analysis; Spectral Intensity   *287*
    13.2. Noise Figure   *293*

14. MULTISTAGE AMPLIFIERS   **298**
    14.1. Cascade Low-Frequency Amplifier Stages   *298*
    14.2. Tuned Multistage Amplifiers   *307*

15. NEGATIVE FEEDBACK   **313**
    15.1. Merits of Negative Feedback   *314*
    15.2. Application of the Feedback Approach to
        Other Circuits   *318*

Appendix A. SEMICONDUCTOR DEVICE THEORY   **323**
    A.1. Basic Laws of Solid-State Electronics   *323*
    A.2. Junction Diodes   *329*
    A.3. Current Flow in *p-n* Junctions   *335*
    A.4. Junction FET   *338*
    A.5. Current Flow in Transistors   *341*
    A.6. High-Frequency Behavior of the *n-p-n* Transistor   *348*

Appendix B. PULSE RESPONSE OF SIMPLE RC-NETWORKS   **352**

SOLUTIONS   **357**

INDEX   **360**

Preface

When teaching a beginning course in electronics, one usually starts with device theory and treats device applications afterwards. In such a case the student feels unhappy because he has to become familiar with a large amount of theoretical material for which he sees no immediate use.

One can prevent this by putting the main emphasis on device applications and treating device theory afterwards. In such a case the student feels unhappy because he must accept too many things without proof.

This book takes an intermediate road. Here the emphasis in the main part of the text is devoted to device applications, and only so such device theory is given as is absolutely essential for understanding these applications. But in addition further discussion of device theory is given in a lengthy appendix, which can either be used as background information for interested students or made an intergral part of the course if the instructor so desires.

Many textbooks give an elaborate discussion of the small-signal theory of FET and transistor applications, so that there is not enough time left to discuss large-signal, nonlinear applications. This becomes more and more a handicap for the student, since most device applications are large-signal and digital, rather than small-signal.

For that reason this book takes a different approach. The small-signal applications are discussed in a single chapter, using a generalized black-box approach. Anybody who has understood what midband response, voltage gain and power gain, lower cutoff frequency, upper cutoff frequency, tuned circuit response, and Miller effect mean can apply this immediately to FETs and transistors. Of course, the student interested in electronics should later take a course in small-signal theory of device applications, but the beginning student gets a very lopsided view of electronics if he is forced to take too much small-signal theory.

About 80% of the device applications are digital applications in which the device is used as a switch or as a generator of waveforms. This should be reflected in a beginning course in electronics. Due to the deemphasis of small-signal theory this book is able to do so. It gives an extensive discussion of large-signal device applications, including applications as switches and as generators of waveforms. Those who have further interest in such applications should take more courses in this field, but an introductory course should certainly treat these topics at some length.

Modern electronic circuits are integrated circuits in most cases. The beginning student should learn to understand such circuits. For that reason this book discusses many of the building blocks used in integrated circuits.

Many textbooks devote much space to concepts that were important in the early days of solid-state device development but have now lost most of their importance. This book has deemphasized these device features.

For example, knowledge about reverse currents and stabilization against variations in I_{CB0} were very important in germanium transistors but are rather insignificant in modern silicon transistors. Therefore this book deemphasizes such topics. As a second example, the usefulness of h- and z-parameters for transistors is still overrated. In many applications one only has to know the parameters h_{fe} and h_{ie} but the parameters h_{re} and h_{oe} are unimportant. Therefore this book gives only as much discussion about h-parameters as is necessary for reading device manuals, and represents the transistor by a simple small-signal equivalent circuit. As a third example, the full Ebers-Moll equations are hardly ever needed for silicon transistors; for that reason this book presents them in a simplified form.

At the request of several reviewers a short chapter on noise was added to the book. Not every instructor may want to teach this chapter. Nevertheless it might serve the function of dispelling the common belief that noise is an advanced topic that cannot be understood by EE juniors.

This book is the outcome of several series of lectures given at the Universities of Minnesota and Florida. Its level has been tested in classroom situations, and it is believed to be neither too elementary nor too advanced.

The book is meant for a two-quarter introductory course in electronics. As such it can be used with considerable flexibility: some sections can be expanded upon and others can be deemphasized. For example, some instructors might want to expand the device theory part of the book somewhat. Other instructors might want to expand the small-signal theory part of the book. Finally, some might want to expand the sections on the large-signal applications of devices. In all these cases the book can serve as a foundation upon which the expansion can take place.

Finally, the book can also be used in a one-quarter introductory course in electronics. In that case the topics discussed in Chapters 2, 4, 5, part of 6, 7, and 8 should form the basis of the course. Again there can be considerable flexibility in which parts of these chapters can be emphasized and which parts can be deemphasized. This is left to the desire of the individual instructor.

The author would like to express his appreciation to all those who contributed to the manuscript in one way or the other. The students who participated in the courses for which this book was developed made a significant contribution. So did many colleagues with whom the author had valuable discussions. Dr. Shu-wei Shen and Mr. Mark Lundstrom read the manuscript and made many valuable suggestions that clarified the text. Mrs. van der Ziel typed the manuscript.

A. VAN DER ZIEL

Introductory Electronics

1

Introduction

Since the aim of this book is to promote understanding of solid-state electronic circuits, including integrated circuits, Section 1.1 can be treated as background material. Section 1.2 is needed for an understanding of integrated resistors, whereas Section 1.3 gives a descriptive background of the most important solid-state devices and can also be treated as background material.

1.1. PROPERTIES OF SEMICONDUCTING MATERIALS

Semiconducting materials are materials that have a conductivity somewhere between a metal and an insulator. Technically the most important semiconducting material is silicon, which is located in the fourth column of the periodic table of elements. It forms cubic crystals in which each silicon atom is surrounded symmetrically (= tetrahedrally) by four silicon atoms to form the so-called diamond structure. This structure comes about because the four outer electrons are used to make covalent bonds with its four neighbors (Fig. 1.1).

Silicon became useful for the electronics industry because the following conditions could be met*:

---

*The same conditions hold for germanium. Since silicon devices have better properties than germanium devices, the latter are hardly used any more.

Fig. 1.1. Atom tetrahedrally surrounded by four neighbors.

1. The material could be purified to a high degree.
2. The material could be put in single-crystal form.
3. The purified material could be doped with controlled amounts of properly chosen impurities.
4. Ohmic contacts could be made to the material (an ohmic contact is a contact that allows free passage of current carriers across the interface).

This is not the place to discuss in detail how these operations are performed; it is sufficient for our present understanding that they *can* be performed. We shall now discuss some properties of single-crystal silicon.

Figure 1.2(a) shows a bar of single-crystal high-resistivity silicon, of, e.g., 20,000-Ω-cm resistivity, to which two ohmic contacts are made. If the

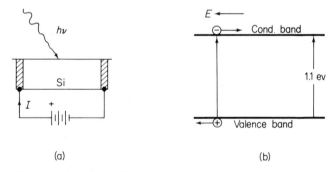

(a)   (b)

Fig. 1.2. (a) Light incident upon high-resistivity silicon gives rise to a photocurrent *I*. (b) Free electron and free hole moving under the influence of an electric field.

sample is irradiated with monochromatic light of varying wavelength, it is found that the conductivity of the material is markedly increased if the energy of the quanta is larger than 1.1 electron volt (eV). This phenomenon is called *photoconductivity;* it is caused by the fact that electrons that were first bound to the atoms are set free in the crystal by the absorption of quanta with an energy > 1.1 eV.

This effect can be interpreted with the help of the so-called energy band model for silicon. The energy levels in the silicon single crystal are concentrated in energy bands, separated by energy ranges (the so-called forbidden gaps) where no energy levels occur. In the pure material one of these bands is completely empty; it is called the *conduction band*. The next lower band is completely filled; it is called the *valence band*. The photoconductivity comes about because electrons are transferred from the valence band to the conduction band by the absorbed quanta; the threshold energy for photoconductivity indicates that the energy difference between the top of the valence band and the bottom of the conduction band is 1.1 eV [Fig. 1.2(b)].

The observed conductivity implies that an electron in the conduction band can move more or less freely through the crystal under the influence of an applied field, except for scattering by collisions with the atoms (Section 1.2a). However, for each electron promoted into the conduction band, there is one electron missing in the valence band; i.e., one electron must be missing from one of the atoms. This missing electron can also move more or less freely from one atom to the next under the influence of an applied field; it behaves in every respect like a positive charge and is called a *hole*. Photoconductivity is therefore caused by the motion of electrons and holes.

By adding impurities of group V of the periodic table of elements (P, As, Sb), one can strongly increase the conductivity of the material, which then becomes proportional to the density of impurity atoms added. This comes about as follows: The impurity atoms, which have *five* outer electrons, take the place of the silicon atoms. Four of the outer electrons are used to make covalent bonds with the four neighboring silicon atoms, leaving a very loosely bound electron behind. In the energy band picture these impurity atoms thus give rise to *occupied* energy levels slightly below the bottom of the conduction band [Fig. 1.3(a)]. Very little energy is needed to bring these

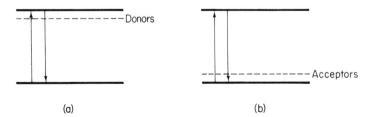

Fig. 1.3. (a) Energy level diagram of *n*-type material. (b) Energy level diagram of *p*-type material.

electrons into the conduction band, leaving positively ionized impurities behind. These impurities are therefore called *donors*, since they "donate" electrons to the conduction band. At room temperature practically all these donor atoms are ionized so that the density of electrons in the conduction band nearly equals the donor density N_d. Since in this type of material the current is carried by (negative) electrons, it is called *n-type* material.

By adding impurities of group III of the periodic table of elements (B, Al, Ga, In), one can also increase the conductivity of the material, which then becomes proportional to the density of impurity atoms added. This comes about as follows: The impurity atoms, which have *three* outer electrons, take the place of the silicon atoms. The outer electrons make covalent bonds with the neighboring silicon atoms, leaving a loosely bound hole behind. In the energy band picture these impurity atoms thus give rise to *unoccupied* energy levels slightly above the top of the valence band [Fig. 1.3(b)]. Very little energy is needed to bring an electron from the valence band to an impurity atom; this process creates a "free" hole and a negatively ionized impurity atom. Since these atoms "accept" electrons from the valence band when they are ionized, they are called *acceptors*. At room temperature practically all the acceptor atoms are ionized so that the density of free holes nearly equals the density N_a of acceptor atoms. Since in this type of material the current is caused by (positive) holes, it is called *p-type* material.

Because of the interaction of carriers with the thermal vibrations of the atoms in the crystal, some electrons in the valence band gain enough energy to be promoted into the conduction band, thereby generating a free electron and a free hole simultaneously. This process is called *thermal generation;* because of it, there will always be some holes in *n*-type material and some electrons in *p*-type material. There is also an opposite process in which an electron and a hole are removed simultaneously under release of energy; this process is called *recombination*. When these two processes balance, an equilibrium situation is established. We shall first investigate this equilibrium for holes in *n*-type material.

Let $g(p)$ be the rate of generation of hole-electron pairs and $r(p)$ the rate of recombination of hole-electron pairs, where p is the hole concentration. Then the rate of change dp/dt of the hole concentration is given by the self-evident equation

$$\frac{dp}{dt} = g(p) - r(p) \tag{1.1}$$

Equilibrium is obtained if $dp/dt = 0$; if $p = p_0$ in that case, Eq. (1.1) yields

$$g(p_0) = r(p_0) \tag{1.1a}$$

A theoretical discussion indicates that $g(p) = g_0(T)$ is independent of the hole density p and that $r(p) = a_0(T)pn$, where n is the electron density. The parameters $g_0(T)$ and $a_0(T)$ are independent of p and n and are functions of the absolute temperature T and the semiconductor material only. Therefore $p_0 n_0$ is a function of the absolute temperature T and some parameters dependent on the material. Since $p = n = n_i$ if the material contains no impurities (so-called *intrinsic* material), we have in equilibrium

$$pn = p_0 n_0 = n_i^2 \qquad (1.2)$$

where n_i is the carrier density of intrinsic material. In silicon $n_i \simeq 10^{10}/\text{cm}^3$ at room temperature ($T = 300°\text{K}$).

In the particular case of holes in n-type silicon material, $n_0 \simeq N_d$, the donor concentration, and hence the equilibrium hole concentration p_n, is

$$p_n = \frac{n_i^2}{N_d} \qquad (1.2a)$$

In the case of electrons in silicon p-type material, $p_0 \simeq N_a$, the acceptor concentration, and hence the equilibrium electron concentration n_p is

$$n_p = \frac{n_i^2}{N_a} \qquad (1.2b)$$

These equations hold as long as N_d and N_a are not close to n_i. Generally, this condition is satisfied.

EXAMPLE: Silicon has $n_i = 10^{10}/\text{cm}^3$. Find the electron concentration in p-type material with $N_a = 10^{15}/\text{cm}^3$.

ANSWER:

$$n_p = \frac{n_i^2}{N_a} = \frac{10^{20}}{10^{15}} = 10^5/\text{cm}^3$$

According to statistical mechanics, n_i^2 can be expressed as

$$n_i^2 = 2\left(\frac{2\pi m_n kT}{h^2}\right)^{3/2} \cdot 2\left(\frac{2\pi m_p kT}{h^2}\right)^{3/2} \exp\left(-\frac{E_g}{kT}\right) \qquad (1.3)$$

where m_n and m_p are the electron and hole masses, respectively, k is Boltzmann's constant, h is Planck's constant, T is the absolute temperature, and E_g is the energy gap between the bottom of the conduction band and the top of the valence band. We thus see that n_i increases very rapidly with increasing temperature.

Equation (1.1) also describes how the hole concentration returns to equilibrium after it has been disturbed slightly. Let this disturbance have been caused by the absorption of a pulse of light at $t = 0$, and let $p = p_0 + \Delta p$ after the disturbance, where $\Delta p \ll p_0$. Making a Taylor expansion of the right-hand side of Eq. (1.1), breaking it off after the second term, and bearing in mind Eq. (1.1a) yields

$$\frac{d \Delta p}{dt} = -\left(\frac{dr}{dp} - \frac{dg}{dp}\right)_{p=p_0} \Delta p = -\frac{\Delta p}{\tau_p} \qquad (1.4)$$

where

$$\frac{1}{\tau_p} = \left(\frac{dr}{dp} - \frac{dg}{dp}\right)_{p=p_0} \qquad (1.4a)$$

The meaning of the parameter τ_p can be understood as follows. Let the disturbance give rise to a change $\Delta p = \Delta p_0$ at $t = 0$; then the solution of Eq. (1.4) is

$$\Delta p(t) = \Delta p_0 \exp\left(-\frac{t}{\tau_p}\right) \tag{1.5}$$

so that the initial disturbance decays exponentially with the time constant τ_p. For that reason τ_p is called the *lifetime of the added holes*. A similar discussion can be held for electrons in p-type material.

We can thus distinguish between *majority* carriers (electrons in n-type material and holes in p-type material) and *minority* carriers (holes in n-type material and electrons in p-type material). We shall see that the operation of solid-state diodes and transistors is based on *injection* and *extraction* of minority carriers.

1.2. SEMICONDUCTOR RESISTORS

1.2a. Ohm's Law for Semiconductors

To understand the conduction process in semiconductors in greater detail, we shall first consider a p-type semiconductor. Let p be the hole concentration, and let each hole have a charge e and a mass m_p. If an electric field E is applied in the X-direction, the force F on a hole is eE and the acceleration a of the hole is $(e/m_p)E$. The holes are accelerated by the electric field, collide with the atoms of the lattice and are scattered in random directions, are accelerated again and scattered again, etc. Let the last collision of a particular hole have occurred at $t = t_0$. Then the velocity v_x in the X-direction at a later instant is

$$v_x(t) = v_{x0} + \frac{e}{m_p}E(t - t_0) \tag{1.6}$$

where v_{x0} is the X-component of the velocity right after the collision.

If we now average over all holes in the material at the instant t and if the average is denoted by an overbar, then

$$\overline{v_x} = \overline{v_{x0}} + \frac{e}{m_p}\overline{E(t - t_0)} = \frac{e}{m_p}E\tau \tag{1.6a}$$

where the first average is taken over all initial velocities and the second over all instants t_0. Since positive and negative values of v_{x0} are equally likely, $\overline{v_{x0}} = 0$. Moreover, $\overline{(t - t_0)}$ represents the average time between collisions; it is denoted by τ. The average drift velocity $\overline{v_x}$ of the holes is thus proportional to the electric field E; the proportionality factor is called the *hole mobility* μ_p and is expressed in square centimeters per volt-second. According to Eq. (1.6a),

$$\overline{v_x} = \mu_p E; \qquad \mu_p = \frac{e}{m_p}\tau \tag{1.6b}$$

Next, the current I is evaluated. If A is the cross-sectional area of the sample, pA is the number of carriers per unit length. Since $\overline{v_x}$ is the average hole drift velocity,

$$I_p = ep A \overline{v_x} = ep\mu_p AE = g_p E \qquad (1.7)$$

where

$$g_p = ep\mu_p A = \sigma_p A \quad \text{and} \quad \sigma_p = ep\mu_p \qquad (1.7a)$$

This is Ohm's law; g_p is the hole conductance per unit length of the sample and σ_p is the hole conductivity of the material. It should be remembered that the hole density p is practically equal to the acceptor density N_a.

We can now repeat the same reasoning for n-type material. Since the electrons have a charge $-e$, they drift in a direction opposite to the electric field E. The average drift velocity may thus be written

$$\overline{v_x} = -\mu_n E \qquad (1.8)$$

where μ_n is the *electron mobility*. Hence the current is

$$I_n = -en A \overline{v_x} = en\mu_n AE = g_n E \qquad (1.9)$$

so that $g_n = en\mu_n A = \sigma_n A$ is the *electron conductance* per unit length of the sample and $\sigma_n = en\mu_n$ is the *electron conductivity* of the material. It should be remembered that the electron density n in n-type material is practically equal to the donor density N_d.

We shall now illustrate these equations as follows:

EXAMPLE 1: Silicon has $\mu_n = 1480 \text{ cm}^2/\text{V-s}$ at room temperature. What is the carrier concentration of $1\ \Omega$-cm of n-type material?

ANSWER:

$$en\mu_n = 1; \qquad n = \frac{1}{e\mu_n} = \frac{1}{1.6 \times 10^{-19} \times 1.48 \times 10^3} = 4.2 \times 10^{15}/\text{cm}^3$$

EXAMPLE 2: If $n_i = 10^{10}/\text{cm}^3$ in silicon at room temperature, what is the hole density in $1\ \Omega$-cm of n-type material at that temperature?

ANSWER:

$$pn = n_i^2; \qquad p = \frac{n_i^2}{n} = \frac{10^{20}}{4.2 \times 10^{15}} = 2.4 \times 10^4/\text{cm}^3$$

We thus see that the equilibrium hole density is quite small.

1.2b. Integrated Resistors

Let a sheet of resistive material have a conductivity σ and a thickness d. Let a rectangular piece of width w and length L be cut out of this sheet and let it be provided with end contacts [Fig. 1.4(a)]. Then the resistance R of this device is

$$R = \frac{L}{\sigma w d} = R_s \frac{L}{w} \qquad (1.10)$$

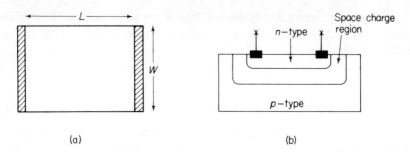

Fig. 1.4. (a) Integrated resistor, top view. (b) Integrated resistor, cross-sectional view. A space-charge region separates the *n*-type island from the *p*-type substrate.

where $R_s = 1/\sigma d$. The parameter R_s is called the *sheet resistance* of the material; it is expressed in ohms per square. That is, if a square is cut out of the sheet and contacts are made to the end, one always obtains the resistance R_s, independent of the size of the square. One can choose R_s by choosing σ and d, and, for a given value of R_s, one can choose the resistance R by adjusting the ratio L/w.

We can now take a chip of *p*-type silicon, diffuse *n*-type impurities in to a depth d, and so make an *n*-type resistance sheet of sheet resistance R_s [Fig. 1.4(b)]. The space-charge region between the *n*-region and the *p*-region isolates the resistor from the other parts of the chip, provided that the proper bias voltage is applied between the *n*-type sheet and the *p*-type substrate. We shall come back to this later (Section 1.3a).

If by suitable masking techniques the *n*-type diffused-in region is given a length L and a width w and ohmic contacts are made to the ends, the resistance R of the device is given by Eq. (1.10). The diffused resistors used in integrated circuits are so designed that the sheet resistance R_s is of the order order of 100–300 Ω/square. By suitable choice of L/w one can easily make integrated resistors with resistance values between 10 and 10,000 Ω.

EXAMPLE: If the diffused-in region has a sheet resistance of 300 Ω/square and the width w is 10 μm, what length L is needed to make a resistor of 6000 Ω?

ANSWER: If L is expressed in micrometers, we have

$$300\frac{L}{w} = 6000; \qquad \frac{L}{w} = 20; \qquad L = 20w = 200 \ \mu\text{m}$$

1.3. SEMICONDUCTOR DEVICES

We are now in a position to understand the structure of the following semiconductor devices:

1. The *p-n* junction diode.
2. The *p-n-p* and *n-p-n* bipolar transistors.
3. The junction gate field effect transistor (JFET).
4. The metal-oxide-semiconductor field effect transistor (MOSFET).

There are many more semiconductor devices available, but the above four are essential for the basic understanding of modern electronics.

1.3a. *p-n* Junction Diode

First we shall discuss the *p-n* diode. It is made by diffusing *n*-type impurities into *p*-type material or *p*-type impurities into *n*-type material. Ohmic contacts are made to the *p*- and *n*-regions and the device is ready for use.*

Often these devices are made on epitaxial material; we shall discuss this for an *n*-type substrate. The epitaxial material consists of a very heavily doped *n*-region, denoted by n^+, onto which a more weakly doped *n*-region of about 5-μm thickness is grown by epitaxial techniques. The *p*-region is then diffused into the more weakly doped *n*-region [Fig. 1.5(a)]. This reduces the series resistance associated with the *n*-region.

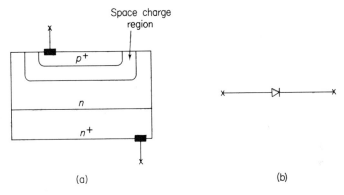

Fig. 1.5. (a) Cross-section of a p^+-*n* junction diode on an n^+-type substrate. (b) Symbol for a diode.

Because of the manufacturing process, there is a (curved) surface in the crystal where the *n*-region goes over into the *p*-region. Near that surface there is a space-charge region, caused by ionized acceptors on the *p*-side of the surface and ionized donors on the *n*-side of the surface; it comes about because electrons were transferred from the *n*-region to the *p*-region. Because of this space-charge region, there is a contact potential difference, known as the diffusion potential V_{dif} between the *n*- and *p*-regions.

---

*Ohmic contacts are contacts that allow free passage of carriers from the one side to the other.

If a voltage V_D is applied to the *p*-region and $V_D > 0$, then the space-charge region contracts, since it now has to accommodate the smaller potential difference $V_{\text{dif}} - V_D$. If $V_D < 0$, then the space-charge region expands since it must now accommodate the larger potential difference $V_{\text{dif}} - V_D$. This voltage dependence of the width of the space-charge region between the *p*- and *n*-regions is the basis for the operation of the junction FET.

For $V_D > 0$ the *p*-region injects holes into the *n*-region, whereas for $V_D < 0$ the *p*-region extracts holes from the *n*-region. Since there are many more holes in the *p*-region than there are in the *n*-region, the current is much smaller for $V_D < 0$ than for $V_D > 0$ and has opposite direction.

What was said for the holes also holds for the electrons. By making the *p*- and *n*-regions such that $N_a \gg N_D$ (so-called p^+-*n* diodes), one can make the hole current much larger than the electron current. If $N_D \gg N_a (n^+$-*p* diodes), practically all the current is carried by electrons.

The bias $V_D > 0$ is called *forward bias* and the bias $V_D < 0$ is called *back bias*. Figure 1.5(b) shows the symbol used for the diode; the triangle points in the direction of easy current flow.

We can now also understand how the substrate of an integrated resistor must be biased. To properly isolate the substrate from the resistor, the resistor-substrate *p-n* diode must be back-biased satisfactorily.

1.3b. *p-n-p* and *n-p-n* Bipolar Transistors

A *p-n-p* transistor is a three-layer device that is made in two diffusion steps. One starts with an epitaxial *p*-layer of about 5-μm thickness on a thicker, heavily doped *p*-region (p^+-region). First an *n*-region is diffused in, then a *p*-region is diffused into part of the *n*-region, and ohmic contacts are made to all three regions [Fig. 1.6(a)]. The diffused-in *p*-region is called the *emitter*, the diffused-in *n*-region is called the *base*, and the p-p^+-region is called the *collector*. *n-p-n* Transistors are constructed in a similar manner. The epitaxial technique is used for discrete transistors because it gives a lower series resistance of the collector region.

In integrated transistors, which are generally of the *n-p-n* variety, this method cannot be used, since one must isolate the different collectors from each other and from the substrate. One starts here with a thin epitaxial *n*-layer on a p^+-type substrate, diffuses p^+-regions in all the way to the substrate to cut out *n*-islands that later become the collector, and then diffuses in the *p*-type base and the *n*-type emitter in the usual manner [Fig. 1.6(b)].

For normal operation of the *p-n-p* transistor the emitter is forward-biased with respect to the base and the collector is back-biased with respect to the base. Consequently, the emitter *injects* holes into the base and the collector *extracts* holes from the base; this explains the names *emitter* and *collector*. The name *base* is a holdover from earlier transistor technology.

In this mode of operation the emitter current I_E flows *into* the emitter,

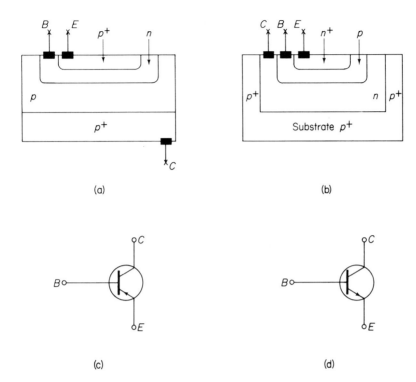

Fig. 1.6. (a) Cross-section of a p^+-n-p transistor on a p^+-substrate.
(b) Integrated n^+-p-n transistor on a p^+-substrate. (c) Symbol for
a p-n-p transistor. (d) Symbol for an n-p-n transistor.

the collector current I_C flows *out of* the collector, and the base current I_B flows *out of* the base. Applying Kirchhoff's laws, we have

$$I_E = I_C + I_B \qquad (1.11)$$

In good p-n-p transistors practically all the current is carried by holes, and practically all the injected holes are collected by the collector, so that $I_C \simeq I_E$ and $I_B \ll I_C$. A very small base current I_B can thus control a much larger collector current I_C; this is the basis for the current amplification properties of the transistor.

The n-p-n transistor is similar to the p-n-p transistor, but all polarities are changed; the emitter current I_E flows *out of* the emitter, the collector current I_C flows *into* the collector, and the base current I_B flows *into* the base. Otherwise the discussion is similar to the p-n-p case. In good n-p-n transistors practically all current is carried by electrons.

Figure 1.6(c) and (d) shows the symbols for the p-n-p and n-p-n transistors, respectively; the arrow indicates the direction of the current flow in the emitter lead.

1.3c. Junction Gate Field Effect Transistor (JFET)

We shall discuss here one particular way in which JFETs can be made. We do so for a device with an *n*-type channel. One starts here with a *p*-type substrate on which a thin region (a few micrometers thick) of *n*-type material is grown by epitaxial techniques. One then diffuses in a p^+-type cylinder of width L that almost reaches to the *p*-substrate, and one connects this region to the substrate. Ohmic contacts are then made to the outer *n*-region, the cylindrical *p*-region, and the inner *n*-region; these contacts are called the *source S*, the *gate G*, and the *drain D*, respectively [Fig. 1.7(a)].

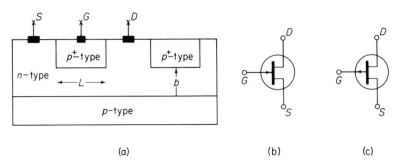

(a)   (b)   (c)

Fig. 1.7. (a) Cross-section of an *n*-channel junction FET on a *p*-type substrate. (b) Symbol for a JFET with *n*-type channel. (c) Symbol for a JFET with *p*-type channel.

First, let no dc bias be applied between D and S, whereas a negative bias V_{GS} is applied between the gate G and the source S. Since the width of the space-charge region between the *p*- and *n*-regions expands if the gate is made more negative, the height b of the *n*-region under the gate decreases if V_{GS} is made more negative and becomes zero at a particular value $V_{GS} = V_P$; this is called the *pinch-off voltage*. The region between D and S thus acts as a voltage-controlled resistor.

If now a positive voltage V_{DS} is applied between D and S and a fixed voltage V_{GS} is applied between G and S, the drain current I_D first increases with increasing V_{DS} until it saturates if $V_{DS} = V_{GS} - V_P$. For $V_{DS} > V_{GS} - V_P$, I_D hardly depends on V_{DS}. The reason for the saturation is that the conducting channel is pinched off at the drain if $V_{GS} - V_{DS} = V_P$ or $V_{DS} = V_{GS} - V_P$.

If V_{GS} is made more negative, the height b of the channel decreases everywhere, and hence the drain current I_D decreases. The voltage V_{GS} can thus control the drain current I_D. Since the gate-channel junction is a back-biased junction that carries hardly any current, the drain current control by V_{GS} is accomplished with practically no current at the gate. It is therefore not surprising that the device can be used as an amplifier.

The above device is a JFET with an n-type channel. One can also make JFETs with a p-type channel. Such a device has opposite polarity and opposite direction of current flow from the n-channel device but otherwise operates along the same principles.

Figure 1.7(b) and (c) shows the symbols used for the two types of JFET; the arrow indicates the direction of easy current flow (forward bias) of the gate junction.

1.3d. Metal-Oxide-Semiconductor Field Effect Transistor (MOSFET)

A MOSFET with an n-type channel is made as follows. One starts with a relatively weak p-type substrate and grows an oxide layer on top of it. One then cuts two holes through the oxide, diffuses n-type impurities into the substrate through these holes, and deposits a metal layer on top of the oxide between the two holes. One then makes ohmic contacts to the two n-regions and to the metallization on top of the oxide. The metallization on top of the oxide is called the gate G, the ohmic contact to the one n-region is called the source S, and the ohmic contact to the other n-region is called the drain D. Voltages V_{GS} and V_{DS} are applied between G and S and D and S, respectively [Fig. 1.8(a)].

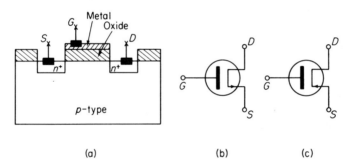

(a)   (b)   (c)

Fig. 1.8. (a) Cross-section of an n-channel MOSFET on a p-type substrate. (b) Symbol for an n-channel MOSFET. (c) Symbol for a p-channel MOSFET.

If first $V_{DS} \simeq 0$ and V_{GS} is varied, one finds that a conducting surface channel occurs if $V_{GS} > V_T$; V_T is called the *turn-on* voltage. By proper preparation of the device and of the gate, V_T can be either positive or negative. The device thus operates for $V_{GS} > V_T$.

Now V_{DS} is gradually increased. If $V_{GS} > V_T$ and V_{GS} is kept constant, the drain current I_D increases with increasing V_{DS} until it saturates at $V_{DS} = V_{GS} - V_T$. For $V_{DS} > V_{GS} - V_T$, I_D hardly depends on V_{DS}. The reason for this behavior is the same as for the JFET; the conducting channel is pinched off at the drain if $V_{GS} - V_{DS} = V_T$ or $V_{DS} = V_{GS} - V_T$.

If V_{DS} is kept constant and V_{GS} is decreased, I_D decreases because the conducting channel becomes less conducting. The voltage V_{GS} can thus be used to control the drain current I_D. The device can therefore be used as an amplifier. Since the gate is isolated from the channel, the gate current is always zero.

MOSFETs with a p-type channel can be made on an n-type substrate. All the voltages and currents have opposite polarity from MOSFETs with an n-type channel, but otherwise the device operates along the same principles. A conducting channel occurs if $V_{GS} < V_T$; V_T is always negative in this case.

Figure 1.8(b) and (c) shows the symbols commonly used for the two types of MOSFETs; the arrow indicates the direction of current flow.

PROBLEMS

1.1. Show that if excess minority carriers can disappear by two processes with time constants τ_1 and τ_2, respectively, then the time constant τ of the combination is given by

$$\frac{1}{\tau} = \frac{1}{\tau_1} + \frac{1}{\tau_2}$$

1.2. (a) Show that the conductivity σ_i of intrinsic material ($n = p = n_i$) is given by $\sigma_i = e(\mu_n + \mu_p)n_i$. (b) If silicon at room temperature has $\mu_n = 1480$ cm²/V-s, $\mu_p = 480$ cm²/V-s, and $n_i = 10^{10}/$cm³, find σ_i.

1.3. n-Type impurities diffused into a p-type substrate give a sheet resistivity of 200 Ω/square. A resistor of 10,000 Ω must be made from this material, and the maximum length of the resistor is 1 mm. Find the width in micrometers.

1.4. How deep is the n-region in Problem 1.3 if the average value of N_d is $10^{16}/$cm³ ?

1.5. In p-type material Ohm's law ceases to be valid if the field strength is higher than 2000 V/cm. Find the velocity of the carriers when this effect begins. $\mu_p = 480$ cm²/V-s.

2

Diode Circuits

A silicon junction diode has a high current for one polarity and a small current for opposite polarity. For that reason it can be idealized by a characteristic of the form

$$I_D = 0 \quad \text{for } V_D < 0; \quad I_D = \infty \quad \text{for } V_D > 0 \qquad (2.1)$$

For $V_D = 0$ the current is positive and limited by the external circuit. A diode with such a characteristic is called an *ideal* diode; it can describe the nonlinear features of most diode circuits. The characteristic of an ideal diode is shown in Fig. 2.1. The symbol for a junction diode was already shown in Fig. 1.5(b).

Fig. 2.1. Characteristic of an ideal diode.

2.1. ACTUAL JUNCTION DIODES AND IDEAL DIODES

2.1a. Determination of the Operating Point

Figure 2.2(a) shows a circuit consisting of a battery V_{BB}, a resistance R, and a diode D. The problem is now how to find the diode current I_D and the diode voltage V_D.

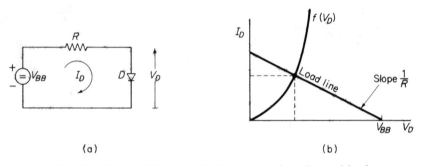

(a)   **(b)**

Fig. 2.2. (a) Actual diode circuit with supply voltage V_{BB} and load resistance R. (b) Graphical determination of the operating point (I_{D0}, V_{D0}) (load-line method).

First, let it be assumed that D is an ideal diode. Then, since $V_{BB} > 0$,

$$V_D = 0, \quad \text{so that } I_D = \frac{V_{BB}}{R} \qquad (2.2)$$

Next, let it be assumed that D is an actual silicon p-n junction diode. As is shown in Section 2.1b, such a diode has a characteristic

$$I_D = I_{rs}\left[\exp\left(\frac{eV_D}{kT}\right) - 1\right] \qquad (2.3)$$

where e is the absolute value of the electron charge, k is Blotzmann's constant, and T is the absolute temperature of the device. The current I_{rs} is called the *reverse saturation current* and is of the order of 10^{-15} A at $T = 300°$K; it increases very rapidly with increasing temperature. The parameter kT/e has the dimension of volt and is often represented by the symbol V_T; $V_T = kT/e = 25.8$ mV at $T = 300°$K.

Before discussing how to find the operating point, we shall ask a few questions that illustrate the meaning of Eq. (2.3).

QUESTION 1: If $I_{rs} = 10^{-15}$ A at $T = 300°$K, at what value of V_D is $I_D = 1$ mA?

ANSWER: $I_D \gg I_{rs}$, or $I_D = I_{rs}\exp(eV_D/kT)$ or $\exp(eV_D/kT) = I_D/I_{rs}$, so that $V_D = (kT/e)\ln(I_D/I_{rs})$ or $V_D = 25.8 \times 10^{-3}\ln 10^{12} = 0.71$ V.

QUESTION 2: If $I_D = I_{D0} = 1$ mA at $V_D = V_{D0} = 0.71$ V at $T = 300°$K, how much must V_D be increased to raise I_D by a factor of 10?

ANSWER: Put $V_D = V_{D0} + \Delta V_D$. Since $I_{D0} = I_{rs} \exp(eV_{D0}/kT)$,

$$I_D = I_{rs} \exp\left[\frac{e(V_{D0} + \Delta V_D)}{kT}\right] = I_{D0} \exp\left(\frac{e \, \Delta V_D}{kT}\right)$$

so that $\exp(e \, \Delta V_D/kT) = I_D/I_{D0} = 10$; $\Delta V_D = (kT/e) \ln 10 = 0.0593$ V. This result is independent of the value of the initial current I_{D0}. In round numbers we can say that at $T = 300°$K $V_{D0} \simeq 0.70$ V at $I_D = 1$ mA and that I_D increases by a factor of 10 if V_D increases by 60 mV. The value $V_D = 0.70$ V is called the *turn-on* voltage of the silicon diode.

There are several ways of determining the operating point (I_D, V_D) for the circuit of Fig. 2.2(a). The graphical method, illustrated in Fig. 2.2(b), will be discussed first. In the first place the current I_D is given by the device characteristic

$$I_D = f(V_D) \tag{2.4}$$

However, I_D is also given by

$$I_D = \frac{V_{BB} - V_D}{R} \tag{2.5}$$

This is a straight line passing through the point $(V_D = V_{BB}, I_D = 0)$ and having a slope $-1/R$. It is called the *load line*. The operating point (I_D, V_D) is thus obtained as the point of intersection between the characteristic (2.4) and the load line (2.5); it can be read directly from the graphs. This so-called load line method of graphical analysis if applicable to any one-port device and can be extended to two-port devices such as FETs and transistors. The graphical method is nothing but a graphical determination of the root(s) of

$$I_D = f(V_D) = \frac{V_{BB} - V_D}{R} \tag{2.6}$$

We now return to the characteristic (2.3) of the *p-n* junction. Since the current flowing through the circuit is of the order of 1 mA and since I_D changes so rapidly with V_D as soon as current flows, the solution of Eq. (2.6) is usually very close to

$$V_D = V_{D0} = 0.70 \text{ V}$$

Therefore in first approximation an actual silicon diode can be represented by an ideal diode with an emf $V_{D0} = 0.70$ V in series. Figure 2.3(a) shows the equivalent circuit; Fig. 2.3(b) shows the idealized characteristic. If this approximation is not sufficient, higher-order approximations are easily made. We shall now investigate this model with the help of examples.

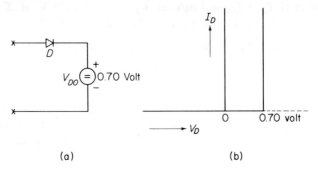

(a)   (b)

Fig. 2.3. (a) Approximate equivalent circuit of an actual diode. (b) Approximate characteristic of an actual diode. The turn-on voltage of the diode is about 0.70 V.

EXAMPLE 1: $V_{BB} = 100$ V, $R = 10^4\ \Omega$, and $I_D = 1.0$ mA at $V_D = 0.70$ V. The ideal diode gives

$$I_D = \frac{V_{BB}}{R} = 10\text{ mA}; \qquad V_D = 0\text{ V}$$

and the approximation of Fig. 2.3(b) gives

$$I_D = I_{D1} = \frac{(V_{BB} - 0.70)}{R} = 9.93\text{ mA}; \qquad V_D = 0.70\text{ V}$$

Insofar as the current is concerned, the ideal diode is quite adequate; the current obtained differs less than 1 % from the value found by the other method. However, the values of V_D are quite different, indicating that the ideal diode method is not a good way of calculating V_D.

We must now make a small correction in V_D, because I_D is not equal to 1.0 mA. Since $I_D \simeq 10$ mA, whereas $V_D = V_{D0} = 0.70$ V at $I_D = I_{D0} = 1.0$ mA and V_D increases by 0.06 V if I_D increases by a factor of 10, in the next approximation we have $V_D = V_{D1} = 0.70 + 0.06 = 0.76$ V. The operating point is thus 9.93 mA, 0.76 V. The approximation of Fig. 2.3(b) needs a small correction in this case.

EXAMPLE 2: $V_{BB} = 10.0$ V, $R = 10^4\ \Omega$, and $I_D = 1.0$ mA at $V_D = 0.70$ V. The ideal diode gives

$$I_D = \frac{V_{BB}}{R} = 1.0\text{ mA}; \qquad V_D = 0\text{ V}$$

and the approximation of Fig. 2.3(b) gives

$$I_D = I_{D1} = \frac{(V_{BB} - 0.70)}{R} = 0.93\text{ mA}; \qquad V_D = 0.70\text{ V}$$

The current is only 7 % off, which is often tolerable. We shall now investigate whether the value of V_D needs a further correction. The actual diode has I_D

$\simeq 1.0$ mA, and since $I_D = 1.0$ mA at $V_D = 0.70$ V, no further correction is needed. The operating point is therefore 0.93 mA, 0.70 V.

EXAMPLE 3: $V_{BB} = 1.00$ V, $R = 10^4$ Ω, and $I_D = 1.0$ mA at $V_D = 0.70$ V. In this case neither the ideal diode approximation nor the approximation of Fig. 2.3(b) is adequate. The latter can be used as a first step in a series of approximations, however. We shall see how this is done.

According to the approximation of Fig. 2.3(b),

$$I_D = I_{D1} = \frac{(V_{BB} - 0.70)}{R} = 0.030 \text{ mA}; \qquad V_D = V_{D0} = 0.70 \text{ V}$$

Since $I_D = I_{D0} = 1.0$ mA at $V_D = V_{D0} = 0.70$ V, we put

$$V_{D1} = V_{D0} + \Delta V_{D1}$$

Because

$$I_{D0} = I_{rs} \exp\left(\frac{eV_{D0}}{kT}\right)$$

we have

$$I_{D1} = I_{D0} \exp\left(\frac{e \Delta V_{D1}}{kT}\right)$$

or

$$\Delta V_{D1} = \frac{kT}{e} \ln\left(\frac{I_{D1}}{I_{D0}}\right) = 25.8 \times 10^{-3} \ln 0.030 = -0.090 \text{ V}$$

Hence

$$V_{D1} = V_{D0} + \Delta V_{D1} = 0.610 \text{ V}$$

For the next approximation we put $V_D = V_{D1} = 0.610$ V into Eq. (2.5). This yields

$$I_D = I_{D2} = \frac{(V_{BB} - V_{D1})}{R} = 0.039 \text{ mA}$$

which differs from the previous value by 30%. We now find the corresponding value V_{D2} for V_D by putting $V_{D2} = V_{D0} + \Delta V_{D2}$, so that

$$I_{D2} = I_{D0} \exp\left(\frac{e \Delta V_{D2}}{kT}\right) \quad \text{or} \quad \Delta V_{D2} = \frac{kT}{e} \ln\left(\frac{I_{D2}}{I_{D0}}\right) = -0.083 \text{ V}$$

so that $V_{D2} = V_{D0} + \Delta V_{D2} = 0.617$ V, which differs from V_{D1} by about 1%. It may thus be assumed that V_{D2} is reasonably accurate. However, we need one further approximation for I_D. Substituting $V_D = V_{D2} = 0.617$ V into Eq. (2.5) yields

$$I_D = I_{D3} = \frac{(V_{BB} - V_{D2})}{R} = 0.383 \text{ mA}$$

which differs from I_{D2} by less than 2%. It may thus be assumed that I_{D3} is reasonably accurate. The operating point is therefore 0.383 mA, 0.617 V. This example illustrates that the series of approximations converges rapidly and gives accurate results in a few steps.

The conclusion to be drawn from these examples is that the ideal diode approximation gives reasonably accurate values for the diode current I_D as long as V_{BB} is larger than 5–10 V, but that it cannot be used to determine the diode voltage V_D. The approximation method of Fig. 2.3(b) gives reasonably accurate values for the diode current I_D and the diode voltage V_D. The latter holds as long as I_D is of the order of 1 mA; otherwise a simple correction must be made for V_D. If V_{BB} is comparable to 0.70 V, the operating point must be determined by a series of successive approximations, as demonstrated in Example 3.

2.1b. Derivation of Eq. (2.3)

Figure 2.4(a) shows the potential distribution in the space-charge region of a *p-n* junction diode for zero external bias. The space-charge gives rise to a contact potential difference V_{dif} between the bulk *n*- and *p*-regions. If a voltage

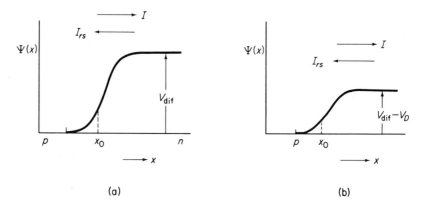

(a)   (b)

Fig. 2.4. (a) Potential distribution in the space-charge region of a diode without bias. (b) Potential distribution in the space-charge region of a diode with bias.

V_D is applied to the *p*-region, the potential difference across the space-charge region changes to $V_{dif} - V_D$ [Fig. 2.4(b)]. There is thus a potential energy barrier of height $E_a = e(V_{dif} - V_D)$ that impedes the flow of hole current.

Holes moving from the *p*- to the *n*-region must climb this potential energy barrier; according to statistical mechanics, the probability that they reach the *n*-region is proportional to $\exp(-E_a/kT)$. Therefore the current flowing from left to right is

$$C_1 \exp\left[\frac{-e(V_{dif} - V_D)}{kT}\right] \tag{2.7}$$

where C_1 is a constant. Holes generated in the *n*-region and moving toward the *p*-region flow downhill; therefore they produce a current I_{rs} from right to left that is independent of the applied bias V_D. Hence the net current is

$$I_D = C_1 \exp\left[\frac{-e(V_{dif} - V_D)}{kT}\right] - I_{rs} \tag{2.8}$$

We now bear in mind that $I_D = 0$ when $V_D = 0$, so that

$$C_1 \exp\left(\frac{-eV_{dif}}{kT}\right) - I_{rs} = 0 \qquad (2.8a)$$

Solving for C_1 and substituting back into I_D yields Eq. (2.3), which may be written in the form

$$I_D = I_{rs} \exp\left(\frac{eV_D}{kT}\right) - I_{rs} \qquad (2.9)$$

Here the first term represents the hole current *injected* into the n-region and the second term the hole current *collected* from the n-region. This hole-injection–hole-collection process is the basis for the operation of p-n-p transistors.

The same reasoning can be applied to electrons, and again Eq. (2.9) is obtained. One must now distinguish between I_{rsp}, the hole current collected from the n-region, and I_{rsn}, the electron current collected from the p-region, and $I_{rs} = I_{rsp} + I_{rsn}$.

The current I_{rsp} should be proportional to p_n, the equilibrium hole concentration in the n-region, and I_{rsn} should be proportional to n_p, the equilibrium electron concentration in the p-region:

$$I_{rsp} = C_p p_n = C_p \frac{n_i^2}{N_d}; \qquad I_{rsn} = C_n n_p = C_n \frac{n_i^2}{N_a} \qquad (2.9a)$$

where C_p and C_n are constants. This indicates that I_{rs} is very strongly temperature-dependent because n_i^2 depends so strongly on temperature. Moreover, in p^+-n diodes practically all current is carried by holes since $N_a \gg N_d$ and hence $I_{rsn} \ll I_{rsp}$, whereas in p-n^+ diodes practically all current is carried by electrons since $N_d \gg N_a$ and hence $I_{rsp} \ll I_{rsn}$.

Finally, I_{rs} is obviously proportional to the junction area A, for if we put two equal diodes in parallel, the total current is doubled.

2.2. CLIPPER CIRCUITS

A clipper circuit is a circuit that transmits only parts of a wave form. Since a diode conducts only for *one* polarity, such devices are well suited as clippers. The basic features of these circuits can be understood by treating the diodes as ideal diodes. Afterwards one can make corrections for the voltage drop (turn-on voltage) in the actual diode.

2.2a. Four Basic Clipper Circuits

Figure 2.5 shows four basic clipper circuits and their respective transfer characteristics. They are obtained by changing the polarity of the diode and the battery.

In Fig. 2.5(a) the ideal diode D does not conduct if $V_i(t) < V_B$, whereas it conducts for $V_i(t) > V_B$. Therefore $V_0(t) = V_B$ for $V_i(t) < V_B$, and $V_0(t) = V_i(t)$ for $V_i(t) > V_B$.

In Fig. 2.5(b) the ideal diode D does not conduct if $V_i(t) > V_B$, whereas it conducts for $V_i(t) < V_B$. Hence $V_0(t) = V_i(t)$ for $V_i(t) < V_B$, and $V_0(t) = V_B$ for $V_i(t) > V_B$.

In Fig. 2.5(c) the ideal diode D does not conduct if $V_i(t) < -V_B$ and conducts for $V_i(t) > -V_B$. Consequently $V_0(t) = -V_B$ for $V_i(t) < -V_B$, and $V_0(t) = V_i(t)$ for $V_i(t) > -V_B$.

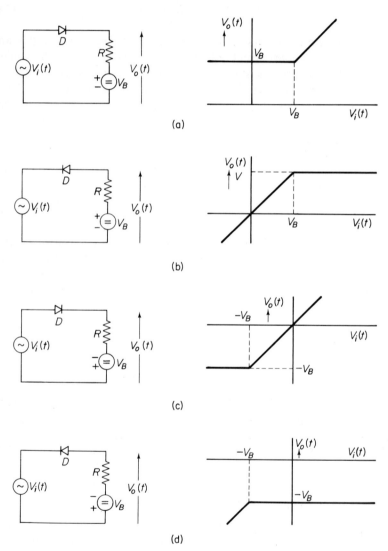

Fig. 2.5. Four possible clipper circuits with their transfer characteristics. (a) Original circuit. (b) Polarity of diode reversed. (c) Polarity of battery reversed. (d) Polarity of diode and battery reversed.

In Fig. 2.5(d) the diode D conducts if $V_i(t) < -V_B$ and does not conduct for $V_i(t) > -V_B$. Therefore $V_0(t) = V_i(t)$ for $V_i(t) < -V_B$, and $V_0(t) = -V_B$ for $V_i(t) > -V_B$.

How are these results altered if we take the voltage drop in the actual diode into account? For the nonconducting part nothing changes: $V_0(t) = V_B$ or $V_0(t) = -V_B$, depending on the polarity of the battery. For the conducting part, however, there is a voltage drop of about 0.70 V in the diode, so that $V_0(t) = V_i(t) - 0.70$ V in Fig. 2.5(a) and (c) and $V_0(t) = V_i(t) + 0.70$ V in Fig 2.5(b) and (d).

2.2b. Applications

As a first application we take the battery charger of Fig. 2.6. If the diode can be considered as ideal, current flows if $V_s(t) > V_B$. Consequently

$$I(t) = \frac{V_s(t) - V_B}{R + R_s} \quad \text{for } V_s(t) > V_B$$
$$= 0 \quad \text{for } V_s(t) < V_B$$

(2.10)

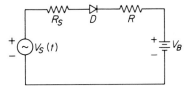

Fig. 2.6. Circuit of a battery charger.

Therefore a current pulse flows through the battery every half-cycle. If $V_s(t) = V_{s0} \cos \omega t$, the maximum current flowing through the battery is $I_{max} = (V_{s0} - V_B)/(R + R_s)$, and the maximum back bias developed across the diode is $-(V_{s0} + V_B)$. The diode must thus be able to stand that forward current and this back bias.

As a second application we consider the diode limiter of Fig. 2.7. It

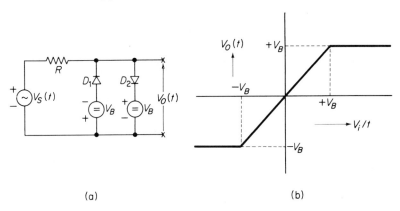

(a)

(b)

Fig. 2.7. (a) Diode limiter circuit. (b) Transfer characteristic.

uses two diodes, each in series with a battery; the two diodes have opposite polarity and so have the two batteries [Fig. 2.7(a)]. The circuit transmits for $-V_B < V_i(t) < V_B$, whereas $V_0(t) = -V_B$ for $V_i(t) < -V_B$ and $V_0(t) = V_B$ for $V_i(t) > V_B$. The transfer characteristic is shown in Fig. 2.7(b). If $V_s(t) = V_{s0} \cos \omega t$ and $V_{s0} > V_B$, the circuit transforms a sinusoidal wave form into a trapezoidal one.

If the two voltages are different, one can transmit a different part of the wave form. Figure 2.8(a) shows a pulse that has noise associated with the base line and with the top. By choosing a circuit that passes the wave form for $V_{B1} < V_i(t) < V_{B2}$, one can completely eliminate the noise on the wave form. Figure 2.8(b) shows the circuit that performs this function.

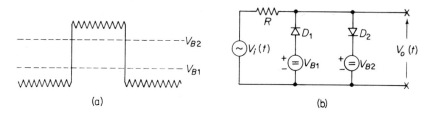

(a)   (b)

Fig. 2.8. (a) Waveform and its proposed clean-up. (b) Circuit providing the clean-up.

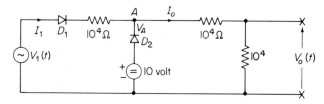

Fig. 2.9. More complicated diode circuit.

A more complicated diode circuit is shown in Fig. 2.9. Three regions of the input and transfer characteristic must be considered here:

1. $V_1 < 10$ V. In this case diode D_2 conducts and diode D_1 is off. This keeps V_A at 10 V, so that

$$I_1 = 0; \qquad V_0 = 10 \times \frac{10^4}{10^4 + 10^4} = 5 \text{ V}$$

and the current flowing through the output is $I_0 = 0.5$ mA; I_0 keeps this value as long as V_A stays at 10 V.

2. $10 < V_1 < 15$ V. In this case D_2 and D_1 both conduct and

$$I_1 = (V_1 - V_A) \times 10^{-4} = (V_1 - 10) \times 10^{-4} \text{ A}$$

so that $I_{D2} = I_0 - I_1 = 0.5$ mA $- I_1$ as long as V_A stays at 10 V.

For $V_1 = 10\ V$ all output current I_0 passes through D_2, whereas for $V_1 = 15$ V $I_1 = 0.5$ mA, and hence $I_{D2} = 0$ so that all of I_0 comes from D_1. The limits of this range are thus clearly established, and the output voltage V_0 stays at 5 V.

3. $V_1 > 15$ V. In this case $V_A > 10$ V and D_2 does not conduct. Consequently

$$I_1 = \frac{V_1}{3 \times 10^4} = 0.333 V_1 \times 10^{-4}\ \text{A}; \qquad V_0 = \frac{1}{3} V_1$$

The input and transfer characteristics are therefore

$$V_1 < 10\ \text{V:} \qquad I_1 = 0 \quad \text{and} \quad V_0 = 5\ \text{V}$$
$$10 < V_1 < 15\ \text{V:} \qquad I_1 = (V_1 - 10) \times 10^{-4}\ \text{A} \quad \text{and} \quad V_0 = 5\ \text{V}$$
$$V_1 > 15\ \text{V:} \qquad I_1 = 0.333 V_1 \times 10^{-4}\ \text{A} \quad \text{and} \quad V_0 = \tfrac{1}{3} V_1$$

The input characteristic thus has two breakpoints, one at $V_1 = 10$ V and one at $V_1 = 15$ V, whereas the transfer characteristic has a breakpoint at 15 V only.

2.2c. Diode Logic Circuits

In digital computers binary arithmetic is used. Closely related to this kind of arithmetic is *Boolean algebra*, which is an algebra of the numbers 0 and 1. We shall see that electronic circuits can perform this kind of algebra.

In Boolean addition $(A + B)$ one has the following rules:

$$0 + 0 = 0$$
$$1 + 0 = 0 + 1 = 1$$
$$1 + 1 = 1$$

In Boolean multiplication $(A \times B = AB)$ the rules are

$$0 \times 0 = 0$$
$$1 \times 0 = 0 \times 1 = 0$$
$$1 \times 1 = 1$$

We further need the operation $\bar{A}$ ("A not") that changes a 0 into a 1 and a 1 into a 0. This is called *inversion*:

$$\bar{0} = 1; \qquad \bar{1} = 0$$

Assigning the words *yes* and *no* to 1 and 0, respectively, the above operations become *logic* operations.

We shall now see how diode circuits can be used to perform the first two operations. First, one has to define 0 and 1 in terms of voltages. In positive logic, 1 is defined as a "high" voltage V_1, say $V_1 > 2.0$ V, and 0 is defined as a "low" voltage V_0, say $V_0 < 0.5$ V.

Addition is performed by the circuit of Fig. 2.10(a), which is called the *or* circuit. If there is no voltage in either channel, $V_0 = 0$ V. If there is an input voltage in *either one* of the channels, the output voltage $V_0(t) = V_i(t) -$ 0.70 V, which is a 1, so that the *or* circuit indeed performs Boolean addition.

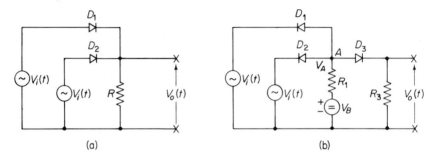

Fig. 2.10. (a) Diode *or* circuit. (b) Diode *and* circuit.

The *and* circuit of Fig. 2.10(b) performs Boolean multiplication. This circuit gives zero output unless there is a voltage $V_i(t)$ in each channel *simultaneously*. For example, if $V_i(t) \geq V_B$ in both channels simultaneously, diodes D_1 and D_2 are turned off and $V_0(t) = (V_B - 0.70)R_3/(R_1 + R_3)$, which is a 1. If there is no voltage $V_i(t)$ in at least one channel, then $V_A = 0.70$ V and $V_0(t) \simeq 0$ V, which is a 0. The *and* circuit thus performs multiplication; the third diode is introduced to make $V_0(t) \simeq 0$ V if $V_i(t) = 0$ V.

Transistor and field effect transistor circuits can be so designed that they transform a 0 into a 1 and a 1 into a 0. They thus act as *inverters*.

Therefore these two diode circuits and the inverter circuit perform all the Boolean operations. All binary operations can be performed by proper combinations of these circuits.

We note that in each diode circuit the 1 output is somewhat smaller than the 1 input. It is therefore not possible to put many of these circuits one behind the other. By using a combination of diode and transistor circuits, this difficulty can be overcome.

2.3. CLAMP CIRCUITS

A clamp circuit is a circuit that shifts the dc level of a wave form and clamps either the top or the bottom of a wave form to a fixed bias. We shall see that a combination of a diode and a capacitor in series can perform this operation.

2.3a. Two Basic Clamp Circuits

The two basic configurations of the clamp circuit are shown in Fig. 2.11(a) and (b); they differ in the polarity of the diode. To operate satisfactorily, the diode must be bypassed by a resistor R such that the RC time constant is

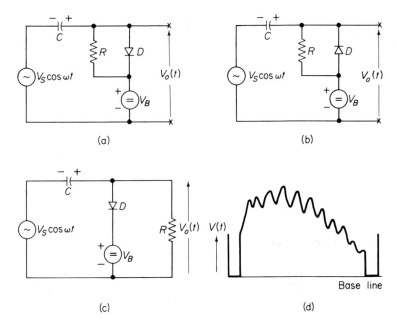

Fig. 2.11. (a) Diode clamp circuit clamping top of waveform. (b) diode clamp circuit clamping bottom of waveform. (c) Clamp circuit that only works above a certain ac voltage. (d) TV waveform, to be restored by a circuit of Fig. 2.11(b).

somewhat larger than the period of the input signal. If the RC time constant is too small, the capacitor C discharges notably during the period of the wave form; if the RC time constant is much too large, the output voltage cannot follow slow variations in the input wave form. If the resistor R is omitted, the circuit either clamps permanently or not at all.

To understand the operation of these circuits, we first turn to Fig. 2.11(a). If $V_s = 0$, the capacitor charges to V_B, with the polarity shown. If the input signal is turned on, so much current flows out of the capacitor C that further discharging is prevented, which means that the voltage across C will be $-V_s + V_B$, so that the voltage developed across the diode D is $-V_s + V_s \cos \omega t$, which is always negative except for $\omega t = 0, 2\pi, \dots$. The output voltage $V_0(t)$ is therefore

$$V_0(t) = V_s \cos \omega t + (-V_s + V_B) = V_B + V_s(-1 + \cos \omega t) \qquad (2.11)$$

this expresses the fact that the *top* of the wave form is clamped to V_B.

We now turn to the circuit of Fig. 2.11(b). If $V_s = 0$, the capacitor charges to V_B, with the polarity shown. If now the input signal is turned on, so much current flows to the capacitor C that further charging is prevented. This is the case if the voltage across C is $V_s + V_B$, so that the voltage developed across the diode D is $-V_s - V_s \cos \omega t$, which is always negative except for $\omega t = \pi, 3\pi, \ldots$. The output voltage $V_0(t)$ is therefore

$$V_0(t) = V_s \cos \omega t + (V_s + V_B) = V_B + V_s(1 + \cos \omega t) \qquad (2.11a)$$

This expresses the fact that the *bottom* of the wave form is clamped to V_B.

It should be noted that the clamping of either the top or the bottom of the wave form depends on the *polarity* of the diode; the polarity of the battery determines whether the wave form is clamped to a positive or negative voltage.

Sometimes the resistor R does not bypass the diode but is connected as shown in Fig. 2.11(c). We shall see that this diode does not clamp at all so long as $V_s < V_B$, for if V_s is first zero, the diode D is back-biased and $V_0(t) = 0$ V, so that the capacitor C is not charged. If now the input voltage is turned on and $V_s < V_B$, the voltage across the diode becomes $-V_B + V_s \cos \omega t$, which is always negative, so that the diode D never conducts. The capacitor C never charges and the circuit does not clamp. The output voltage thus follows the input wave form.

What happens if $V_s > V_B$? In this case the capacitor C is charged to the voltage $-V_s + V_B$ with the polarity shown and hence

$$V_0(t) = V_s \cos \omega t - V_s + V_B = V_B + V_s(-1 + \cos \omega t)$$

which corresponds to (2.11). The top of the wave form is therefore clamped to V_B. The difficulty for $V_s < V_B$ is avoided if the resistor R is connected across D.

The circuit of Fig. 2.11(b) is used in TV circuits as a *dc restorer*. The idea behind it is as follows. The TV signal generated by a television pickup tube is a wave form referred to a fixed base line [Fig. 2.11(d)]; the height $V(t)$ is proportional to the light intensity of the various parts of the scene. If this signal is passed through an ac amplifier, the dc base line is not transmitted; hence it must be restored before the amplified signal can be applied to a picture tube. The circuit of Fig. 2.11(b) accomplishes this.

2.4. RECTIFIERS

A rectifier transforms an ac voltage into a dc voltage. Any *clipper* circuit that clips the top half of the wave form can perform this function, but it will be seen that much better results are obtained if the output load resistance is bypassed by a suitably chosen capacitor C; this turns the circuit into a *clamp* circuit.

2.4a. Rectifiers Without Output Capacitor

As a first example we take the *half-wave rectifier* shown in Fig. 2.12(a). Assuming ideal diode operation, the output voltage is therefore

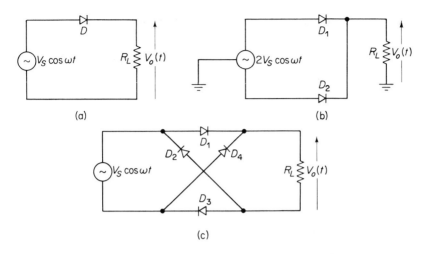

Fig. 2.12. (a) Half-wave rectifier; the diode passes current half the time. (b) Full-wave rectifier. (c) Bridge-type full-wave rectifier.

$$V_0(t) = V_s \cos \omega t \qquad \text{for } \cos \omega t > 0$$
$$= 0 \qquad \text{for } \cos \omega t < 0 \qquad (2.12)$$

This gives a dc voltage,

$$V_{av} = \frac{1}{2\pi} \int_{-\pi}^{\pi} V_0(t)\, d(\omega t) = \frac{V_s}{\pi} \qquad (2.12a)$$

and a large ac ripple component, mostly of the frequency ω.

As a second example we take the *full-wave* rectifier shown in Fig. 2.12(b). Here a voltage $2V_s \cos \omega t$, obtained from the output of a center-tapped transformer, is applied in balance (push-pull) to two diodes D_1 and D_2 and the output is taken from the load resistance R_L between the two diodes and ground. The output voltage is

$$V_0(t) = V_s |\cos \omega t| \qquad (2.13)$$

This gives a dc voltage,

$$V_{av} = \frac{1}{2\pi} \int_{-\pi}^{\pi} V_0(t)\, d(\omega t) = \frac{2V_s}{\pi} \qquad (2.13a)$$

and a large ac ripple component, chiefly of the frequency 2ω.

As a third example we take the bridge rectifier of Fig. 2.12(c). In this circuit the diodes D_1 and D_3 conduct during one half-cycle and the diodes D_2 and D_4 conduct during the other half-cycle. The output voltage is thus the same as in the previous case.

2.4b. Rectifiers with Output Capacitor (Peak Rectifiers)

If a capacitor C is connected in parallel with the load resistor R_L of rectifier circuit, the capacitor charges up to the voltage amplitude V_s and the dc output voltage is about equal to V_s. The circuit thus obtained is therefore called a *peak rectifier*. Figure 2.13(a) shows a half-wave rectifier improved in that

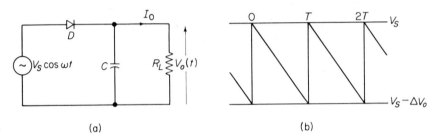

(a)   (b)

Fig. 2.13. (a) Half-wave rectifier with capacitive output filter. (b) Waveform of the voltage developed across the capacitor.

manner. If the capacitor C is sufficiently large, it is charged to the voltage V_s and stays that way, so that $V_0(t) = V_s$ in first approximation. The maximum back bias across the diode is now $2V_s$ and the diode must be so chosen that it can stand this.

If the current I_0 drawn by the load resistance R_L is appreciable, the capacitor C is charged to the voltage V_s when the input voltage reaches its peak and is partly discharged during the remainder of the full cycle. Assuming I_0 to be constant, $V_0(t)$ will drop linearly from the value V_s at $t = 0$ to the value $V_s - \Delta V_0$ at $t = T = 1/f$, after which the charge-discharge cycle repeats itself [Fig. 2.13(b)]. Consequently the average output voltage drops to $V_s - \frac{1}{2}\Delta V_0$.

The value of ΔV_0 is easily calculated, for

$$I_0 = -C\frac{dV_0}{dt} = C\frac{\Delta V_0}{T} \tag{2.14}$$

since $-dV_0/dt = \Delta V_0/T$ because V_0 varies linearly with time during the period $T = 1/f$ of the wave form. Hence

$$\Delta V_0 = \frac{I_0 T}{C} = \frac{I_0}{fC} \tag{2.14a}$$

which is small if I_0 is small and C large. The output wave form is

$$V_0(t) = V_s - \Delta V_0 \frac{t}{T} \qquad \text{for } 0 < t < T \tag{2.15}$$

and periodically repeated outside that interval [Fig. 2.13(b)].

EXAMPLE: If $V_s = 20$ V, $f = 60$ Hz, and $R_L = 1000\ \Omega$, find the value of C that reduces the output ripple to 1 V peak to peak.

ANSWER: $V_0 = V_s = 20$ V at the peak, so that $I_0 = 20$ mA. Hence from (2.14a)

$$C = \frac{I_0}{f\,\Delta V_0} = \frac{0.02}{60 \times 1} = 333\ \mu\text{F}$$

Up to now we have assumed that the diode could be treated as an ideal diode. Actually, the diode has a voltage drop of about 0.70 V when it is on. Therefore the actual dc output voltage is $V_s - \frac{1}{2}\Delta V_0 - 0.70$ V.

Figure 2.14(a) shows a full-wave rectifier. The capacitor C again clamps

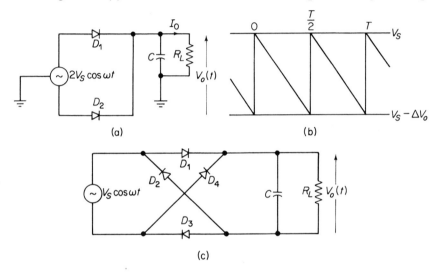

(a)   (b)

(c)

Fig. 2.14. (a) Full-wave rectifier with capacitive output. (b) Waveform of the voltage developed across the capacitor. (c) Bridge-type rectifier with capacitive output.

the output voltage to V_s, but the capacitor is now charged twice during each cycle, once at $t = 0$ and then at $t = T/2$, where $T = 1/f$ is the period of the wave form. Consequently

$$V_0(t) = V_s - \Delta V_0 \frac{2t}{T} \qquad \text{for } 0 < t < T/2 \qquad (2.16)$$

and periodically repeated outside that interval [Fig.2.14(b)]. Here

$$\Delta V_0 = \frac{I_0 T}{2C} = \frac{I_0}{2fC} \qquad (2.17)$$

since the discharge interval is now $T/2$ rather than T as in the previous case. For the same current I_0 and the same value of the capacitor C the output voltage thus has half the ripple of that in the half-wave rectifier and the ripple voltage has twice the frequency.

The maximum back bias developed across each diode is again $2V_s$. If the voltage drop of 0.70 V during the *on* period of the diode is taken into account, the dc output voltage bacomes $V_s - \frac{1}{2}\Delta V_0 - 0.70$ V.

We finally turn to the bridge rectifier with the capacitive output shown in Fig. 2.14(c). Again the capacitor C is charged twice during each cycle so that Eq. (2.16) and (2.17) are valid. However, there are two differences with the full-wave rectifier. First, the back bias across each diode is now V_s instead of $2V_s$, so that the requirements that the diodes have to meet are less stringent than in the previous cases. For example, let D_1 and D_3 be conducting and let the input voltage have a peak value V_s. If D_3 is ideal, no voltage is developed across D_3, and hence the back voltage developed across D_2 is $V_s - 0 = V_s$. Second, two diodes in series are "on" at the same time so that the output voltage is two diode voltage drops less than $V_s - \frac{1}{2}\Delta V_0$, corresponding to $V_s - \frac{1}{2}\Delta V_0 - 1.40$ V.

2.4c. Voltage Doubler

By combining a clamper circuit with a peak rectifier circuit as shown in Fig. 2.15(a), one obtains a *voltage doubler* circuit, i.e., a rectifier circuit that gives

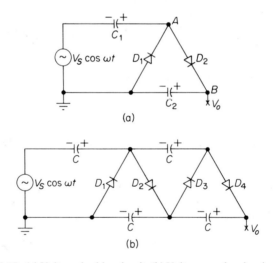

(a)

(b)

Fig. 2.15. (a) Voltage doubler circuit. (b) Voltage quadrupler circuit.

a dc output voltage equal to *twice* the ac input amplitude. The clamper circuit develops a dc voltage V_s across C_1 with the polarity shown, so that a voltage $V_s(1 + \cos \omega t)$ is developed between point A and ground. To prevent the flow of current through D_2, the capacitance C_2 must be charged to the voltage $2V_s$. The dc output voltage for zero current is therefore $2V_s$.

The same idea can be used to make a voltage quadrupler. Figure 2.15(b) shows the circuit.

2.4d. ac Filters

Rectifier circuits with capacitive output show a sawtooth-type voltage ripple of peak-to-peak amplitude ΔV_0 and frequency f (for a half-wave rectifier) or frequency $2f$ (full-wave rectifier or bridge rectifier). To remove this ripple, two types of filters are used: the RC-filter and the LC-filter.

Figure 2.16(a) shows an RC-filter. If V_s is the dc input voltage of the filter, the dc output voltage is $V_{00} = V_s - I_0R$, where I_0 is the current drawn

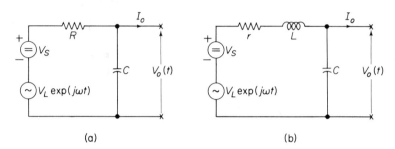

(a)   (b)

Fig. 2.16. (a) RC-filter used in rectifier circuits for low load currents. (b) LC-filter used in rectifier circuits for high load currents.

by the load. For good operation of the filter one must require $I_0R \ll V_s$, so that this type of filter is used only if a small current I_0 is drawn. The ac output voltage has an amplitude of

$$V_0 = V_L \frac{1/(j\omega C)}{R + 1/(j\omega C)} = \frac{V_L}{1 + j\omega CR}; \quad \left|\frac{V_0}{V_L}\right| = \frac{1}{(1 + \omega^2 C^2 R^2)^{1/2}} \quad (2.18)$$

For good filtering one must require $\omega^2 C^2 R^2 \gg 1$; the attenuation per filtering stage is then ωCR.

Figure 2.16(b) shows an LC-filter. If the inductor has a series resistance r and the dc input voltage is V_s, then the dc output voltage is $V_{00} = V_s - I_0r$. Since r can be relatively small, this filter can handle considerably larger dc currents than the previous one. The ac output voltage has an amplitude of

$$V_0 = V_L \frac{1/(j\omega C)}{j\omega L + 1/(j\omega C)} = \frac{V_L}{1 - \omega^2 LC}; \quad \left|\frac{V_0}{V_L}\right| = \frac{1}{|1 - \omega^2 LC|} \quad (2.19)$$

For good filtering one must require $\omega^2 LC \gg 1$; the attenuation per filtering stage is then $\omega^2 LC$.

EXAMPLE: In an RC-filter $V_s = 20$ V, $I_0 = 1$ mA, $f = 60$ Hz, the allowed dc voltage drop in R is 5 V, and the required attenuation factor of the filter is 10. Find the values of R and C.

ANSWER: $I_0R = 5$ V, or $R = 5000\ \Omega$. $2\pi fCR = 10$, $C = 10/(2\pi fR) = 5\ \mu$F.

2.5. FURTHER RECTIFIER CIRCUITS

2.5a. ac Diode Voltmeter

In an ac diode voltmeter the ac voltage is first transformed into a dc voltage by a diode rectifier and then this dc voltage is measured by a dc voltmeter.

Figure 2.17(a) shows a simple rectifier circuit that performs this func-

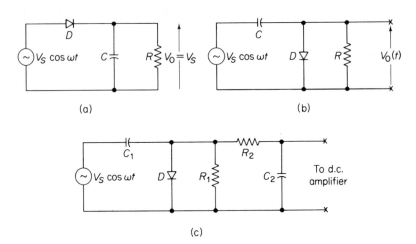

(a)   (b)

(c)

Fig. 2.17. (a) One form of diode voltmeter circuit. (b) Another form of diode voltmeter circuit. (c) Circuit (b) with $R_2 C_2$ output filter.

tion. If it is assumed that the diode D is an ideal diode and that the capacitor C is so chosen that $\omega CR \gg 1$, the full ac voltage is developed across the diode and the current flows in infinitely short pulses. The dc output voltage is V_s and hence the dc diode current is $I_0 = V_s/R$. We now make a Fourier analysis of the diode current $I(t)$ by writing

$$I(t) = I_0 + I_1 \cos \omega t + I_2 \cos 2\omega t + \cdots \tag{2.20}$$

where

$$I_0 = \frac{1}{2\pi} \int_{-\pi}^{\pi} I(t)\, d(\omega t); \quad I_1 = \frac{1}{\pi} \int_{-\pi}^{\pi} I(t) \cos \omega t\, d(\omega t); \ldots \tag{2.21}$$

If the current flows in very sharp pulses, $\cos \omega t$ is still nearly unity when $I(t)$ has already dropped to zero. We may thus replace $\cos \omega t$ by unity and then find $I_1 \simeq 2I_0$. The ac input conductance of the rectifier circuit is therefore

$$g = \frac{I_1}{V_s} = \frac{2I_0}{V_s} = \frac{2}{R} \tag{2.22}$$

so that the input resistance of the circuit is $R/2$.

This circuit cannot be used if the signal source has no dc path, because the dc current must pass through the signal source. The circuit of Fig. 2.17(b) remedies this situation. Here the capacitor C separates the signal source from the diode for dc. If C is so chosen that $\omega CR \gg 1$, the full input signal $V_s \cos \omega t$ is developed across the diode D. Applying the same reasoning as before, the input conductance is found to be

$$g = \frac{2}{R} + \frac{1}{R} = \frac{3}{R} \tag{2.23}$$

The first term comes from the diode and corresponds to Eq. (2.22); the second term comes from the fact that the resistance R is directly in parallel with the signal source.

The output voltage $V_0(t)$ now contains not only the dc term V_s but also the ac signal $V_s \cos \omega t$. The latter must be filtered out; this is done by a $C_2 R_2$-filter for which $\omega C_2 R_2 \gg 1$. The dc voltage thus obtained is fed into a dc amplifier and the amplified dc signal is fed into the indicating instrument [Fig. 2.17(c)].

The diode voltmeter must be so designed that the input resistance does not load the signal source too heavily.

2.5b. Detector for Amplitude-Modulated Signals

In most radio and TV transmission systems the signal to be transmitted is used to amplitude-modulate a signal of carrier frequency ω_i. This carrier frequency is easily transmitted and carries the modulating signal with it. In a receiver the modulated signal must be demodulated to recover the transmitted information.

The circuit of Fig. 2.17(a) can be used for the demodulation. The modulated signal can be written as

$$\begin{aligned} V_i(1 &+ m \cos \omega_p t) \cos \omega_i t \\ &= V_i \cos \omega_i t + \tfrac{1}{2} m V_i [\cos (\omega_i + \omega_p)t + \cos (\omega_i - \omega_p)t] \end{aligned} \tag{2.24}$$

where $m < 1$; it contains a *carrier* signal of frequency ω_i and side-band signals of frequencies $\omega_i + \omega_p$ and $\omega_i - \omega_p$, respectively. Figure 2.18 shows the demodulation circuit in greater detail. It must be so designed that $\omega_i CR \gg 1$, so that the full modulated signal appears across the diode, whereas $\omega_p CR \ll 1$, so that the capacitance acts as an open circuit for the frequency

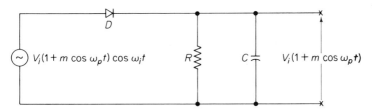

Fig. 2.18. Demodulator circuit that rectifies a modulated signal and produces at the output a dc signal plus the low-frequency modulation.

ω_p. If these two conditions are met, the output voltage is equal to the input amplitude, or

$$V_0(t) = V_i(1 + m \cos \omega_p t) \tag{2.25}$$

so that the modulating signal $mV_i \cos \omega_p t$ has been recovered.

It is sometimes necessary, especially if ω_i is not too much larger than ω_p, to use a more complicated filter that has a flatter passband for the modulation frequencies and that cuts off more sharply at higher frequencies. Care must also be taken that the input conductance g, given by Eq. (2.22), does not load the signal source too heavily.

PROBLEMS

2.1. In the circuit of Fig. 2.2(a), $V_{BB} = 1.40$ V, $R = 5000 \, \Omega$. The diode has $V_D = V_{D0} = 0.70$ V at $I_D = I_{D0} = 1.0$ mA. Find the voltage V_D and the current I_D by successive approximation.

2.2. The battery charger of Fig. 2.6 has $V_s(t) = 50 \cos 2\pi f t$ and $R_s = 0.50 \, \Omega$, whereas $V_B = 12.0$ V. (a) Find R so that the maximum current is 30 A. (b) Find the maximum back bias that the diode must be able to withstand. (c) Find the part of the cycle during which the diode is carrying current. (d) Find the average current passing through the battery. (e) Find how long it takes to charge the fully discharged battery if its capacity is 75 Ah. *Hint for (d):* $I_{av} = (1/2\pi) \int_{-\pi}^{\pi} I(x) \, dx$, where $x = \omega t$ is the variable.

2.3. (a) In Problem 2.2, find the average power supplied by $V_s(t)$ during the charging of the battery. (b) How many kilowatthours are needed to charge the battery? *Hint for (a):* $P_{av} = (1/2\pi) \int_{-\pi}^{\pi} P(x) \, dx$, where $x = \omega t$ is the variable.

2.4. Design a limiter circuit of Fig. 2.7 that transforms the signal $V_s(t) = 20 \cos \omega t$ V into a trapezoidal wave form with 10.0-V peak-to-peak amplitude, bearing in mind that each diode has a turn-on voltage of 0.70 V.

2.5. Plot the voltage transfer characteristic of the circuit of Fig. 2.19 and plot the

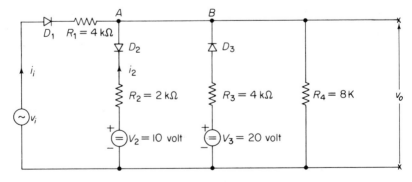

Fig. 2.19.

input current as a function of the input voltage. Locate especially the break-points in these characteristics. Neglect the turn-on voltage of the diodes.

2.6. In the *or* circuit of Fig. 2.10(a), $V_i(t) = 5.0$ V. Find V_0 if the devices have a turn-on voltage of 0.7 V.

2.7. In the *and* circuit of Fig. 2.10(b), $V_B = 5.0$ V and $V_i(t) = 5.0$ V in each channel. Find V_0 if $R_1 = 5$ kΩ and $R_3 = 15$ kΩ. The diodes have a turn-on voltage of 0.7 V.

2.8. In the circuit of Fig. 2.11(c), $V_s = 10$ V. (a) Find the output wave form for $V_B = 5, 10,$ and 15 V, respectively. You may neglect the turn-on voltage of the diode. (b) Show that the circuit clamps the top of the wave form to V_B if $V_B < 10$ V.

2.9. If $V_s = 20$ V, find the maximum back bias that the diodes must withstand (a) in Fig. 2.12(a), (b), and (c); (b) in Figs. 2.13(a) and 2.14(a) and (c). You may neglect the turn-on voltage of the diodes in either case.

2.10. If $V_s = 20$ V and the diodes have a turn-on voltage of 0.7 V, find the dc output voltage of the rectifier circuits of Figs. 2.13(a) and 2.14(a) and (c) if the capacitance C is sufficiently large.

2.11. The rectifier circuits of Figs. 2.13(a) and 2.14(a) and (c) are driven by a 60-Hz signal and $V_s = 20$ V in each case. (a) Calculate the value of the capacitance C for which the output ripple has a peak amplitude of 1 V when the load draws a current I_0 of 100 mA. (b) Taking into account the turn-on voltage of 0.7 V of each diode, calculate the average output voltage of each circuit.

2.12. The voltage quadrupler circuit of Fig. 2.15(b) has $V_s = 10^4$ V, the frequency is 10,000 Hz, and the output current is 1.0 mA. Find the output voltage and the value of the capacitance C needed to keep the peak-to-peak output ripple at 20 V.

2.13. (a) The output voltage of a half-wave rectifier driven by a 60-Hz signal has a peak-to-peak output ripple ΔV_0. Find the 60-Hz component of this ripple. (b) Design an *LC*-filter with $L = 10$ H that attenuates this 60-Hz component by a factor of 100.

2.14. (a) The output voltage of a full-wave rectifier driven by a 60-Hz signal has a peak-to-peak output ripple ΔV_0. Find the 120-Hz component of this ripple. (b) Design an *LC*-filter with $L = 10$ H that attenuates this 120-Hz component by a factor of 100.

2.15. A full-wave rectifier driven by a 60-Hz signal produces a peak voltage of 20 V with a 1-V peak-to-peak ripple. This signal is passed through an *RC*-filter that reduces the 120-Hz ripple by a factor of 10. Design the *RC*-filter so that the dc output voltage is 14.5 V if a dc current of 1.0 mA is drawn from the filter.

2.16. Two filter stages of the type mentioned in Problem 2.15 are connected in cascade. (a) Find the dc output voltage if a dc current of 1.0 mA is drawn. (b) Find the 120-Hz component at the output. (c) By connecting a capacitor

C' between input and output, you can reduce the 120-Hz component to practically zero (see Fig. 2.20). Find C'.

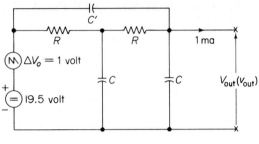

Fig. 2.20.

2.17. In the ac voltmeter of Fig. 2.17(c), $R_1 = R_2 = 10^6 \ \Omega$. (a) Design the filter so that the ac output amplitude at 10 Hz is 1 % of the dc voltage supplied to the dc amplifier. (b) Find the input impedance of the diode voltmeter as seen by V_s.

2.18. In the diode rectifier of Fig. 2.18, $R = 1000 \ \Omega$, $f_i = 10$ MHz, and $f_p = 100$ kHz. Find the value of C that gives a good compromise so that the modulated HF signal appears across the diode and the modulation across R.

2.19. (a) Show that the current in Fig. 2.21 flowing through the meter M consists

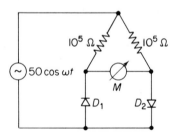

Fig. 2.21.

of a rectified sine wave (full-wave rectification) and calculate the peak current. (b) Determine the dc current passing through the meter.

2.20. The resistors in Fig. 2.21 are replaced by capacitors having the same absolute value of the impedance at the frequency ω as the resistors of Fig. 2.21. (a) Show that this circuit acts as a full-wave rectifier. (b) Draw the wave form of the ac current and compare it with the wave form of the applied voltage. (c) Show that the peak current and the dc current flowing through the meter are the same as in Problem 2.19.

3

Further Applications
of the *p-n* Diode

3.1. SMALL-SIGNAL APPLICATIONS OF THE p-n DIODE

3.1a. Small-Signal Capacitance of the Back-Biased Diode

The space-charge region of a back-biased *p-n* junction diode has a charge $-Q$ on the *p*-side of the junction due to ionized acceptors and a charge $+Q$ on the *n*-side of the junction due to ionized donors. Since the width d of the space-charge region depends on the voltage V_D, the charge $-Q$ depends on the voltage V_D, say $-Q = f(V_D)$. A small change ΔV_D in applied voltage produces a corresponding change ΔQ in the stored charge, where

$$\Delta Q = \frac{d(-Q)}{dV_D}\Delta V_D = C\Delta V_D; \qquad C = -\frac{dQ}{dV_D} \qquad (3.1)$$

so that it a makes sense to call C the small-signal capacitance of the back-biased diode. It should be noted that Q decreases with increasing V_D, so that C has a positive value.

Often C can be written in the form

$$C = \frac{C_0}{(V_{\text{dif}} - V_D)^n} \qquad (3.2)$$

where V_{dif} is the diffusion potential or contact potential and $V_{\text{dif}} - V_D$ the potential difference across the space-charge region. The parameter C_0 is the value of the capacitance for 1-V potential difference across the space-charge region. It should be noted that V_D is the voltage applied to the p-region and that V_D is negative for a back-biased diode.

In diodes in which the acceptor concentration N_a is constant in the p-region and the donor concentration N_d is constant in the n-region, one finds $n = \frac{1}{2}$, but for specially designed diodes with properly chosen impurity profiles, n can have a larger value, e.g., $n = 2$.

The reason for this behavior is the following. Let the junction have an area A and let ϵ be the relative dielectric constant of the material. Let d be the width of the space-charge region and let the applied voltage V_D increase by a small amount ΔV_D; then the width of the space-charge region shrinks slightly. One can represent this by saying that a charge $+\Delta Q$ has been added to the p-side edge of the space-charge region and a charge $-\Delta Q$ has been added to the n-side edge of the space-charge region. The two charges $+\Delta Q$ and $-\Delta Q$ are thus a distance d apart and can be considered as charges on a planar capacitor of plate area A and plate distance d and filled with a dielectric of relative dielectric constant ϵ [Fig. 3.1(a)]. Consequently

$$C = \frac{\Delta Q}{\Delta V_D} = \frac{\epsilon \epsilon_0 A}{d} \tag{3.3}$$

where $\epsilon_0 = 8.85 \times 10^{-14}$ F/cm is the electric conversion factor for MKS units, usually called the *dielectric constant of free space*. (We put ϵ_0 in farads per centimeter so that all dimensions can be expressed in centimeters.) Furthermore, one may write

$$d = d_0(V_{\text{dif}} - V_D)^n \tag{3.4}$$

where d_0 is the width of the space-charge region for a 1-V potential difference

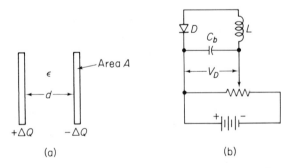

(a)   (b)

Fig. 3.1. (a) Diode space-charge region presented as a capacitor of electrode area A, electrode distance d, filled with a material of dielectric constant ϵ. (b) Back-biased diode used as a voltage-dependent tuning element, tuning an inductance L to the desired frequency.

across that region. The parameter C_0 may thus be expressed as

$$C_0 = \frac{\epsilon\epsilon_0 A}{d_0} \tag{3.4a}$$

A calculation given in Appendix A shows that if the acceptor concentration N_a is constant in the p-region and the donor concentration N_d is constant in the n-region, then

$$d = \left[\frac{2\epsilon\epsilon_0(V_{dif} - V_D)(N_a + N_d)}{N_a N_d}\right]^{1/2} = d_0(V_{dif} - V_D)^{1/2} \tag{3.5}$$

where

$$d_0 = \left[\frac{2\epsilon\epsilon_0(N_a + N_d)}{eN_a N_d}\right]^{1/2} \tag{3.5a}$$

For a p^+-n diode $N_a \gg N_d$, so that d_0 reduces to

$$d_0 = \left(\frac{2\epsilon\epsilon_0}{eN_d}\right)^{1/2} \tag{3.5b}$$

The value of d_0 is of the order of a few tenths of 1 μm (10^{-4} cm). Furthermore, it can be shown that if d_p and d_n are the widths of the p-part and the n-part of the space-charge region, respectively, then

$$d_p = d\frac{N_d}{N_a + N_d}; \qquad d_n = d\frac{N_a}{N_a + N_d} \tag{3.5c}$$

so that the space-charge region extends mainly into the region of the smallest impurity concentration.

We shall illustrate this with a few examples:

EXAMPLE 1: Silicon has $\epsilon = 12$. Find d_0 in a p^+-n diode with $N_d = 10^{16}/\text{cm}^3$.

ANSWER:

$$d_0 = \left(\frac{2\epsilon\epsilon_0}{eN_d}\right)^{1/2} = \left(\frac{2 \times 12 \times 8.85 \times 10^{-14}}{1.60 \times 10^{-19} \times 10^{16}}\right)^{1/2}$$
$$= 0.37 \times 10^{-4} \text{ cm} = 0.37 \ \mu\text{m}$$

EXAMPLE 2: What is the value of C_0 for Example 1 if the junction area $A = 10^{-4}$ cm^2?

ANSWER:

$$C_0 = \frac{\epsilon\epsilon_0 A}{d_0} = \frac{12 \times 8.85 \times 10^{-14} \times 10^{-4}}{0.37 \times 10^{-4}} \text{ F} = 2.8 \text{ pF}$$

The small-signal capacitance of a p-n junction diode can be used to tune a tuned circuit by adjusting the voltage V_D. This is illustrated by the following example [Fig. 3.1(b)].

EXAMPLE 3: Voltage-tune a 10^{-3}-H coil to 1 MHz by means of the voltage-dependent small-signal capacitance of a p-n junction diode of the form

$$C = \frac{100}{(0.90 - V_D)} \text{pF}$$

SOLUTION: The required value of C is

$$C = \frac{1}{\omega^2 L} = \frac{1}{4\pi^2 \times 10^{12} \times 10^{-3}} = 25 \text{ pF}$$

Therefore

$$\frac{100}{(0.90 - V_D)} = 25; \qquad 0.90 - V_D = 4; \qquad V_D = -3.10 \text{ V}$$

p-n Junction diodes usable as voltage-dependent capacitors are commercially available under the name *varactor* diodes.

3.1b. Zener Diode

If a very large negative bias ($V_D \ll 0$) is applied to a *p-n* junction, avalanche breakdown occurs in which the electrons and holes crossing the junction generate an avalanche of hole-electron pairs that are collected by the *p*- and *n*-regions, respectively. The resulting device current must be controlled by the external circuit; the voltage V_b at which breakdown occurs is very well defined and is a characteristic of the diode in question.

While every diode shows breakdown when sufficiently back-biased, special diodes, known as Zener diodes, are commercially available that have a very well-defined value of V_b. The breakdown voltage V_b can be controlled by proper choice of the impurity concentrations N_a and N_d. The symbol commonly used for a Zener diode is shown in Fig. 3.2(a).

Zener diodes are used for voltage control purposes. The circuit is shown in Fig. 3.2(b). If $V_{BB} > V_b$, the Zener diode breaks down and keeps the output voltage at V_b, independent of a possible fluctuation in V_{BB}. Figure 3.2(c) shows the characteristic and the load line.

As is seen by inspection of the circuit, the current I flowing through the

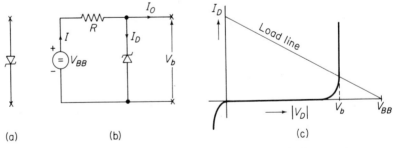

Fig. 3.2. (a) Symbol of a Zener diode. (b) Zener diode control circuit that keeps the output voltage constant. (c) Operation of the Zener diode control circuit explained with the help of the load line method.

resistance R is

$$I = \frac{V_{BB} - V_b}{R} \tag{3.6}$$

If the output current is I_0, the current $I - I_0$ flows through the diode. We must thus require $I_0 < I$. Actually, Zener diodes cease to operate properly if the device current drops below a value I_{min}, which is usually 1–2 mA. This means that the maximum current I_0 that can be drawn from the circuit is

$$I_0 = I - I_{min} \tag{3.6a}$$

If the load current I_0 is removed, the full current I flows through the diode. The diode must thus be able to dissipate the power $P_{max} = IV_b$.

3.1c. Small-Signal Impedance of the Forward-Biased Diode

For $I_D \gg I_{rs}$, the forward-biased diode ($V_D > 0$) has a characteristic

$$I_D = I_{rs} \exp\left(\frac{eV_D}{kT}\right) \tag{3.7}$$

A small change ΔV_D in applied voltage produces a corresponding change ΔI_D in current:

$$\Delta I_D = I_{rs} \exp\left[\frac{e(V_{D0} + \Delta V_D)}{kT}\right] - I_{rs} \exp\left(\frac{eV_{D0}}{kT}\right) = I_D\left[\exp\left(\frac{e\,\Delta V_D}{kT}\right) - 1\right] \tag{3.8}$$

If now $\Delta V_D \ll kT/e$, the last form in brackets may be written as $e\,\Delta V_D/kT$, since $\exp(x) - 1 \simeq x$ for small x. Consequently

$$\Delta I_D = g_d \,\Delta V_D = \frac{eI_D}{kT}\,\Delta V_D \tag{3.8a}$$

The parameter g_d thus represents the small-signal conductance of the diode. Instead of g_d we may also speak of the small-signal resistance $R_d = 1/g_d$. Consequently

$$g_d = \frac{eI_D}{kT} = \frac{dI_D}{dV_D}; \qquad R_d = \frac{kT}{eI_D} = \frac{dV_D}{dI_D} \tag{3.8b}$$

Hence the junction diode behaves as a current-controlled resistor. Since $kT/e = 25.8$ mV at $T = 300°$K,

$$R_d = \frac{25.8}{I_D}\,\Omega \tag{3.8c}$$

where I_d is in milliamperes.

The expression for the small-signal resistance holds only if $(e\,\Delta V_D)/kT < 1$, so that the signal must be less than about 10 mV. That limits the usefulness of this current-controlled resistor.

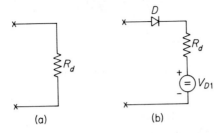

Fig. 3.3. (a) Small-signal equivalent circuit of a diode. (b) Full equivalent circuit of a diode near its quiescent operating point.

Figure 3.3(a) shows the ac equivalent circuit of the diode. If a diode has an operating point (I_{D0}, V_{D0}), then for voltages V_D close to V_{D0} the diode can be represented by the equivalent circuit shown in Fig. 3.3(b), where

$$V_{D1} = V_{D0} - I_{D0}R_d \qquad (3.8d)$$

The circuit is useful only for a very limited voltage range for which $|V_D - V_{D0}| < 10$ mV, since $I_{D0}R_d = kT/e = 25.8$ mV.

3.1d. Diode-Type Voltage Regulator Circuit

Figure 3.4(a) shows a diode-type voltage regulator circuit that is sometimes used in integrated circuits to stabilize voltages. Here V_1 is an external voltage applied to the integrated circuit. The control circuit consists of a resistor R in series with m diodes (in our example, $m = 6$). Consequently the current

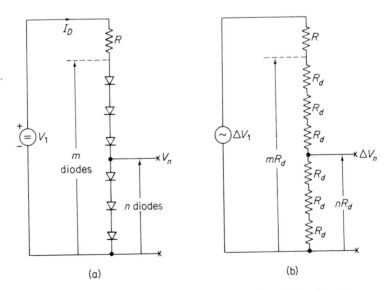

Fig. 3.4. (a) Integrated diode voltage control circuit keeping the output voltage equal to an integral number of times the turn-on voltage of each diode. (b) Small signal equivalent circuit of (a).

I_D flowing in the circuit, since each diode has a voltage of about 0.70 V, is

$$I_D = \frac{V_1 - 0.70m}{R} \tag{3.9}$$

If n diodes in series are used to supply the controlled voltage (in our example, $n = 3$) the controlled voltage is

$$V_n = 0.70n \text{ V} \tag{3.10}$$

If now V_1 shows a small fluctuation ΔV_1, then the corresponding fluctuation ΔV_n in V_n is found by treating the diode as a small-signal resistance $R_d = kT/eI_D$ [Fig. 3.4(b)]. Consequently

$$\Delta V_n = \Delta V_1 \frac{nR_d}{R + mR_d} \tag{3.11}$$

EXAMPLE: If $V_1 = 10$ V, $m = 6$ and $n = 4$, $R = 2000\ \Omega$, and $\Delta V_1 = 1.0$ V, find ΔV_n.

ANSWER: $I_D = (10 - 4.2)/2000 = 2.9$ mA; $R_d = 25.8/I_D = 9\ \Omega$. Thus

$$\Delta V_n = 1.0 \times \frac{4 \times 9}{2000 + 6 \times 9} = 18 \text{ mV}$$

This is small enough to make the use of R_d feasible.

3.2. PULSE RESPONSE OF THE p-n JUNCTION DIODE

3.2a. Small-Signal Admittance of the Diode at High Frequencies

At higher frequencies the small-signal impedance of the diode becomes complex. It is found that the equivalent circuit can be represented by a small-signal capacitance C_d in parallel with the small-signal resistance R_d (Fig. 3.5). The capacitance C_d is usually much larger than the small-signal capacitance C of the space-charge region; it is caused by the storage of holes in the n-region in a p^+-n diode (or by the storage of electrons in the p-region in an n^+-p diode).

This can be understood as follows. The holes injected into the n-region of a p^+-n diode are removed by recombination with electrons. One can now introduce a lifetime τ for the injected holes; as a consequence the charge stored in the n-region is finite. Let it be denoted by Q_s; then the steady-state recombination current is Q_s/τ. However since all the injected holes will

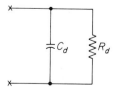

Fig. 3.5. High frequency equivalent circuit of a diode.

recombine eventually, this must be equal to the steady-state current I_D, so that

$$I_D = \frac{Q_s}{\tau} \quad \text{or} \quad Q_s = I_D\tau \tag{3.12}$$

The small-signal storage capacitance C_s is given by

$$C_s = \frac{dQ_s}{dV_D} = \tau\frac{dI_D}{dV_D} = \frac{\tau}{R_d} \quad \text{or} \quad C_s R_d = \tau \tag{3.13}$$

The product $C_s R_d$ is thus independent of the current and equal to the carrier lifetime τ. If we equate C_s to C_d, we have

$$C_d R_d = \tau \tag{3.13a}$$

independent of the current*.

3.2b. Transient Behavior of the *p-n* Junction Diode

Figure 3.6(a) shows a circuit for the testing of the pulse response of diodes. At $t = 0$ the voltage $V(t)$ is raised from zero to V_1; we assume for the sake of simplicity that $V_1 \gg 0.70$ V, the turn-on voltage of the diode. It is then found that $V_D(t)$ rises very rapidly, in a time equal to about 0.1τ, to the final value, which is of the order of 0.70 V. This is the *turn-on transient* [Fig. 3.6(b)]. If, on the other hand, the circuit has been on for a long time with $V(t)$ at V_1 and $V(t)$ is turned off at $t = 0$, the output voltage $V_D(t)$ drops almost linearly

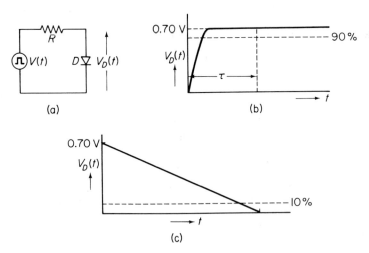

(a)   (b)

(c)

Fig. 3.6. (a) Pulse voltage $V(t)$ applied to a diode. (b) Turn-on transient. (c) Turn-off transient.

*Actually, for reasons that cannot be explained here, $C_d = \frac{1}{2}C_s$. Due to this, all expressions $\exp(-t/\tau)$ in Section 3.2b become $\exp(-2t/\tau)$; the transients are thus speeded up by a factor of two by this effect.

from about 0.70 V to 0 V in a time large in comparison with τ [Fig. 3.6(c)]. This is the *turn-off transient*. These observations require an explanation.

If we try to write a differential equation for $V_D(t)$, we run into considerable difficulties, since the diode characteristic is so strongly nonlinear. We must therefore search for a variable that changes the problem into a linear one. We saw in the previous section that the stored hole charge Q_s in a p^+-n diode varied linearly with the current I_D. It thus seems worthwhile to investigate whether $Q_s(t)$ is a suitable variable.

If a current $I(t)$ is passed through a p^+-n diode, the part dQ_s/dt of this current is used to provide for the buildup of the stored charge $Q_s(t)$ and the part Q_s/τ gives the recombination current. The time dependence of the stored charge should thus be given by the differential equation

$$\frac{dQ_s}{dt} + \frac{Q_s}{\tau} = I(t) \tag{3.14}$$

which is a simple extension of Eq. (3.12). The model introduced here is called the *charge control model* of the diode. We see that the differential equation is linear, as required.

We shall now solve this equation for the turn-on transient. Since the circuit is turned on at $t = 0$, we have as the initial condition $Q_s(t) = 0$ at $t = 0$. Next we evaluate $I(t)$. Assuming that $V_1 \gg 0.70$ V, we have

$$I(t) = \frac{V_1 - V_D(t)}{R} \simeq \frac{V_1}{R} = I_1 \quad \text{for } t > 0 \tag{3.15}$$

if $V_D(t) \ll V_1$. It is thus seen by inspection that the solution of Eq. (3.14) under the above initial condition can be written as

$$Q_s(t) = I_1\tau\left[1 - \exp\left(-\frac{t}{\tau}\right)\right] \tag{3.16}$$

since this solution satisfies the initial condition as well as the differential equation. $Q_s(t)$ thus rises gradually from the value $Q_s(0) = 0$ to the value $Q_s(\infty) = I_1\tau$ for $t \gg \tau$. Compare also Appendix B.

We now solve for the turn-off transient. We assume that $V(t) = V_1$ has been on for $t < 0$ and is turned off at $t = 0$. Then the initial condition is

$$Q_s(t) = I_1\tau \quad \text{for } t = 0 \tag{3.17}$$

If we neglect the small current $-V_D(t)/R$ flowing in the external circuit, Eq. (3.14) becomes

$$\frac{dQ_s}{dt} + \frac{Q_s}{\tau} = 0 \tag{3.17a}$$

The solution of this equation is

$$Q_s(t) = I_1\tau \exp\left(-\frac{t}{\tau}\right) \tag{3.18}$$

because it satisfies Eq. (3.17a) and (3.17). Compare also Appendix B.

There are several ways of speeding up the response. One way consists of reducing the lifetime τ of the injected carriers. Diodes with a very short lifetime are commercially available or can be made by connecting a transistor as a diode (Section 7.1 d).

There is also a circuit arrangement that speeds up the response. It consists of connecting a capacitance C in parallel with the resistance R and choosing C such that

$$CR = C_d R_d = \tau \tag{3.19}$$

(Fig. 3.7). This speeds up the response because the circuit acts as a simple voltage divider, which is most easily seen from the frequency response of the circuit:

$$V_D = V \frac{R_d/(1 + j\omega C_d R_d)}{R/(1 + j\omega CR) + R_d/(1 + j\omega C_d R_d)} = V \frac{R_d}{R + R_d} \tag{3.19a}$$

because $CR = C_d R_d = \tau$. Since this is independent of frequency, $V_D(t)$ follows $V(t)$ instantaneously.

Knowing the stored charge $Q_s(t)$, we can evaluate the diode voltage $V_D(t)$ as follows. Since $I_D = I_{rs} \exp(eV_D/kT)$, we have, if we multiply by τ and equate $Q_{rs} = I_{rs}\tau$,

$$Q_s = Q_{rs} \exp\left(\frac{eV_D}{kT}\right); \qquad \exp\left(\frac{eV_D}{kT}\right) = \frac{Q_s}{Q_{rs}} \tag{3.20}$$

or

$$V_D(t) = \frac{kT}{e} \ln\left[\frac{Q_s(t)}{Q_{rs}}\right] \tag{3.21}$$

This relationship holds as long as $Q_s(t) \gg Q_{rs}$.

First, we shall consider the turn-on transient. Substituting Eq. (3.16) into (3.21), we obtain

$$V_D(t) = \frac{kT}{e} \left\{\ln\left(\frac{I_1}{I_{rs}}\right) + \ln\left[1 - \exp\left(-\frac{t}{\tau}\right)\right]\right\} \tag{3.22}$$

We now observe that $(kT/e) \ln(I_1/I_{rs})$ is of the order of 0.70 V and that $(kT/e) \ln[1 - \exp(-t/\tau)]$ is of the order of -0.06 for $t/\tau = 0.1$.* Therefore

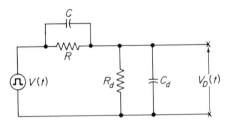

Fig. 3.7. Pulsed diode circuit to which a capacitor C is added to speed up the response.

*Here we have made use of the fact that $1 - \exp(-x) \simeq x$ for small x and have substituted $kT/e = 25.8$ mV at $T = 300°K$.

the diode voltage $V_D(t)$ rises very quickly, in a time of the order of 0.1τ, to 90% of the final value, in agreement with Fig. 3.6(b).

Next we shall consider the turn-off transient. Substituting Eq. (3.18) into (3.21) and bearing in mind that the latter equation is applicable only if $Q_s(t) \gg Q_{rs}$, we have

$$V_D(t) = \frac{kT}{e}\left[\ln\left(\frac{I_1}{I_{rs}}\right) - \frac{t}{\tau}\right] \tag{3.23}$$

holding for $t/\tau < \ln(I_1/I_{rs})$. This decreases linearly from the initial value of about 0.70 V and reaches zero at $t = \tau \ln(I_1/I_{rs})$. Since $I_1/I_{rs} \simeq 10^{12}$ ($I_1 \simeq$ 1 mA; $I_{rs} \simeq 10^{-12}$ mA), the turn-off transient takes a time much larger than τ, in agreement with Fig. 3.6(c).

Actually, the current $-V_D(t)/R$ cannot be ignored; it has the tendency of speeding up the response considerably. Nevertheless, it remains true that $V_D(t)$ varies approximately linearly with time, with a sharp drop-off to zero at the end, and that the turn-off transient lasts a time larger than τ.

The turn-off response can be speeded up by switching from $+V_1$ to $-V_2$ at $t = 0$ rather than from $+V_1$ to 0 V. Since $V_D(t) \simeq 0.70$ V when the circuit is turned off, we have

$$I(t) = -\frac{V_2 + V_D(t)}{R} \simeq -\frac{V_2}{R} = -I_2 \tag{3.24}$$

if $V_2 \gg V_D(t)$. Equation (3.14) thus becomes

$$\frac{dQ_s}{dt} + \frac{Q_s}{\tau} = -I_2 \qquad \text{for} \quad t > 0, \quad Q_s(t) > 0 \tag{3.25}$$

with $Q_s(t) = I_1\tau$ at $t = 0$. As is seen by inspection, the solution is

$$Q_s(t) = I_1\tau \exp\left(-\frac{t}{\tau}\right) - I_2\tau\left[1 - \exp\left(-\frac{t}{\tau}\right)\right] \tag{3.26}$$

as long as $Q_s(t) > 0$. The output voltage $V_D(t)$ is zero if $Q_s(t) \simeq Q_{rs}$, but since Q_{rs} is so small, we have in good approximation that $V_D(t)$ is zero if $Q_s(t)$ is zero, or

$$(I_1 + I_2)\tau \exp\left(-\frac{t}{\tau}\right) - I_2\tau = 0 \tag{3.27}$$

so that

$$\exp\left(\frac{t}{\tau}\right) = \frac{I_1 + I_2}{I_2}, \qquad t = \tau \ln\left(\frac{I_1 + I_2}{I_2}\right) \tag{3.28}$$

If I_2 is of the order of I_1, the turn-on time is reduced to a value of about τ. This speedup in the turn-off response comes about because the current I_2 removes the stored charge more rapidly.

PROBLEMS

3.1. In a silicon p^+-n junction ($N_a \gg N_d$), $V_{\text{dif}} = 0.80$ V and $N_d = 10^{15}/\text{cm}^3$. Find d_0 and the width d of the space-charge region as a function of the applied voltage V_D. Silicon has $\epsilon = 12$.

3.2. For the diode of Problem 3.1, find the capacitance as a function of the applied voltage V_D for a junction area of 10^{-3} cm^2.

3.3. A silicon diode that is especially designed has a capacitance

$$C = \frac{270}{0.90 - V_D}\,\text{pF}$$

It is connected in parallel to a tuned circuit that has an external parallel capacitance of 20 pF. The maximum value of V_D is 0.00 V. (a) Design the tuning coil so that the tuned circuit is tuned at 550 kHz at $V_D = 0.00$ V. (b) Find the voltage V_D for which the circuit is tuned at 1600 kHz.

3.4. A Zener diode having a breakdown voltage of 10 V is used in the voltage control circuit of Fig. 3.2(b). The Zener diode operates properly if the diode current is larger than 1.0 mA. (a) If V_{BB} varies between 25 and 35 V, choose R such that the load current I_0 is 10 mA under worst-case conditions. (b) If the load current is suddenly removed, find the dissipation in the Zener diode under worst-case conditions. (c) If V_{BB} is now kept constant at 30 V, find the range of load current I_0 for which the Zener diode stays in the range of allowed operations if R is as in part (a).

3.5. A silicon p-n diode at $T = 300°$K, having $I_D = 1.0$ mA at $V_D = V_{D0} = 0.70$ V, is used as a small-signal resistor. Find the range of V_D values for which the small-signal resistance varies between 50 and 5000 Ω.

3.6. A silicon diode voltage regulator circuit as shown in Fig. 3.4(a) is used to make a stabilized voltage of about 3.50 V. If $V_1 = 12.50$ V, find (a) the resistance R_d so that the diode current is 3.0 mA when no current is drawn from the voltage regulator—you may assume that $V_D \simeq 0.70$ V for each diode—how many diodes are needed? (b) the actual output voltage when no current is drawn from the circuit if each diode has a diode voltage of 0.70 V at $I_D = 1$ mA, (c) the voltage developed across the diodes if the voltage regulator supplies a current of 2.7 mA to a load, and (d) the voltage ripple in part (b) if V_D varies between 12.0 and 13.0 V for zero load current.

3.7. Find the storage capacitance in a silicon diode carrying a diode current of 1.0 mA if the carrier lifetime is 10^{-7} s.

3.8. In the circuit of Fig. 3.7 the diode has a lifetime of 10^{-7} s. The resistance R in series with the diode is 2000 Ω. Find the capacitance C that must be put in parallel with R to eliminate the turn-on transient.

3.9. In the circuit of Fig. 3.6 the silicon diode has $\tau = 10^{-7}$ s and $I_D = I_{D0} = 1.0$ mA at $V_D = V_{D0} = 0.70$ V. (a) If a pulse of 6.0 V is applied to the diode, find the resistance R for which the maximum current is 1.0 mA. (b) Find the time interval for which the turn-on voltage transient has risen from 0 to 90% of the final value.

3.10. (a) In the circuit of Problem 3.9 (Fig. 3.6), $V(t)$ switches from $+6.0$ to -6.0 V. Calculate I_1 and I_2 accurately by not ignoring V_D, and then calculate the turn-off transient. (b) Repeat the calculation if $V(t)$ switches from 6.0 to 0 V.

3.11. In the diode of Problem 3.9, $V(t)$ switches from 0 to 6.0 V. The diode starts drawing current for $V_D(t) \simeq 0.70$ V. Calculate the turn-on delay time of the diode if the input capacitance C is 5.0 pF when the diode is not drawing current. The turn-on delay time is defined as the time it takes for $V_D(t)$ to reach the value of 0.70 V.

4

Linear
Amplifiers

Before discussing linear amplifiers involving transistors and FETs, it seems appropriate to give the general framework into which these individual amplifiers can be fitted. To that end we introduce the concepts of voltage gain, current gain, power gain, upper cutoff frequency, lower cutoff frequency, and the like.

An amplifier is strictly linear only for small signals. The general approach given in this chapter, insofar as it applies to amplifiers, is therefore a *small-signal approach*. If the signals are larger, nonlinear effects, such as second and higher harmonic distortion, become important. Such complications should be avoided. Several techniques are available that can reduce the distortion; the most important one is negative feedback. These effects and their reduction are discussed in subsequent chapters of the book.

It would also be worthwhile to discuss the general framework of large-signal applications, especially digital applications, but this can be done properly only after the characteristics of the devices have been discussed in detail.

4.1. LINEAR AMPLIFIER

A simple passive, linear two-port network can be represented either as a π-network [Fig. 4.1(a)] or as a T-network [Fig. 4.1(b)] without any controlled sources, i.e., sources that depend linearly on the input voltage v_i or on the input current i_i.

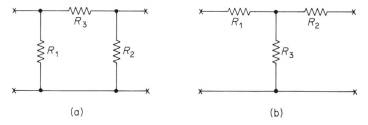

Fig. 4.1. (a) Equivalent π-network of a passive resistive circuit. (b) Equivalent T-network of a passive resistive circuit.

A linear amplifier can also be represented as a π- or T-network, but in this case one must either add a controlled current source i_0 in parallel with the output or a controlled voltage source e_0 in series with the output. The simplest linear amplifier thus consists *only* of such a controlled source. This is called an *ideal linear amplifier*. In it the other elements of the π- or T-network are ignored.

Actual amplifiers and ideal amplifiers differ in that in the former the effects of some of the elements of the equivalent π- or T-network cannot be ignored.

4.1a. Ideal Amplifier

There are four basic forms of the ideal amplifier:

1. The controlled current source i_0 is proportional to the input current i_i [Fig. 4.2(a)]:

$$i_0 = \beta i_i \tag{4.1}$$

   The factor β is called the *current amplification factor of the device* and the amplifier is called an *ideal current amplifier*. For proper amplification it is necessary that $\beta > 1$. We shall see that this representation is useful in transistors.

2. The controlled current source i_0 is proportional to the input voltage v_i [Fig. 4.2(b)]:

$$i_0 = g_m v_i \tag{4.2}$$

   The factor g_m has the dimension of a conductance and is called the *transconductance* of the amplifier. We shall see that this representation is useful for both transistors and FETs.

3. The controlled voltage source e_0 is proportional to the input current i_i [Fig. 4.2(c)]:

$$e_0 = R_m i_i \tag{4.3}$$

   The factor R_m has the dimension of a resistance and is called the

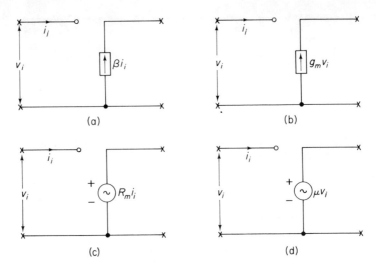

Fig. 4.2. Four basic amplifier diagrams with a controlled source at the output. (a) Controlled current source proportional to the input current (current amplifier). (b) Controlled current source proportional to the input voltage. (c) Controlled EMF proportional to the input current. (d) Controlled EMF proportional to the input voltage (voltage amplifier).

transresistance of the amplifier. This representation was once used in discussing transistor circuits, but this is no longer the case.

4. The controlled voltage source e_0 is proportional to the input voltage v_i [Fig. 4.2(d)]:

$$e_0 = \mu v_i \tag{4.4}$$

The factor μ is called the *voltage amplification factor* of the amplifier and the amplifier is called an *ideal voltage amplifier*. For proper amplification it is necessary that $\mu > 1$. We shall see that this representation is useful in some transistor and FET circuits.

Often the output of one amplifier is connected to the input of the next amplifier and so forth. This is called the *cascade* connection.

Figure 4.3(a) shows a cascade connection of three voltage amplifiers with voltage amplification factors μ_1, μ_2, and μ_3. It is easily seen by inspection that the open-circuit output voltage v_4 is given by

$$v_4 = \mu_1 \mu_2 \mu_3 v_1 \tag{4.5}$$

so that a very large *voltage gain* v_4/v_1 is obtained if all μs are much larger than unity. As a matter of fact, one can obtain an arbitrary voltage gain by connecting a sufficient number of voltage amplifiers in cascade.

Figure 4.3(b) shows a cascade connection of three current amplifiers

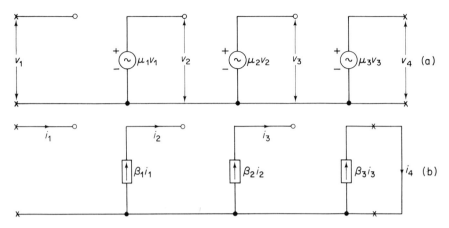

Fig. 4.3. (a) Several voltage amplifiers connected in cascade. (b) Several current amplifiers connected in cascade.

with current amplification factors β_1, β_2, and β_3. It is easily seen by inspection that the short-circuit output current is given by

$$i_4 = \beta_1 \beta_2 \beta_3 i_1 \qquad (4.5a)$$

so that a very large *current gain* i_4/i_1 is obtained if all the βs are much larger than unity. As a matter of fact, one can obtain an arbitrary current gain by connecting a sufficient number of current amplifiers in cascade.

It is common practice to express the voltage gain of a voltage amplifier in *decibels* (dB) with the help of the following definition:

$$\text{voltage gain in dB} = 10\left(10\log\left|\frac{v_{\text{out}}}{v_{\text{in}}}\right|^2\right) = 20\left(10\log\left|\frac{v_{\text{out}}}{v_{\text{in}}}\right|\right) \qquad (4.6)$$

where v_{in} is the input voltage amplitude and v_{out} the output voltage amplitude.

The advantage of using decibels is that the voltage gains of a number of stages in cascade become additive instead of multiplicative. For example, if we have n identical voltage amplifier stages in cascade, each having a voltage gain of G_{v1} dB, then the voltage gain of the complete amplifier in decibels is

$$G_{vn} = nG_{v1} \qquad (4.6a)$$

EXAMPLE 1: How many amplifier stages, each having a voltage gain of 20 dB, are needed to obtain a total voltage gain of 120 dB?

ANSWER: $n = 120/20 = 6$.

In the same way the current gain of a current amplifier is expressed in decibels with the help of the following definition:

$$\text{current gain in dB} = 10\left(10\log\left|\frac{i_{\text{out}}}{i_{\text{in}}}\right|^2\right) = 20\left(10\log\left|\frac{i_{\text{out}}}{i_{\text{in}}}\right|\right) \qquad (4.6b)$$

where i_{in} is the input current amplitude and i_{out} the output current amplitude.

4.1b. Actual Amplifiers

Actual amplifiers differ from ideal ones in that they have more network elements in their equivalent circuits. For example, amplifiers have an input resistance R_i for short-circuited output. This is most easily seen in Fig. 4.2(a) and (c), where the current i_i is associated with the voltage v_i; in these cases $R_i = v_i/i_i$. Also, amplifiers have an output resistance R_0. This is most easily seen from Fig. 4.2(c) and (d), for if there were no resistance in series with the EMF e_0, the short-circuit output current would be infinite.

Finally, there is often a feedback admittance between output and input; this feedback can be removed by neutralization. We shall generally ignore it, except in Section 4.4, where the effect of a feedback capacitance is explored.

If we ignore the feedback effect, the four amplifier circuits of Fig. 4.2(a)–(d) must be extended to the circuits shown in Fig. 4.4(a)–(d). Not only do the new circuits give a better description of what goes on in the amplifier, but in addition the four circuits become interchangeable.

For example, if we compare Fig. 4.4(a) and (b), we see that

$$\beta i_i = \frac{\beta}{R_i} v_i = g_m v_i \quad \text{or} \quad g_m = \frac{\beta}{R_i} \tag{4.7}$$

If we compare Fig. 4.4(a) and (c), we see that

$$e_0 = R_m i_i = \beta i_i R_0 \quad \text{or} \quad R_m = \beta R_0 \tag{4.8}$$

If we compare Fig. 4.4(a) and (d), we see that

$$e_0 = \mu v_i = \beta i_i R_0 = \beta \frac{R_0}{R_i} v_i \quad \text{or} \quad \mu = \beta \frac{R_0}{R_i} \tag{4.9}$$

These results are summarized as follows:

$$\mu = g_m R_0 = \beta \frac{R_0}{R_i} = \frac{R_m}{R_i} \tag{4.10}$$

$$\beta = g_m R_i = \mu \frac{R_i}{R_0} = \frac{R_m}{R_0} \tag{4.11}$$

$$g_m = \frac{\mu}{R_0} = \frac{\beta}{R_i} = \frac{R_m}{R_0 R_i} \tag{4.12}$$

$$R_m = g_m R_0 R_i = \mu R_i = \beta R_0 \tag{4.13}$$

Circuit 4.4(b) is generally applicable and makes it possible to describe FET and transistor circuits by the same formalism—hence its popularity. Circuit 4.4(a) is of value in those transistor circuits where the current gain aspect of the device is utilized. Circuit 4.4(d) is of value in those FET or transistor circuits where the voltage gain aspect of the device is utilized.

Usually one requires an actual voltage gain or $\mu > 1$. However, a circuit with $\mu = 1$ can still be useful. Figure 4.5 shows such a circuit; it has a high input resistance $R_i = 10^6 \ \Omega$ and a low output resistance $R_0 = 100 \ \Omega$.

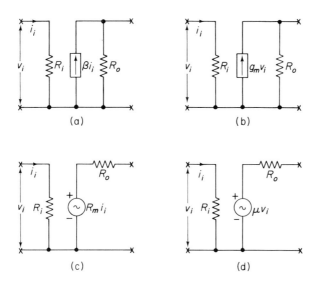

Fig. 4.4. The four basic amplifier diagrams with input and output resistance. (a) Controlled current source proportional to the input current (current amplifier). (b) Controlled current source proportional to the input voltage. (c) Controlled EMF proportional to the input current. (d) Controlled EMF proportional to the input voltage (voltage amplifier).

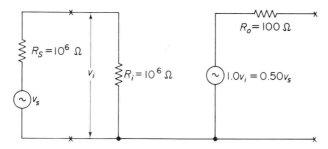

Fig. 4.5. Circuit transforming from a high to a low impedance level.

The circuit is connected to a signal source with an internal resistance $R_s = 10^6 \, \Omega$. The circuit transforms the signal v_s at the impedance level of $10^6 \, \Omega$ to a signal $0.50v_s$ in series with $100 \, \Omega$. This could never have been achieved by any passive network such as a transformer. We shall see that this circuit corresponds to the FET source follower.

We shall now investigate how the actual and the ideal amplifiers compare. First, the voltage amplifier is considered [Fig. 4.6(a)]. As is seen by

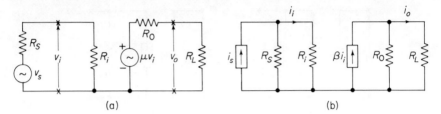

Fig. 4.6. (a) The voltage amplifier of Fig. 4.4(d) with signal source and output load. (b) The current amplifier of Fig. 4.4(a) with signal source and output load.

inspection,

$$v_i = v_s \frac{R_i}{R_s + R_i}; \qquad v_0 = \mu v_i \frac{R_L}{R_0 + R_L} = \mu \frac{R_i}{R_s + R_i} \frac{R_L}{R_0 + R_L} v_s \qquad (4.14)$$

so that the voltage gain v_0/v_s is smaller than μ. However, if $R_s \ll R_i$ and $R_0 \ll R_L$, the voltage gain approaches μ and the actual amplifier performance approaches that of the ideal one.

The corresponding current amplifier is shown in Fig. 4.6(b). As is seen by inspection,

$$i_i = i_s \frac{R_s}{R_s + R_i}; \qquad i_0 = \beta i_i \frac{R_0}{R_0 + R_L} = \beta \frac{R_s}{R_s + R_i} \frac{R_0}{R_0 + R_L} i_s \qquad (4.15)$$

so that the current gain i_0/i_s is smaller than β. However, if $R_s \gg R_i$ and $R_0 \gg R_L$, the current gain approaches β, so that the actual amplifier performance approaches that of the ideal one.

4.1c. Voltage Gain in Cascaded Amplifiers

Figure 4.7 shows the cascade connection of two actual amplifiers. The voltage gain of the first amplifier is

$$\frac{v_2}{v_1} = g_m R_0 \frac{R_1}{R_0 + R_1} \qquad (4.16)$$

and the voltage gain of the second amplifier is

$$\frac{v_3}{v_2} = g_m R_0 \frac{R_L}{R_0 + R_L} \qquad (4.17)$$

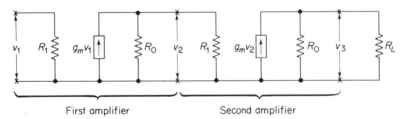

First amplifier   Second amplifier

Fig. 4.7. Two amplifier stages connected in cascade.

so that the voltage gain of the combination is

$$\frac{v_3}{v_1} = \frac{v_3}{v_2}\frac{v_2}{v_1} = (g_m R_0)^2 \frac{R_1}{R_0 + R_1}\frac{R_L}{R_0 + R_L} \qquad (4.18)$$

In a similar way, for n amplifiers in cascade, in which the output is connected to a load resistance R_L, we have

$$\frac{v_{n+1}}{v_1} = (g_m R_0)^n \left(\frac{R_1}{R_0 + R_1}\right)^{n-1} \frac{R_L}{R_0 + R_L} \qquad (4.19)$$

Note that $g_m R_0$ equals the amplification factor μ of each amplifier and that the total voltage gain approaches μ^n for $R_1 \gg R_0$ and $R_L \gg R_0$.

4.2. POWER CONSIDERATIONS

Sometimes it is more convenient to look at an amplifier from a power rather than a voltage point of view, because an amplifier can have a voltage gain but a power loss. It would then have been better to match the signal source directly to the load by means of a matching transformer.

4.2a. Matching; Available Power

Figure 4.8(a) shows a signal source of EMF v_s and internal resistance R_s connected to a variable load resistance R_L. The voltage developed across R_L is

$$v_L = v_s \frac{R_L}{R_s + R_L} \qquad (4.20)$$

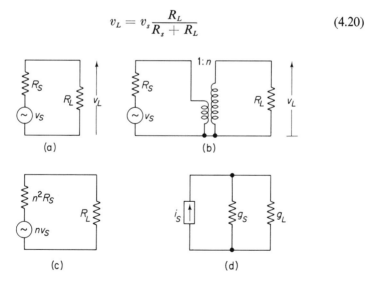

Fig. 4.8. (a) Signal source directly connected to a load resistance. (b) Signal source connected to a load resistance by means of a matching transformer of turns-ratio n. (c) Equivalent circuit of (b). (d) Current source connected to a load conductance.

so that the power fed into R_L is

$$P = \frac{1}{2}\frac{|v_L|^2}{R_L} = \frac{1}{2}|v_s|^2\frac{R_L}{(R_s + R_L)^2} \qquad (4.21)$$

Considered as a function of R_L, this has a maximum value if $R_L = R_s$. The load is then said to be *matched* to the source and the power thus obtained is called the *available power*, P_{av}, of the source:

$$P_{av} = \frac{1}{8}\frac{|v_s|^2}{R_s} \qquad (4.22)$$

Often one must deal with a *fixed* source and a *fixed* load. Matching can then be obtained by an ideal (= lossless) transformer of turns ratio n [Fig. 4.8(b)] such that

$$n^2 R_s = R_L \qquad (4.23)$$

(matching condition). This, in turn, can be represented by the equivalent circuit of Fig. 4.8(c). The available power is now

$$P_{av} = \frac{1}{8}\frac{|nv_s|^2}{n^2 R_s} = \frac{1}{8}\frac{|v_s|^2}{R_s} \qquad (4.24)$$

which corresponds to Eq. (4.22).

Figure 4.8(d) shows a signal source consisting of a current generator i_s and a conductance g_s in parallel to which a load conductance g_L is connected. In that case for $g_L = g_s$ we have

$$P = P_{av} = \frac{1}{8}\frac{|i_s|^2}{g_s} \qquad (4.25)$$

4.2b. Power Gain of an Amplifier

Intuitively one would define the power gain G of an amplifier as

$$G = \frac{\text{power fed into load}}{\text{power fed into input}} \qquad (4.26)$$

This gives misleading numbers, however, unless the signal source is matched to the input. For a considerable mismatch between source and input the amplifier power should be compared with what would be obtained if the source were directly matched to the load. This leads to the following definition:

$$G = \frac{\text{power fed into load}}{\text{power available at source}} \qquad (4.27)$$

For $G < 1$ the amplifier could better be replaced by a lossless matching transformer, whereas for $G > 1$ the amplifier is useful, even if badly mismatched at the source.

As an example, consider the amplifier of Fig. 4.9. As is seen by inspec-

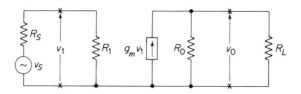

Fig. 4.9. Circuit of Fig. 4.4(b) connected to a source and an output load.

tion,

$$v_1 = v_s \frac{R_1}{R_s + R_1}; \qquad v_0 = g_m v_1 \frac{R_0 R_L}{R_0 + R_L} \qquad (4.28)$$

$$P_{av} = \frac{1}{8} \frac{|v_s|^2}{R_s}; \qquad P_{load} = \frac{1}{2} \frac{|v_0|^2}{R_L} \qquad (4.29)$$

so that

$$G = \frac{P_{load}}{P_{av}} = 4 \frac{R_s}{R_L} \left| \frac{v_0}{v_s} \right|^2 = 4 g_m^2 R_s R_L \left(\frac{R_1}{R_s + R_1} \right)^2 \left(\frac{R_0}{R_0 + R_L} \right)^2 \qquad (4.30)$$

If the load is matched to the output ($R_L = R_0$), the power gain is called the *available power gain*, denoted by G_{av}:

$$G = G_{av} = g_m^2 R_0 R_s \left(\frac{R_1}{R_s + R_1} \right)^2 \qquad (4.31)$$

If, in addition, the source is matched to the input ($R_s = R_1$), the power gain is called the *maximum power gain*, denoted by G_{max}:

$$G_{max} = \frac{1}{4} g_m^2 R_0 R_1 = \frac{1}{4} \mu \beta = \frac{1}{4} \beta^2 \frac{R_0}{R_1} \qquad (4.31a)$$

If $G_{max} < 1$, the circuit is not useful under any condition.

It is common practice to express the power gain of an amplifier in decibels with the help of the following definition:

$$\text{power gain in dB} = 10 \,^{10}\!\log \frac{P_{load}}{P_{av}} \qquad (4.32)$$

where P_{load} is the power fed into the load and P_{av} the power available at the source.

It is easily seen that the power gain in decibels is related to the voltage gain in decibels, for let G_p be the power gain in decibels and G_v the voltage gain. Then

$$G_p = 10 \left(^{10}\!\log \frac{P_{load}}{P_{av}} \right) = 10 \left\{ ^{10}\!\log \left[4 \frac{R_s}{R_L} \left| \frac{v_0}{v_s} \right|^2 \right] \right\} = G_v + 10 \left[^{10}\!\log \left(\frac{4R_s}{R_L} \right) \right]$$

$$(4.32a)$$

where v_0 is the output voltage, v_s the source voltage, R_L the load impedance, and R_s the source impedance.

EXAMPLE: To illustrate the concept of power gain, we shall compare two amplifiers each with the load matched to the output ($R_0 = R_L$). One amplifier has a bad mismatch at the input, whereas the second amplifier is matched at the input.

For both amplifiers $R_s = 2 \times 10^8\ \Omega$ and $R_L = 10^4\ \Omega$. Because of the matching condition at the output, $R_0 = R_L$ in each case. Both amplifiers have the same transconductance $g_m = 10^{-2}$ mho, but the first amplifier has $R_1 = 10^4\ \Omega$, whereas the second has $R_1 = 2 \times 10^8\ \Omega$ ($R_1 = R_s$).

Substituting into Eq. (4.31), we have for the first amplifier

$$G = G_{av} = 10^{-4} \times 10^4 \times 2 \times 10^8 \left(\frac{10^4}{2 \times 10^8}\right)^2 = \frac{1}{2}$$

so that it is not useful. For the second amplifier we find

$$G_{av} = G_{max} = \tfrac{1}{4} \times 10^{-4} \times 10^4 \times 2 \times 10^8 = \tfrac{1}{2} \times 10^8$$

which is very large.

This is not an artificial example, for the first amplifier corresponds to a transistor amplifier and the second to an FET amplifier. It was chosen to demonstrate the disastrous effect of a bad mismatch at the input and the importance of matching source and input.

4.3. FREQUENCY RESPONSE OF LOW-FREQUENCY AMPLIFIERS

The frequency responses of FET and transistor low-frequency amplifiers follow the same general pattern, and for that reason this pattern is now discussed. We do so for a cascade connection of several amplifier stages, each containing an active element (Fig. 4.10).

First, one wants to separate the output of the one stage from the input of the next stage dc-wise to allow greater flexibility in adjusting the bias of the two stages; to that end the two stages are connected by means of a coupling capacitor C_b. Then one wants to bias the output of the one stage and the input of the next stage properly. This is represented schematically by a supply voltage V_{B1} in series with a resistance R_L (load resistance) and a supply voltage V_{B2} in series with the resistance R_b (bias resistance), respectively. The active

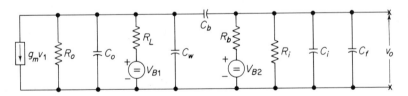

Fig. 4.10. Full interstage network of a low-frequency amplifier.

elements used in the stages have an output resistance R_0 and an input resistance R_i, respectively. The signal transfer properties of the device are represented by the current generator $g_m v_1$, where v_1 is the input voltage of the first stage and g_m the transconductance of the active element.

At high frequencies one must take into account that there are some capacitances associated with the circuit. Generally there is an output capacitance C_0 in parallel with the output resistance R_0 of the one stage and an input capacitance C_i in parallel with the input resistance R_i of the next stage. In addition there may be a Miller effect capacitance C_f in parallel with C_i (see Section 4.4). Finally, there is some wiring capacitance C_w associated with R_L and R_b.

The capacitances affect the frequency response of the circuit and influence the pulse response. We shall now consider these effects in some detail.

4.3a. Midband and High-Frequency Response

If the capacitance C_b is properly chosen, it acts as a short circuit for most frequencies amplified by the cascade connection. Therefore its effect can be ignored, and the interstage network can be represented by an equivalent network, shown in Fig. 4.11(a). It consists of the current generator $g_m v_1$, a resistance R_{par}, and a capacitance C_{par} in parallel, where

$$\frac{1}{R_{par}} = \frac{1}{R_0} + \frac{1}{R_L} + \frac{1}{R_b} + \frac{1}{R_i}; \qquad C_{par} = C_0 + C_w + C_i + C_f \qquad (4.33)$$

This, in turn, can be replaced by the equivalent circuit of Fig. 4.11(b), consisting of an EMF $-g_m R_{par} v_1$ in series with the resistance R_{par} and the capacitance C_{par}.

We first turn to the ac frequency response. According to Fig. 4.11 (b),

$$v_0 = -g_m R_{par} v_1 \frac{1/(j\omega C_{par})}{R_{par} + 1/(j\omega C_{par})} = -\frac{g_m R_{par}}{1 + j\omega C_{par} R_{par}} v_1 = \frac{g_{v0}}{1 + jf/f_1} v_1$$

$$(4.34)$$

where $\omega = 2\pi f$ and

$$g_{v0} = -g_m R_{par}, \qquad 2\pi f_1 C_{par} R_{par} = 1 \qquad (4.34a)$$

The parameter g_{v0} is called the *midband gain*, and f_1 is called the *upper cutoff frequency;* it is the frequency at which the absolute value of the gain, $|v_0/v_1|$, has dropped to $1/\sqrt{2}$ times the midband value $|g_{v0}|$. We see that $|v_0/v_1|$ decreases with increasing frequency so that the capacitance C_{par} indeed sets a high-frequency limit to the gain.

We see that the product

$$g_{v0} f_1 = \frac{g_m}{2\pi C_{par}} \qquad (4.34b)$$

is independent of R_L. It is called the *gain-bandwidth product* of the amplifier stage in cascaded connection. One can thus trade gain against bandwidth.

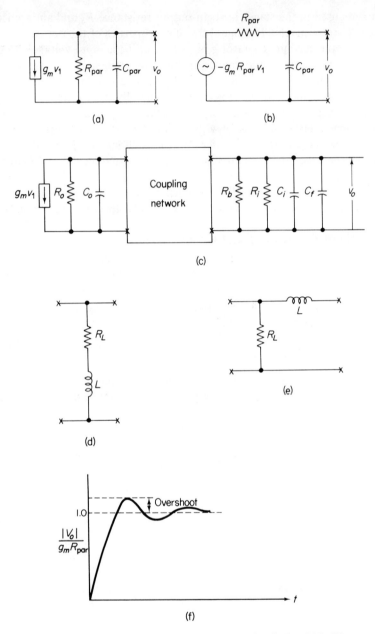

Fig. 4.11. (a) High-frequency equivalent circuit of Fig. 4.10. (b) Alternate equivalent circuit of (a). (c) General coupling network for producing improved frequency response. (d) Shunt-peaked load circuit. (e) Series-peaked load circuit. (f) Relative pulse response showing overshoot.

If we define $g_v(f) = v_0/v_1$ as the high-frequency voltage gain, we see that the relative frequency response $|g_v(f)/g_{v0}|$ has the value

$$\left| \frac{g_v(f)}{g_{v0}} \right| = \frac{1}{(1 + f^2/f_1^2)^{1/2}} \tag{4.34c}$$

If the load resistance R_L is replaced by a more complicated network containing inductances, an improved performance of the interstage network is possible in two respects:

1. The upper cutoff frequency f_1 is again defined so that

$$\left| \frac{g_v(f)}{g_{v0}} \right| = \frac{1}{2}\sqrt{2} \quad \text{at } f = f_1 \tag{4.34d}$$

and the circuit has a larger gain-bandwidth product $|g_{v0}f_1|$ than the circuit of Fig. 4.10.

2. The relative frequency response can be so designed that for $0 < f < f_1$

$$1 \geqq \left| \frac{g_v(f)}{g_{v0}} \right| > \frac{1}{(1 + f^2/f_1^2)^{1/2}} \quad \text{and} \quad \left| \frac{g_v(f)}{g_{v0}} \right| < \frac{1}{(1 + f^2/f_1^2)^{1/2}}$$

for $f > f_1$. That is, the circuit gives a flatter response for $f < f_1$ and a steeper drop for $f > f_1$ than the circuit of Fig. 4.10. The ideal frequency response

$$\left| \frac{g_v(f)}{g_{v0}} \right| = 1 \quad \text{for } f < f_1; \qquad \left| \frac{g_v(f)}{g_{v0}} \right| = 0 \quad \text{for } f > f_1 \tag{4.34e}$$

cannot be fully achieved, however.

It is shown in Chapter 6 that a considerable improvement is possible by proper choice of the network replacing R_L. Figure 4.11(c) shows the schematic diagram of such a circuit. Figure 4.11(d) and (e) show two particular simple cases. In Fig. 4.11(d) a suitably chosen inductance is put in series with R_L; this is called *shunt peaking*. In Fig. 4.11(e) a suitably chosen inductance is put in series with the input of the next stage; this is called *series peaking*. Both circuits can give an improved frequency response, as will be seen in Chapters 6 and 8.

We shall now investigate the pulse response by replacing v_1 by the unit step function $u(t)$. This yields (see Appendix B)

$$v_0(t) = -g_m R_{par} \left[1 - \exp\left(-\frac{t}{\tau_1}\right) \right] \tag{4.35}$$

where $\tau_1 = C_{par} R_{par}$ is the time constant of the RC-circuit. We note that τ_1 is related to f_1; as a matter of fact,

$$2\pi\tau_1 = \frac{1}{f_1} \tag{4.35a}$$

We also note that

$$|g_{vo}|f_1 = \frac{g_m}{2\pi C_{\text{par}}} = \frac{|g_{vo}|}{2\pi\tau_1} \tag{4.36}$$

so that an improved high-frequency response (high value of f_1) and an improved pulse response (short value of τ_1) can be obtained by sacrificing gain.

The rise time of the pulse response can be defined as the time it takes for the response to reach 90% of its maximum value. According to Eq. (4.35),

$$1 - \exp\left(-\frac{t}{\tau_1}\right) = 0.90; \quad \exp\left(-\frac{t}{\tau_1}\right) = 0.10; \quad \exp\left(\frac{t}{\tau_1}\right) = 10$$

so that

$$t = t_r = \tau_1 \ln 10 = 2.3\tau_1 \tag{4.37}$$

We shall now illustrate the discussion with a few examples.

EXAMPLE 1: An interstage network has $C_{\text{par}} = 20\,\text{pF}$ and the amplifying device has $g_m = 10^{-2}$ mhos. Design the network so that $f_1 = 4\,\text{MHz}$ and evaluate the midband gain per stage.

ANSWER:

$$R_{\text{par}} = \frac{1}{2\pi f_1 C_{\text{par}}} = \frac{1}{2\pi \times 4 \times 10^6 \times 20 \times 10^{-12}} = 2000\,\Omega$$

$$g_m R_{\text{par}} = 10^{-2} \times 2000 = 20$$

EXAMPLE 2: Find the rise time of the amplifier stage of Example 1 when used as a pulse amplifier.

ANSWER:

$$t_r = 2.3\tau_1 = 2.3 R_{\text{par}} C_{\text{par}} = 2.3 \times 2 \times 10^3 \times 2 \times 10^{-11}$$
$$= 9.2 \times 10^{-8}\,\text{s}$$

If R_L is replaced by a more complicated network containing inductances, a decrease in rise time can be achieved. In some cases, however, this improvement is bought at the price of a considerable *overshoot*, in which $|v_0(t)|$ rises above the value $g_m R_{\text{par}}$ [Fig. 4.11(f)]. This problem should be avoided by proper circuit design. The overshoot is usually expressed in percent.

4.3b. Low-Frequency Response

At low frequencies the effect of all the parallel capacitances can be ignored, but the effect of the coupling capacitance C_b becomes important. Figure 4.12(a) shows the resulting equivalent circuit of the interstage network and Fig. 4.12(b) shows an equivalent version of it. Here

$$\frac{1}{R_1} = \frac{1}{R_0} + \frac{1}{R_L}; \quad \frac{1}{R_2} = \frac{1}{R_b} + \frac{1}{R_i}; \quad \frac{1}{R_{\text{par}}} = \frac{1}{R_1} + \frac{1}{R_2} \tag{4.38}$$

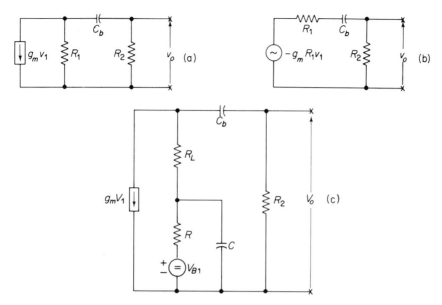

Fig. 4.12. (a) Low-frequency equivalent circuit of Fig. 4.10. (b) Alternate equivalent circuit of (a). (c) Circuit with improved low-frequency response.

We first turn to the ac frequency response. From Fig. 4.12(b)

$$v_0 = -g_m R_1 v_1 \frac{R_2}{R_1 + 1/(j\omega C_b) + R_2}$$

$$= -g_m \frac{R_1 R_2}{R_1 + R_2} v_1 \frac{1}{1 - j/[\omega C_b(R_1 + R_2)]} = \frac{g_{v0}}{1 - jf_2/f} v_1 \quad (4.39)$$

where g_{v0} is again the midband gain and f_2 is the *lower cutoff frequency*, defined by

$$2\pi f_2 C_b(R_1 + R_2) = 1 \quad (4.39a)$$

This frequency is so defined that $|v_0/v_1| = 1/\sqrt{2}$ at $f = f_2$.

An improved frequency response can be obtained by replacing R_L in Fig. 4.10 by a more complicated network that does not alter the midband response but improves the low-frequency response. A particular simple case occurs if $R_1 \simeq R_L$ and $R_L \ll R_2$, so that $f_2 = (2\pi C_b R_2)^{-1}$. One then adds the RC parallel combination shown in Fig. 4.12(c) in series with R_L, where R and C are so chosen that $CR \gg C_b R_2$. This does not change the midband frequency response, since $\omega CR \gg 1$ for midband frequencies, but it improves the low-frequency response. A calculation that will be omitted here shows that if the capacitance C is so chosen that

$$C_b R_2 = C \frac{R_L R}{R_L + R} \quad (4.39b)$$

then

$$v_0 = \frac{g_{v0}}{1 - j/(\omega CR)} v_1 = \frac{g_{v0}}{1 - jf_2'/f} v_1 \tag{4.39c}$$

We see that $|v_0/g_{v0}v_1| = \frac{1}{2}\sqrt{2}$ at the new lower cutoff frequency f_2'

$$f_2' = \frac{1}{2\pi CR} \ll \frac{1}{2\pi C_b R_2} = f_2 \tag{4.39d}$$

since we had chosen $CR \gg C_b R_2$.

An additional advantage of the circuit is that it decouples the supply voltage V_{B1} from the interstage network and thereby decreases feedback through the internal impedance of this supply voltage.

Next we turn to the pulse response by replacing v_1 by the unit step function $u(t)$. This yields (compare Appendix B)

$$v_0(t) = g_{v0} \exp\left(-\frac{t}{\tau_2}\right) \tag{4.40}$$

where

$$\tau_2 = C_b(R_1 + R_2) = \frac{1}{2\pi f_2} \tag{4.40a}$$

so that τ_2 and f_2 are related. We see that the response reaches its maximum at $t = 0$ and then sags. For $t/\tau_2 \ll 1$, Eq. (4.40) may be written as

$$v_0(t) = g_{v0}\left(1 - \frac{t}{\tau_2}\right) \tag{4.40b}$$

The sag is usually expressed in percent and is written as

$$\text{sag} = \frac{100t}{\tau_2} \% \tag{4.41}$$

It is often given in percent per millisecond.

If one compares Eqs. (4.39) and (4.39c), one observes that they differ only in the lower cutoff frequency. Therefore the circuit of Fig. 4.12(c) gives a smaller sag, since the sag changes from $200\pi t f_2$ to $200\pi t f_2' \%$, according to Eqs. (4.41) and (4.40a).

We shall illustrate this with the help of a few examples.

EXAMPLE 3: If $R_1 = 2000\ \Omega$ and $R_2 = 10^6\ \Omega$, find C_b so that the lower cutoff frequency is 10 Hz. (This corresponds to an FET amplifier.)

ANSWER:

$$C_b = \frac{1}{2\pi f_2(R_1 + R_2)} \simeq \frac{1}{2\pi \times 10 \times 10^6} = 1.6 \times 10^{-2}\ \mu\text{F}.$$

EXAMPLE 4: Repeat the calculation of Example 3 for $R_1 = 1000\ \Omega$ and $R_2 = 2000\ \Omega$. (This corresponds to a transistor amplifier.)

ANSWER:

$$C_b = \frac{1}{2\pi f_2(R_1 + R_2)} = \frac{1}{2\pi \times 10 \times 3 \times 10^3} = 5\ \mu\text{F}$$

EXAMPLE 5: Find the sag of the amplifiers in Examples 3 and 4.

ANSWER:

$$\text{sag} = \frac{100t}{\tau_2} = 200\pi t f_2 = 200\pi \times 10^{-3} \times 10$$

$$= 6.3\%$$

4.4. MILLER EFFECT

Up to now we have neglected the effect of a possible feedback between output and input of an amplifier stage. The most common feedback is that a capacitance C_{oi} exists between output and input. We shall show that this gives rise to an apparent capacitance

$$C_f = C_{oi}(1 + |g_{vo}|) \tag{4.42}$$

in parallel with the input, where g_{vo} is the midband gain of the amplifier stage. This effect is called the *Miller effect*.

To prove Eq. (4.42), we represent the interstage network by the resistance R_{par} defined by Eq. (4.33) and apply an EMF to the input; since the network parameters R_i and C_i do not influence the feedback effect, we omit them here. We thus obtain the equivalent network of Fig. 4.13(a), which can be replaced by the network of Fig. 4.13(b). The input current i_1 is therefore

$$i_1 = \frac{v_1 + g_m R_{par} v_1}{R_{par} + 1/(j\omega C_{oi})} = v_1(1 + |g_{vo}|)\frac{j\omega C_{oi}}{1 + j\omega C_{oi} R_{par}} \tag{4.43}$$

For most frequencies of practical interest in the circuit, $\omega C_{oi} R_{par} \ll 1$; Eq. (4.43) may then be rewritten as

$$i_1 \simeq v_1(1 + |g_{vo}|)j\omega C_{oi} = v_1 j\omega C_f \tag{4.43a}$$

which proves Eq. (4.42).

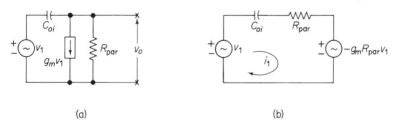

(a)   (b)

Fig. 4.13. (a) Equivalent circuit demonstrating Miller effect. (b) Alternate equivalent circuit for calculating Miller effect.

4.5. AMPLIFIER WITH TUNED INTERSTAGE NETWORK

At high frequencies the capacitance C_{par} of an interstage network between two amplifying devices shunts the resistance R_{par} and this results in a very low voltage gain, or even a voltage loss, according to Section 4.3a. To remedy the situation, the interstage network is tuned to the center frequency f_0 of the frequency band that must be amplified. Figure 4.14(a) shows such an arrangement. Here R_c is the tuned circuit impedance and C the tuned circuit capacitance; we assume that the feedback capacitance in the device between interstages is negligible. The resistance R_b is a bias resistance; in some cases it is not needed, in some other cases its effect is negligible.

4.5a. Midband Gain; Bandwidth

The circuit of Fig. 4.14(a) can be simplified to the circuit of Fig. 4.14(b) where all the resistances have been lumped into a resistance R_{par} and all capacitances into a capacitance C_{par}:

$$\frac{1}{R_{\text{par}}} = \frac{1}{R_0} + \frac{1}{R_c} + \frac{1}{R_b} + \frac{1}{R_i}; \qquad C_{\text{par}} = C_0 + C + C_i \qquad (4.44)$$

Since the circuit is tuned at the frequency f_0, we have

$$\omega_0^2 L C_{\text{par}} = 1 \qquad (4.45)$$

where $\omega_0 = 2\pi f_0$. The output voltage at the center frequency is thus

$$v_0 = -g_m R_{\text{par}} v_i \qquad (4.46)$$

so that the voltage gain at the midband frequency is

$$g_v = g_{v0} = -g_m R_{\text{par}} \qquad (4.46a)$$

This is called the midband gain. At all other frequencies

$$v_0 = \frac{-g_m v_1}{1/R_{\text{par}} + 1/(j\omega L) + j\omega C_{\text{par}}} = -\frac{g_m R_{\text{par}} v_1}{1 + j(\omega/\omega_0 - \omega_0/\omega)\omega_0 C_{\text{par}} R_{\text{par}}} \qquad (4.47)$$

where we replaced $1/L$ by $\omega_0^2 C_{\text{par}}$. As long as $\Delta\omega = \omega - \omega_0$ is small in comparison with ω_0, we may put

$$\frac{\omega}{\omega_0} - \frac{\omega_0}{\omega} \simeq \frac{2\Delta\omega}{\omega_0} \qquad (4.47a)$$

so that

$$v_0 \simeq -\frac{g_m R_{\text{par}}}{1 + 2j\,\Delta\omega C_{\text{par}} R_{\text{par}}} v_1 \qquad (4.48)$$

from which the voltage gain $g_v = v_0/v_1$ follows.

We observe that $|v_0/v_1|$ is equal to $g_m R_{\text{par}}$ for $\Delta\omega = 0$ and that it has dropped to $\frac{1}{2}\sqrt{2}$ times the midband value when

$$2\Delta\omega C_{\text{par}} R_{\text{par}} = \pm 1 \qquad (4.49)$$

Putting $2\Delta f = B$, where $\Delta f = \Delta\omega/(2\pi)$, we thus have

$$2\pi B C_{\text{par}} R_{\text{par}} = 1 \qquad (4.49a)$$

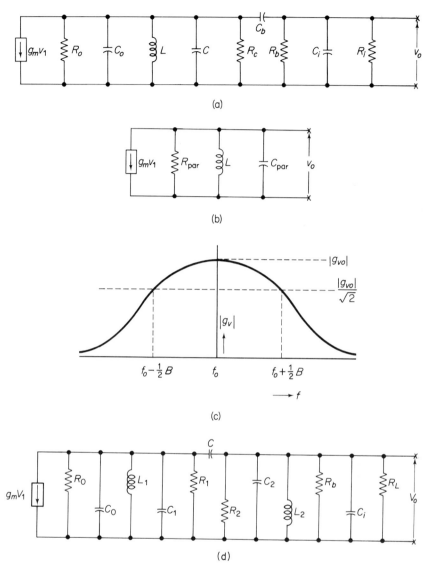

Fig. 4.14. (a) Full interstage network of a tuned amplifier. (b) Equivalent circuit of (a). (c) Frequency response of the circuit (a) showing the half-power points. (d) Interstage network with double-tuned circuit.

The parameter B is called the *bandwidth* of the tuned circuit. Its meaning is that at the frequencies $f_0 \pm \frac{1}{2}B$ the absolute value of v_0/v_1 has dropped to $\frac{1}{2}\sqrt{2}$ times the midband value. The frequencies $f_0 \pm \frac{1}{2}B$ are therefore called the *half-power points*. The situation is illustrated in Fig. 4.14(c).

Combining Eqs. (4.46a) and (4.49a), we have

$$|g_{v0}|B = \frac{g_m}{2\pi C_{par}} \tag{4.50}$$

This expression is called the *gain-bandwidth* product of the amplifier stage. One can thus again trade gain against bandwidth as in the low-frequency amplifier (Section 4.3a).

It is again possible to replace the tuning circuit by a more complicated network. On simple solution is to use the double-tuned circuit shown in Fig. 4.14(d). By proper dimensioning of the circuit one can obtain a flatter response within the passband and a sharper drop in response outside the passband. Here the passband is again defined as the frequency range between the two half-power points as in the single-tuned circuit case.

Another possibility arises here that does not exist in the low-frequency amplifier. Here one can tune the subsequent interstage networks to different frequencies within the passband B. By proper design the amplifier can have a flatter frequency response inside the passband and a steeper drop in response outside the passband. Such an amplifier is called a *stagger-tuned* amplifier; it is especially suitable if one wants a broad-band response. We refer to Chapter 14 for details.

EXAMPLE 1: An 30-MHz interstage network has $C_{par} = 20$ pF and each device has $g_m = 10^{-2}$ mhos. Find the value of R_{par} that gives a bandwidth of 4 MHz, the midband gain, and the inductance L needed to tune the interstage network to 30 MHz.

ANSWER:

$$2\pi B C_{par} R_{par} = 1$$

$$R_{par} = \frac{1}{2\pi B C_{par}} = \frac{1}{2\pi \times 4 \times 10^6 \times 20 \times 10^{-12}} = 2000 \; \Omega$$

$$g_m R_{par} = 10^{-2} \times 2000 = 20$$

$$\omega^2 LC = 1; \quad L = \frac{1}{4\pi^2 \times 9 \times 10^{14} \times 20 \times 10^{-12}} = 1.4 \; \mu H$$

4.5b. Effect of a Feedback Capacitance C_{oi} Between Output and Input

Up to now we ignored the effect of the feedback capacitance C_{oi} between output and input. In some cases this is allowed; for example, in the FET cascode circuit, C_{oi} can be made quite small. In other cases the effect of the capacitance C_{oi} is so large that the circuit may break into oscillations.

This is illustrated in Fig. 4.15(a), which shows schematically two interstage networks coupled by means of an amplifiying device having a transconductance g_m and a feedback capacitance C_{oi}. In Fig. 4.15(a) the tuning of the interstage network is represented by reactances X_1 and X_2, which can each vary independently from $-\infty$ to $+\infty$.

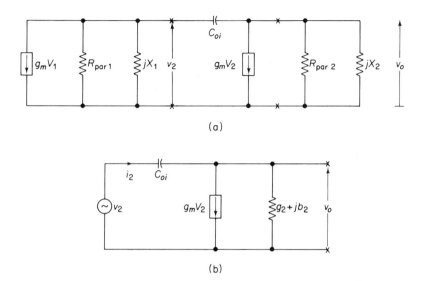

(a)

(b)

Fig. 4.15. (a) Equivalent circuit of a tuned amplifier stage with feedback capacitance C_{oi}. (b) Equivalent circuit used to calculate the input admittance of the amplifier under (a).

If C_{oi} were zero, tuning would be obtained if $X_1 = X_2 = 0$ at $f = f_0$ and the two interstage networks would not interact. Because C_{oi} is not zero, an interaction occurs that can lead to instability in the following manner: One first tunes X_2 so that v_0 is a maximum. One then tunes X_1 so that v_0 is further increased. One now readjusts X_2 so that v_0 again attains a maximum value and this process can be repeated ad infinitum. In some cases this tuning process converges in that v_0 reaches an absolute limit; the circuit is then perfectly stable. In other cases the tuning process diverges, in that v_0 grows without limits and then the circuit will break into oscillations. A calculation shows that the condition for stability may be written as

$$\tfrac{1}{2}\omega C_{oi}g_m R_{\text{par}1}R_{\text{par}2} < 1 \tag{4.51}$$

We have purposely chosen the parallel resistances $R_{\text{par }1}$ and $R_{\text{par }2}$ in Fig. 4.15(a) to be different. If both are equal to R_{par}, then the midband gain is $g_m R_{\text{par}}$, so that Eq. (4.51) may be written as

$$(g_m R_{\text{par}})^2 < \frac{2g_m}{\omega C_{oi}} = \frac{g_m}{\pi f C_{oi}} \tag{4.51a}$$

The feedback thus limits the midband voltage gain that can be obtained in this manner.

EXAMPLE 2: $g_m = 10^{-2}$ mho, $C_{oi} = 2$ pF, and $f = 10^7$ Hz. Then

$$\frac{g_m}{\pi f C_{oi}} = \frac{10^{-2}}{\pi \times 10^7 \times 2 \times 10^{-12}} = 160 \quad \text{or} \quad g_m R_{\text{par}} = 12.5$$

In a cascode circuit such as is discussed in Chapter 6, one alternately makes R_{par} small and large. For example, one chooses $R_{par\,1}$ such that $g_m R_{par\,1} = 1$, so that the part of the circuit associated with $R_{par\,1}$ has a voltage gain of unity. According to Eq. (4.51), one can then make $R_{par\,2}$ much larger so that the part of the circuit associated with $R_{par\,2}$ has the large midband voltage gain $g_m R_{par\,2}$. The combination is now stable if the condition

$$g_m R_{par\,2} < \frac{2g_m}{\omega C_{oi}} \tag{4.51b}$$

is satisfied.

This illustrates only one property of the cascode circuit. An additional feature is that the circuit associated with $R_{par\,1}$ is untuned and that the part of the amplifier with a tuned output impedance $R_{par\,2}$ has a very small value of C_{oi}. Both features result in a further improvement of the stability of the circuit.

We shall finally prove Eq. (4.51). To that end we turn to Fig. 4.15(b), with the help of which we evaluate the input admittance of the feedback circuit of Fig. 4.15(a). We replace the input circuit by an EMF v_2, calculate the input current i_2, and define the input admittance $Y_{in} = i_2/v_2$. We replace $R_{par\,2}$ by $g_2 = 1/R_{par\,2}$ and jX_2 by jb_2, where $b_2 = -1/X_2$.

We now have from Fig. 4.15(b)

$$i_2 = (v_2 - v_0)j\omega C_{oi} \tag{4.52}$$

$$v_0(g_2 + jb_2) = -g_m v_2 + j\omega C_{oi}(v_2 - v_0) \tag{4.53}$$

so that if $\omega C_{oi} \ll g_m$,

$$v_0 = -\frac{g_m v_2}{g_2 + jb}; \qquad i_2 = j\omega C_{oi} v_2\left(1 + \frac{g_m}{g_2 + jb}\right) \tag{4.54}$$

where $b = b_2 + \omega C_{oi}$. Hence

$$Y_i = \frac{i_2}{v_2} = j\omega C_{oi} + \frac{j\omega C_{oi}g_m}{g_2^2 + b^2}(g_2 - jb) = g_{in} + jb_{in} \tag{4.55}$$

The imaginary part of Y_i can be removed by tuning; the real part,

$$g_{in} = g_m \omega C_{oi}\frac{b}{g_2^2 + b^2} \tag{4.55a}$$

has a minimum value $(g_i)_{min}$ if $b = -g_2$. Consequently

$$(g_i)_{min} = -\tfrac{1}{2}g_m \omega C_{oi} R_{par\,2} \tag{4.56}$$

Stability is assured if

$$(g_i)_{min} + \frac{1}{R_{par\,1}} > 0 \quad \text{or} \quad \frac{1}{2}g_m \omega C_{oi} R_{par\,1} R_{par\,2} < 1$$

which corresponds to Eq. (4.51).

Circuits exist that reduce or eliminate the feedback through the capacitance C_{oi}. They are called *neutralization circuits*, since they neutralize the effect of the feedback capacitance C_{oi}. One way of achieving this is to tune the capacitance C_{oi} to the center frequency f_0 of the passband; this is acceptable for amplifiers tuned at a fixed frequency but is unacceptable if the amplifier must be tuned to different frequencies.

A particular simple solution is possible in tuned push-pull amplifiers. Here two parallel amplifier stages with common ground are used; the input signals are applied to the two inputs in opposite phase, and the output signals are taken from the two outputs in opposite phase (push-pull operation). One can now achieve complete neutralization over a wide frequency range by connecting an external capacitance C_{oi} between the output of each amplifier stage and the input of the other amplifier stage. An example of this method of neutralization is shown in Fig. 9.6(a).

PROBLEMS

4.1. A voltage amplifier stage has $\mu = 100$ and $R_i = R_0 = R_s = R_L = 2000\ \Omega$. What is the voltage gain of this amplifier stage in decibels?

4.2. A voltage amplifier consists of three amplifier stages connected in cascade. Each has $g_m = 50$ millimhos, $R_0 = \infty$, and $R_1 = R_L = 2000\ \Omega$. Each interstage network is loaded down by a load resistance $R'_L = R_L = 2000\ \Omega$. (a) If v_0 is the output voltage developed across R_L and v_1 the voltage applied to the first stage, show that the voltage gain of this amplifier is

$$\left|\frac{v_0}{v_1}\right| = \frac{1}{4}(g_m R_L)^3$$

(b) Express this voltage gain in decibels.

4.3. In the voltage amplifier of Problem 4.2 a signal source v_s with resistance $R_s = R_1 = 2000\ \Omega$ in series is connected to the input of the amplifier. (a) Show that in this case the voltage gain of the amplifier is

$$\left|\frac{v_0}{v_s}\right| = \frac{1}{8}(g_m R_L)^3$$

(b) Express this voltage gain in decibels. (c) Express the power gain of the amplifier in decibels and compare with the result in part (b).

4.4. The amplifier of Fig. 4.16 is connected to a 10-Ω load resistance by means of a matching transformer of turns ratio n. (a) Find the turns ratio n needed

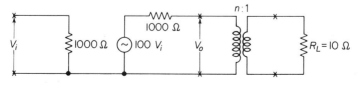

Fig. 4.16.

for matching. (b) Find the rms input voltage needed to supply 100 mW of power to the output load.

4.5. The amplifier of Fig. 4.17 is directly connected to a 10-Ω load resistance. Find the rms input voltage needed to supply 100 mW to the output load.

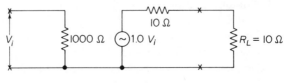

Fig. 4.17.

4.6. The amplifier of Fig. 4.16 is connected directly to the input of the amplifier of Fig. 4.17 and the output of that amplifier is directly connected to the 10-Ω load. Find the rms input voltage needed to supply 100 mW to the output load.

4.7. Two amplifiers of the type shown in Fig. 4.16 are directly connected in cascade, and the output of the second amplifier is directly connected to the input of the amplifier shown in Fig. 4.17. That amplifier, in turn, is connected directly to a 10-Ω load resistance. (a) Find the rms input voltage needed to supply 100 mW to the output load. (b) If a signal source v_s with internal resistance $R_s = 1000 \ \Omega$ is connected to the input of this amplifier combination, find the rms value of v_s needed to supply 100 mW to the output load resistance.

4.8. The circuit of Fig. 4.18 is, among other things, used to reduce the output ripple of rectifier power supplies. R is so chosen that the Zener diode D operates properly. $V_b = 10$ V. (a) In this particular application the potentiometer is so adjusted that the input voltage of the amplifier is zero when $V_1 = 50$ V. Find R_2. (b) Express the output ripple ΔV_0 in terms of the input ripple ΔV_i and the amplification μ.

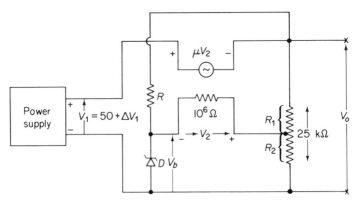

Fig. 4.18.

4.9. In the circuit of Fig. 4.18 the potentiometer is so adjusted that R_2 is increased above the value used in Problem 4.8. (a) Find the dc output voltage V_0 in terms of the ratio $r = R_2/(R_1 + R_2)$ for $0.20 < r < 1$. (b) Express this result numerically if $\mu = 100$, $V_1 = 50$ V. (c) Express the ac output voltage ripple in terms of r.

4.10. A number of wide-band amplifiers are connected in cascade and the interstage networks are so designed that an upper cutoff frequency f_1 and a lower cutoff frequency f_2 are obtained. Evaluate the half-power points of the overall response for n amplifiers connected in cascade.

4.11. A transistor amplifier has $R_0 = \infty$, $R_i = 2000\,\Omega$, $R_b = 5000\,\Omega$, and $C_{par} = 50$ pF. (a) Find R_L so that the upper cutoff frequency f_1 of an individual stage is 10 MHz. (b) For that particular value of R_L, find C_b so that the lower cutoff frequency is 10 Hz. (c) If $g_m = 50$ millimhos, find the midband gain.

4.12. An FET amplifier has $R_0 = \infty$, $R_i = \infty$, $R_b = 1.0$ MΩ, and $C_{par} = 30$ pF. (a) Find R_L so that the upper cutoff frequency is 10 MHz. (b) For that particular value of R_L, find C_b so that the lower cutoff frequency is 10 Hz. (c) If $g_m = 10$ millimhos, find the midband gain.

4.13. An FET amplifier stage has $C_{oi} = C_{dg} = 2.0$ pF and $C_i = C_{gs} = 5.0$ pF. The transconductance is 5×10^{-3} mho and the load resistance R_d in the output is $2000\,\Omega$. $R_0 = \infty$. Find the input capacitance of the amplifier due to the Miller effect.

4.14. A tuned FET amplifier is so designed that the feedback capacitance between output and input is negligible (such a circuit is called a cascode circuit or FET tetrode circuit). $R_0 = R_i = R_b = \infty$, $C_{par} = 20$ pF, and $R_{par} = R_c$. (a) Design the circuit so that the interstage network has a bandwidth of 4 MHz. (b) Evaluate the midband gain of the circuit if $g_m = 10^{-2}$ mho. (c) Choose the tuning inductance so that the interstage network is tuned at 30 MHz.

5

Simple FET
Circuits

5.1. JUNCTION FET AND MOSFET

5.1a. Characteristic of the Junction FET with an n-Type Channel

The junction FET with an n-type channel was discussed in Section 1.3c. It consists of a voltage-controlled resistance between two ohmic contacts, the source S and the drain D. The height b of the conducting channel between D and S is controlled by the voltage V_{GS} applied between a junction gate G and the source S. We shall consider here a slightly different geometry from the one used in Section 3.1c in that the voltage-controlled resistance consists of a thin slab of semiconductor with two (internally or externally) connected p-regions diffused into each side.

If at first the voltage V_{DS} between drain and source is kept zero and a voltage V_{GS} is applied between G and S, then the height b of the channel decreases if V_{GS} is made more negative and becomes zero for $V_{GS} = V_P$. For that reason V_P is called the *pinch-off* voltage or *turn-on* voltage. Here $V_P < 0$ [Fig. 5.1(a)]. If now a voltage $V_{DS} > 0$ is applied between D and S and V_{GS} is so chosen that the channel is nowhere pinched off, then current flows; the width b of the channel now decreases by going from source to drain, and the voltage V in the channel, taken with respect to the source, increases with increasing distance to the source. If we choose an X-axis through the middle of the channel, call the length of the channel L, and put

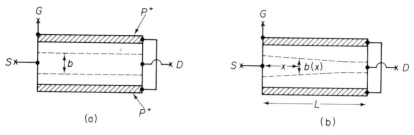

Fig. 5.1. (a) Cross-section of *n*-channel JFET with zero drain bias, showing shape of conducting channel. (b) Cross-section of *n*-channel JFET with positive drain bias, showing shape of conducting channel.

the origin of the coordinate system at S, then $V(x) = 0$ at $x = 0$ (contact S) and $V(x) = V_{DS}$ at $x = L$ (contact D). The potential difference between the gate and the channel at x is $V_{GS} - V(x)$. Therefore the height b of the channel is a function $b(x)$ of x; $b(x)$ is largest at $x = 0$ and smallest at $x = L$. For $V_{GS} - V_{DS} = V_P$ the channel height at $x = L$ is zero; the channel is then said to be *pinched off* at the drain. The channel is nowhere pinched off if $V_{GS} - V_{DS} > V_P$ or $V_{DS} < V_{GS} - V_P$ [Fig. 5.1(b)].

The height $b(x)$ of the channel as a function of the voltage difference $V_{GS} - V(x)$ is a complicated function of x. It is usually simplified by assuming that $b(x)$ decreases linearly with increasing $V(x)$. Since we know that $b(L) = 0$ for $V(L) = V_{GS} - V_P$, we may write

$$b(x) = b_0 [V_{GS} - V_P - V(x)] \tag{5.1}$$

where b_0 is a constant that need not concern us here. We see indeed that $b(x)$ decreases linearly with increasing $V(x)$ and that it is zero if $V(x) = V_{GS} - V_P$.

Now if σ is the conductivity of the channel and w its width, then the conductance $g(x)$ per unit length at x may be written as

$$g(x) = \sigma w b(x) \quad \text{or} \quad g(V) = \sigma w b_0 [V_{GS} - V_P - V(x)] \tag{5.2}$$

where we have switched over to $V(x)$ as a new variable.

If we now split the channel up into small sections Δx and look at a particular section in greater detail, we see that the section Δx has a resistance $\Delta R = \Delta x / g(V)$; if the voltage developed across the section is ΔV, the current I_D is

$$I_D = \frac{\Delta V}{\Delta R} = \frac{g(V)}{\Delta x} \Delta V \quad \text{or} \quad I_D \Delta x = g(V) \Delta V \tag{5.3}$$

Since there is no charge stored in the channel, I_D is independent of x. Integrating over the channel length L, we obtain

$$I_D \int_0^L dx = I_D L = \int_0^{V_{DS}} g(V) \, dV = \sigma w b_0 \left[(V_{GS} - V_P) V_{DS} - \frac{1}{2} V_{DS}^2 \right] \tag{5.4}$$

as is found by substituting Eq. (5.2). Consequently

$$I_D = K[2(V_{GS} - V_P)V_{DS} - V_{DS}^2] \qquad (5.5)$$

where

$$K = \frac{1}{2}\frac{\sigma b_0 w}{L} \qquad (5.5a)$$

In practical devices K is of the order of 1 mA/V^2, but it can be made much larger or smaller by proper choice of w/L. We finally observe that I_D flows *into* the drain.

The important things to remember from this derivation are that the characteristic is quadratic and that the constant K varies as w/L; to obtain large values of K, one should make devices with a large channel width w and a short channel length L.

Since Eq. (5.2) was used in deriving the expression for I_D, this implies that the channel is nowhere cut off, except possibly at the drain. This means that Eq. (5.5) is correct only for $V_{DS} \leq V_{GS} - V_P$. To find out what happens beyond that point, we first apply a small fluctuation ΔV_{DS} and note the change ΔI_D in I_D. Obviously

$$\Delta I_D = \frac{\partial I_D}{\partial V_{DS}}\Delta V_{DS}; \qquad \frac{\partial I_D}{\partial V_{DS}} = 2K(V_{GS} - V_P - V_{DS}) \qquad (5.6)$$

so that I_D increases with increasing V_{DS}, reaching its maximum value,

$$I_D = K(V_{GS} - V_P)^2 \qquad (5.7)$$

at $V_{DS} = V_{GS} - V_P$. Experimentally one finds that I_D remains practically constant for $V_{DS} \geq V_{GS} - V_P$, so that Eq. (5.7) remains correct for $V_{DS} \geq V_{GS} - V_P$.

To sum up our results, we thus write

$$I_D = K[2(V_{GS} - V_P)V_{DS} - V_{DS}^2] \qquad \text{for } 0 \leq V_{DS} \leq V_{GS} - V_P \qquad (5.8)$$
$$= K(V_{GS} - V_P)^2 \qquad \text{for } V_{DS} \geq V_{GS} - V_P \qquad (5.8a)$$

The first region of the I_D, V_{DS} characteristic is called the *non-pinch-off* region and the second region is called the *pinch-off* region; it is used in linear amplifiers. The non-pinch-off region is important in switching circuits involving FETs, especially the region where $V_{DS} \ll V_{GS} - V_P$. We may then ignore the V_{DS}^2 term in Eq. (5.8) and obtain

$$I_D = 2K(V_{GS} - V_P)V_{DS} = \frac{V_{DS}}{R_{\text{on}}} \qquad (5.9)$$

so that the device becomes a linear resistor of value

$$R_{\text{on}} = \frac{V_{DS}}{I_D} = \frac{1}{2K(V_{GS} - V_P)} \qquad (5.9a)$$

This is called the *ohmic region* of the FET characteristic.

We finally note that the small-signal drain conductance is

$$g_d = \frac{\partial I_D}{\partial V_{DS}} = 2K(V_{GS} - V_P - V_{DS}) \qquad (5.10)$$

It varies from the value $g_{d0} = 2K(V_{GS} - V_P) = 1/R_{on}$ for $V_{DS} = 0$ to zero value at pinch off ($V_{DS} = V_{GS} - V_P$).

Here we have discussed the restrictions on V_{DS}. What, if any, are the restrictions on V_{GS}? First, one wants a conducting channel, and this means that $V_{GS} > V_P$. Second, one must avoid gate current. Now for $V_{GS} > 0$, the *p-n* junction between gate and channel is forward-biased near the source. In silicon JFETs the gate current is less than 1 μA if $V_{GS} < 0.50$ V. Assuming that such a current is acceptable, we must require $V_P \leq V_{GS} \leq 0.50$ V. In some high-impedance circuits one might want a somewhat lower maximum value of V_{GS}.

The device is usually characterized by the pinch-off voltage V_p and the pinch-off current I_{DSS} at $V_{GS} = 0$. Since

$$I_{DSS} = KV_P^2 \quad \text{or} \quad K = \frac{I_{DSS}}{V_P^2} \qquad (5.11)$$

the constant K can be evaluated from the data. We shall always characterize the device by the values of K and V_P. Different samples of the same device type differ in the characteristics because the values of K and V_P are different.

EXAMPLE 1: An *n*-channel JFET has $I_{DSS} = 9$ mA and $V_P = -3$ V. Find K.

ANSWER:

$$K = \frac{I_{DSS}}{V_P^2} = \frac{9}{9} = 1.0 \text{ mA/V}^2$$

Figure 5.2(a) shows the (I_D, V_{DS}) characteristics for different values of V_{GS}. The pinch-off regime occurs for $V_{DS} \geq V_{GS} - V_P$. Figure 5.2(b) shows the (I_D, V_{GS}) characteristic in the pinch-off region.

If one looks at the characteristic more carefully, one finds that I_D increases slightly with increasing V_{DS} in the pinch-off region because the length L' of the conducting channel, which has the value L at pinch off, decreases slowly with the increasing value of V_{DS}. In the region $L' < x < L$ the field becomes very large and the current flow in that region is caused by this field [Fig. 5.2(c)].

Figure 5.2(d) shows the symbol commonly used for the *n*-channel JFET. The arrow indicates the direction of gate current flow for a forward-biased gate.

5.1b. Characteristic of the *p*-Channel JFET

The discussion of the *p*-channel JFET is quite similar to the *n*-channel JFET. The only difference is that the carriers responsible for the current flow have

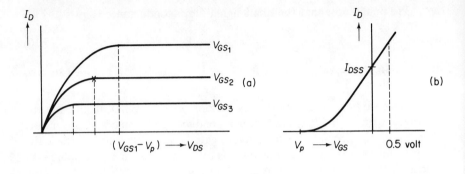

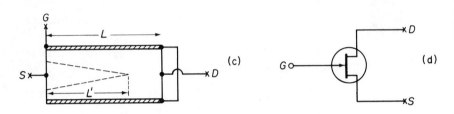

Fig. 5.2. (a) (I_D, V_{DS}) characteristic of n-channel JFET. (b)(I_D, V_{GS}) characteristic of n-channel JFET in pinch-off region. (c) Shape of conducting channel when drain voltage is very large. (d) Symbol for n-channel JFET.

opposite polarity; the channel is p-type, the gate regions are n-type, all voltages have opposite polarity, and all currents flow in the opposite direction.

Since the gate regions are n-type, the height of the conducting channel is decreased if V_{GS} is made more positive; consequently the pinch-off voltage $V_P > 0$. Since for $V_{GS} < 0$ the gate-channel junction is forward-biased near the source, the useful range of values for V_{GS} in a silicon p-channel JEFT is $-0.50 \text{ V} \leq V_{GS} \leq V_P$.

In addition, the drain bias V_{DS} is now negative, and the drain current I_D flows *out of* the device. If we take as standard direction of positive current flow the direction *into* the drain from the outside, we must write

$$I_D = -K[2(V_{GS} - V_P)V_{DS} - V_{DS}^2] \qquad \text{for } V_{GS} - V_P \leq V_{DS} \leq 0 \qquad (5.12)$$

$$= -K(V_{GS} - V_P)^2 \qquad \text{for } V_{DS} \leq V_{GS} - V_P \qquad (5.12a)$$

The latter region is again the *pinch-off* region of the characteristic. The device can again be characterized by the pinch-off voltage V_P and the pinch-off value I_{DSS} of I_D for $V_{GS} = 0$:

$$I_{DSS} = -KV_P^2; \qquad K = -\frac{I_{DSS}}{V_P^2} \qquad (5.12b)$$

Here we use characterization by K and V_P, however.

The (I_D, V_{DS}) characteristics of this type of device are shown in Fig. 5.3(a); the pinch-off characteristics (I_D, V_{GS}) are shown in Fig. 5.3(b). The symbol commonly used for a p-channel JFET is shown in Fig. 5.3(c). The arrow again shows the direction of gate current flow for a forward-biased gain junction.

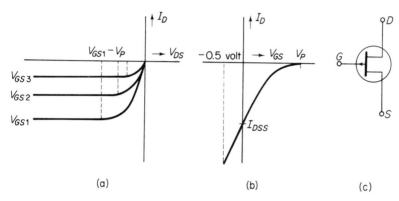

(a)   (b)   (c)

Fig. 5.3. (a) (I_D, V_{DS}) characteristic of p-channel JFET. (b) (I_D, V_{GS}) characteristic of p-channel JFET in pinch-off region. (c) Symbol for p-channel JFET.

5.1c. *n*-Channel MOSFET

An *n*-channel MOSFET is made on a p-type substrate. Two strongly n-type regions at a distance L are diffused into the substrate and an oxide is grown or deposited over the region between the two n^+-regions. A metal layer is deposited on this oxide and ohmic contacts are made to the two n-regions and to the metal layer. The two n-regions form the source S and the drain D and the metal on top of the oxide forms the gate G. The thickness of the oxide layer is of the order of 1000 Å (10^{-5} cm) (Fig. 5.4).

If now a sufficiently large positive voltage V_{GS} is applied between the gate and the source and the voltage V_{DS} between gate and source is kept zero, a negative mobile charge is induced in the region under the oxide that is proportional to V_{GS}. As a consequence a conducting channel is formed be-

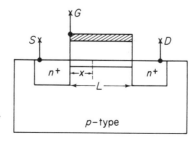

Fig. 5.4. Cross-section of *n*-channel MOSFET.

tween gate and source and the conductance of the channel increases linearly with increasing V_{GS}. The channel disappears if V_{GS} drops below the value V_T, which is therefore called the *turn-on* voltage. For most n-channel silicon MOSFETs there is already a channel present at $V_{GS} = 0$ V, indicating that $V_T < 0$ in those cases; by special design one can make devices with $V_T > 0$. In either case the charge per unit length, Q, in the channel may be written as

$$Q = -C_{ox}w(V_{GS} - V_T) \qquad (5.13)$$

where C_{ox} is the capacitance per unit area between gate and channel and w the width of the conducting channel.

If now a voltage V_{DS} is applied between drain and source, current flows. A voltage $V(x)$ is developed in the channel; the voltage between gate and channel at x is $V_{GS} - V(x)$, so that the expression for the charge per unit length, $Q(x)$, in the channel may be written as

$$Q(x) = -C_{ox}w[V_{GS} - V_T - V(x)] \qquad (5.14)$$

Let it first be assumed that $V_{DS} \leq V_{GS} - V_T$; then $Q(x)$ is nowhere zero. For $V_{DS} = V_{GS} - V_T$, $Q(x)$ is zero at $x = L$ (drain); the channel is then said to be *pinched off* at the drain. If $u_d(x) = \mu_n \, dV/dx$ is the average drift velocity at x, the current at x is

$$I_D = -Q(x)u_d(x) = -\mu_n Q(x) \frac{dV}{dx} = g(V) \frac{dV}{dx} \qquad (5.15)$$

where

$$g(V) = -\mu_n Q(x) = \mu_n C_{ox}w[V_{GS} - V_T - V(x)] \qquad (5.15a)$$

is the conductance per unit length of the channel. The mobility μ_n is called the *surface* mobility; it is somewhat smaller than the mobility in bulk material.

Since there can be no charge stored anywhere in the channel, I_D is independent of x. Therefore, if we write Eq. (5.15) as

$$I_D \, dx = g(V) \, dV \qquad (5.15b)$$

and integrate from source to drain, we obtain

$$I_D L = \int_0^{V_{DS}} g(V) \, dV = \mu_n C_{ox}w \left[(V_{GS} - V_T)V_{DS} - \frac{1}{2}V_{DS}^2 \right] \qquad (5.16)$$

so that

$$I_D = \frac{\mu_n C_{ox}w}{2L}[2(V_{GS} - V_T)V_{DS} - V_{DS}^2] = K[2(V_{GS} - V_T)V_{DS} - V_{DS}^2] \qquad (5.17)$$

where

$$K = \frac{\mu_n C_{ox}w}{2L} \qquad (5.17a)$$

Since the expression for $g(V)$ holds for $V_{DS} \leq V_{GS} - V_T$, Eq. (5.17) is valid for $0 \leq V_{DS} \leq V_{GS} - V_T$. For $V_{DS} = V_{GS} - V_T$, I_D has the value

$$I_D = K(V_{GS} - V_T)^2 \tag{5.18}$$

Experimentally one finds that I_D is practically independent of V_{DS} for $V_{DS} \geq V_{GS} - V_T$, so that I_D is given by Eq. (5.18) for $V_{DS} \geq V_{GS} - V_T$. This is called the *pinch-off* mode of operation of the device; it is the mode of operation used in linear amplifiers.

To illustrate this further, we evaluate the small-signal output conductance g_d of the device. From Eq. (5.17)

$$g_d = \frac{\partial I_D}{\partial V_{DS}} = 2K(V_{GS} - V_T - V_{DS}) \tag{5.19}$$

which has a maximum value of

$$g_{d0} = 2K(V_{GS} - V_T) \tag{5.19a}$$

for $V_{DS} = 0$ and reaches zero value at $V_{DS} = V_{GS} - V_T$. Therefore I_{DS}, considered as a function of V_{DS}, increases with increasing V_{DS}, attaining a maximum value at $V_{DS} = V_{GS} - V_T$.

In pulse applications the device is used chiefly in the region of small V_{DS}. The term V_{DS}^2 in Eq. (5.17) can then be ignored and

$$I_D \simeq 2K(V_{GS} - V_T)V_{DS} = \frac{V_{DS}}{R_{on}} \tag{5.20}$$

where

$$R_{on} = \frac{1}{g_{d0}} = \frac{1}{2K(V_{GS} - V_T)} \tag{5.20a}$$

Because of the linear relationship between I_D and V_{DS}, this is called the *ohmic* regime; the device there has the output resistance R_{on}.

We note that the characteristic for the n-channel MOSFET is formally equivalent to that of the n-channel JFET, therefore the characteristics for the non-pinch-off and the pinch-off regions are again given by Fig. 5.2(a) and (b), respectively. Both the constants K and V_T can be much better controlled in the MOSFET than in the JFET. There is one further difference between the JFET and the MOSFET, in that the gate is completely isolated from the channel, so that the gate current is zero under all conditions. There is therefore no restriction on V_{GS}.

A device with $V_T < 0$ can operate in two modes:

1. $V_T < V_{GS} < 0$; this is called the *depletion mode*.
2. $V_{GS} > 0$; this is called the *enhancement mode*.

An n-channel device with $V_T > 0$ only has a conducting channel for $V_{GS} > V_T > 0$; therefore it can operate only in the enhancement mode.

We characterized the device by the parameters K and V_T. Device manuals characterize depletion mode devices by the drain current I_{DSS} under pinch-off conditions at $V_{DS} = 0$. This means that

$$I_{DSS} = KV_T^2; \qquad K = \frac{I_{DSS}}{V_T^2} \tag{5.21}$$

Devices that can operate only in the enhancement mode are characterized by V_T and the device current I_{D1} under pinch-off condition at $V_{GS} = V_{GS1} > V_T$, so that

$$I_{D1} = K(V_{GS1} - V_T)^2 \quad \text{or} \quad K = \frac{I_{D1}}{(V_{GS1} - V_T)^2} \tag{5.22}$$

The parameter K is again of the order of 1 mA/V² but can be made much smaller or larger if necessary.

It is again found that I_D increases slightly with increasing V_{DS} beyond pinch off. The reason is the same as for the JFET: beyond pinch off the length L' of the conducting channel decreases slowly with increasing V_{DS} [see Fig. 5.2(c)].

The symbol for a MOSFET with an n-type channel is shown in Fig. 5.5(a). The arrow indicates the direction of positive current flow. In some cases the substrate is directly connected to the source; in other cases the

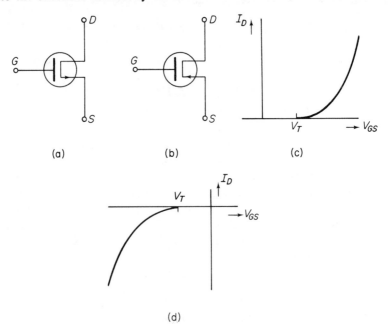

Fig. 5.5. (a) Symbol for n-channel MOSFET. (b) Symbol for p-channel MOSFET. (c) (I_D, V_{GS}) characteristic of enhancement-mode n-channel MOSFET. (d) (I_D, V_{GS}) characteristic of enhancement-mode p-channel MOSFET.

substrate is supplied with a separate lead, so that it can be biased separately. We did not show this in Fig. 5.5(a).

Devices that can operate in the enhancement mode only have certain advantages in direct-coupled amplifiers (Section 5.3) and in switching devices (Section 6.6) over devices that can operate both in the depletion and the enhancement modes. This is one of the reasons integrated MOSFET circuits use depletion mode devices.

5.1d. *p*-Channel MOSFET

The *p*-channel MOSFET is made on an *n*-type substrate. The source S and the drain D are introduced in the same way as in the previous case, but they are now p^+-type. The gate G is again deposited on top of an oxide layer deposited or grown between S and D. If the gate voltage V_{GS} is sufficiently negative and $V_{DS} = 0$, a mobile hole charge is induced in the surface region under the oxide, giving rise to a conducting channel between S and D; the conductance of the channel varies linearly with V_{GS} and disappears for $V_{GS} > V_T$, where V_T is the *turn-on* voltage of the channel. For normal operation we must thus require $V_{GS} < V_T$. Generally $V_T < 0$, so that MOSFETs with a *p*-type channel can operate only in the enhancement mode.

Otherwise the theory of the *n*-channel device can be applied to the *p*-channel device, but the voltages V_{GS} and V_{DS} are now negative and I_D flows *out of* the drain. Since the external drain current is counted positive if it flows into the drain, we must count I_D negative. We thus have

$$I_D = -K[2(V_{GS} - V_T)V_{DS} - V_{DS}^2] \quad \text{for } V_{GS} - V_T \leq V_{DS} \leq 0 \quad (5.23)$$

$$= -K(V_{GS} - V_T)^2 \quad \text{for } V_{DS} \leq V_{GS} - V_T \quad (5.24)$$

The latter mode of operation is again called the *pinch-off* mode of operation. The *ohmic* mode of operation is the operation for small V_{DS}, in which case

$$I_D \simeq -2K(V_{GS} - V_T)V_{DS} = \frac{V_{DS}}{R_{on}} \quad (5.25)$$

where

$$R_{on} = -\frac{1}{2K(V_{GS} - V_T)} \quad (5.25a)$$

The symbol for a MOSFET with a *p*-type channel is shown in Fig. 5.5(b); again the arrow indicates the direction of positive current flow. Figure 5.5(c) and 5.5(d) show the (I_D, V_{GS}) characteristics for enhancement mode *n*-channel and *p*-channel MOSFETs at pinch off, respectively.

5.1e. Breakdown in JFETs and MOSFETs

Since JFETs have a space-charge region between the gate and the conducting channel and MOSFETs have a space-charge region between the substrate and the conducting channel as well as between substrate and drain, and the *p-n* junctions thus formed are back-biased when the device is operated in the pinch-off mode, one would expect avalanche breakdown to occur when $|V_{DS}|$ is sufficiently large. This breakdown can actually be observed.

It is often masked, however, by another type of breakdown occurring in the region between the conducting channel and the drain. In that type of breakdown hole-electron pairs are formed also, but the minority carriers are quickly removed by the gate (in JFETs) or by the substrate (in MOSFETs), so that only the *majority* carriers multiply. Although not as objectionable as avalanche breakdown, it is still an undesirable effect, since it gives rise to an appreciable increase in drain current with increasing V_{DS} and to an appreciable gate current (in JFETs) or substrate current (in MOSFETs).

Since all these breakdown effects should be avoided, $|V_{DS}|$ should not be raised above a maximum value which is a characteristic limitation for any particular type of device.

5.2. SMALL-SIGNAL OPERATION OF THE FET

5.2a. *n*-Channel JFET as a Small-Signal Amplifier

Let an *n*-channel JFET be operated in the pinch-off mode ($V_{DS} \geq V_{GS} - V_P$) so that

$$I_D = K(V_{GS} - V_P)^2 \tag{5.26}$$

We now apply a dc bias V_{GS0} and a small ac signal $v_1 \cos \omega t$ to the gate and ask for the dc and ac drain current. We have [Fig. 5.6(a)]

$$
\begin{aligned}
I_D &= K(V_{GS0} - V_P + v_1 \cos \omega t)^2 \\
&= K(V_{GS0} - V_P)^2 + 2K(V_{GS0} - V_P)v_1 \cos \omega t + Kv_1^2 \cos^2 \omega t \\
&= I_{D0} + g_m v_1 \cos \omega t + \tfrac{1}{2}Kv_1^2 + \tfrac{1}{2}Kv_1^2 \cos 2\omega t \tag{5.27}
\end{aligned}
$$

where

$$I_{D0} = K(V_{GS0} - V_P)^2; \qquad g_m = 2K(V_{GS0} - V_P) \tag{5.27a}$$

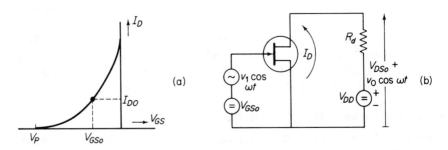

Fig. 5.6. (a) (I_D, V_{DS}) characteristic in pinch-off region with operating point. (b) Amplifier circuit, demonstrating dc and small ac signal operation.

The current I_{D0} is called the *quiescent current*, the parameter g_m is called the *transconductance* of the device, the term $\frac{1}{2}Kv_1^2$ is called the *rectified current*, and the term $\frac{1}{2}Kv_1^2 \cos 2\omega t$ is called the *second harmonic current*. The last two terms are negligible if v_1 is sufficiently small, but the second harmonic current is important since it determines the largest ac voltage that can be tolerated without objectionable second harmonic distortion. The ratio

$$d_2 = \frac{\frac{1}{2}Kv_1^2}{g_m v_1} = \frac{\frac{1}{2}Kv_1}{2K(V_{GS0} - V_P)} = \frac{v_1}{4(V_{GS0} - V_P)} \tag{5.28}$$

is called the *second harmonic distortion;* it is usually expressed in percent. The larger $V_{GS0} - V_P$, the smaller the distortion for a given value of v_1, or the larger the amplitude v_1 that can be tolerated for a given distortion.

We shall now illustrate this with the help of examples.

EXAMPLE 1: A JFET with an *n*-type channel has $K = 1.0$ mA/V^2 and $V_p = -4$ V. Find the quiescent current I_{D0} and the transconductance g_m at $V_{GS0} = -2$ V.

ANSWER:

$$I_{D0} = 10^{-3} \times (-2 + 4)^2 = 4.0 \text{ mA}$$

$$g_m = 2 \times 10^{-3}(-2 + 4) = 4.0 \text{ millimhos}$$

EXAMPLE 2: In the JFET in Example 1 one requires a maximum harmonic distortion of 1 %. When is the maximum amplitude v_1 that can be tolerated?

ANSWER:

$$d_2 = \frac{v_1}{4(V_{GS0} - V_P)} = 0.01$$

$$v_1 = 0.04(V_{GS0} - V_P) = 0.04(-2 + 4) = 0.08 \text{ V}$$

We shall see that this is two orders of magnitude better than for transistors.

Figure 5.6(a) shows how the quiescent current I_{D0} can be determined graphically from the characteristic.

We now put a load resistor R_d in the drain circuit, determine the dc operating voltage V_{DS0}, and require that the device operates in the pinch-off region, i.e., $V_{DS0} \geq V_{GS} - V_P$. We now turn to Fig. 5.6(b) and see by inspection that

$$V_{DS0} = V_{DD} - I_{D0}R_d \geq V_{GS0} - V_P \tag{5.29}$$

or that

$$I_{D0}R_d \leq V_{DD} - V_{GS0} + V_P; \qquad R_d \leq \frac{V_{DD} - V_{GS0} + V_P}{I_{D0}} \tag{5.29a}$$

That is, the requirement that the device must be operating at pinch off limits the value of R_d that can be tolerated. We shall illustrate this result with the help of examples.

EXAMPLE 3: If $V_{DD} = 20$ V, $V_{GS0} = -2$ V, $V_P = -4$ V, and $I_{D0} = 4$ mA, find the maximum value of R_d that can be tolerated.

ANSWER:

$$(R_d)_{\max} = \frac{V_{DD} - V_{GS0} + V_P}{I_{D0}} = \frac{20 + 2 - 4}{4 \times 10^{-3}} = 4500 \ \Omega$$

Actually one does not want so large a value of R_d, since one must be able to exchange devices without the risk of having the new device operating outside the pinch-off region. For that reason one wants V_{DS0} considerably larger than $V_{GS0} - V_P$. In that case the required value of R_d is

$$R_d = \frac{V_{DD} - V_{DS0}}{I_{D0}} \tag{5.30}$$

EXAMPLE 4: If in the circuit in Example 3 $V_{DD} = 20$ V, at $I_{D0} = 4$ mA one requires $V_{DS0} = 8$ V. Find the value of R_d required.

ANSWER:

$$R_d = \frac{V_{DD} - V_{DS0}}{I_{D0}} = \frac{20 - 8}{4 \times 10^{-1}} = 3000 \ \Omega$$

EXAMPLE 5: The operating point is $V_{GS0} = -2$ V and $V_P = -4$ V. The JFET of Example 4 is replaced by a JFET of the same type but with a larger value of K. What value of I_{D0} can be tolerated so that the device stays within the pinch-off region?

ANSWER: The minimum value of V_{DS0} is $-2 + 4 = 2$ V. Therefore

$$I_{D0} = \frac{V_{DD} - V_{DS0}}{R_d} = \frac{20 - 2}{3000} = 6 \ \text{mA}$$

Next we evaluate the ac drain voltage $v_0 \cos \omega t$. As is seen by inspecting Fig. 5.6(b),

$$v_0 \cos \omega t = -(g_m v_1 \cos \omega t) R_d = -g_m R_d v_1 \cos \omega t \tag{5.31}$$

The ratio

$$g_v = \frac{v_0}{v_1} = -g_m R_d \tag{5.31a}$$

is called the *voltage gain* of the device; the minus sign of g_v indicates that the output voltage is 180 degrees out of phase with respect to the input voltage.

EXAMPLE 6: Find the voltage gains in Examples 3 and 4.

ANSWER: In Example 3, $g_m = 4.0$ millimhos and $R_d = 4500 \ \Omega$, so that $g_v = -18$. In Example 4, $g_m = 4.0$ millimhos and $R_d = 3000 \ \Omega$, so that $g_v = -12$. One can thus achieve greater tolerance in the drain current I_{D0} by sacrificing a small amount of gain. We also see how the minimum drain voltage requirement limits the voltage gain that can be achieved.

We shall now investigate how the gain can be increased by a change in the gate bias conditions. To that end we take Eq. (5.31a) and substitute for g_m and R_d. According to Eq. (5.27a),

$$g_m = 2K(V_{GS0} - V_P); \qquad I_{D0} = K(V_{GS0} - V_P)^2$$

and according to Eq. (5.30), $R_d = (V_{DD} - V_{DS0})/I_{D0}$. Consequently, if V_{DD} and V_{DS0} are kept constant,

$$g_v = g_m R_d = 2K(V_{GS0} - V_P)\frac{V_{DD} - V_{DS0}}{K(V_{GS0} - V_P)^2} = 2\frac{V_{DD} - V_{DS0}}{(V_{GS0} - V_P)} \qquad (5.32)$$

so that a larger voltage gain can be obtained if $V_{GS0} - V_P$ is made smaller. This larger voltage gain is bought at the price of considerably more distortion, however.

EXAMPLE 7: In Example 4, $V_{DD} = 20$ V and $V_{DS0} = 8$ V. If $V_P = -4$ V and V_{GS0} is changed from -2 to -3.5 V, how much is the voltage gain changed? What is the new load resistance R_d? What value of amplitude v_1 now produces 1% distortion?

ANSWER: Since $V_{GS0} - V_P$ changes from $-2 + 4 = 2$ V to $-3.5 + 4 = 0.5$ V, $|g_v|$ is increased by a factor of 4 and I_{D0} is decreased by a factor of 16, so that R_d is increased by a factor 16. Finally, because of Eq. (5.28), the allowed value of v_1 is decreased by a factor of 4, so that the value of v_1 that gives 1% distortion is reduced to 0.02 V.

The story of the drain bias can be illustrated graphically by the load line method (Fig. 5.7). Since V_{GS0} is given, $I_D = I_{D0}$ for $V_{DS} > V_{GS0} - V_P$. We now draw the I_D, V_{DS} characteristic that has $I_D = I_{D0}$ at pinch off. Next we draw the *load line* given by

$$I_D = \frac{V_{DD} - V_{DS}}{R_d} \qquad (5.33)$$

Where the load line and the characteristic meet, we find the quiescent drain voltage V_{DS0} (operating point B). Also shown is the point $A(I_{D0}, V_{GS0} - V_P)$, where the device is just pinched off. To have the device operating in the pinch-off region, one must thus require that operating point B lie to the right

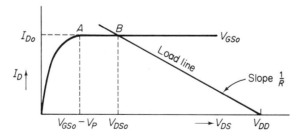

Fig. 5.7. (I_D, V_{DS}) characteristic with load line demonstrating proper choice (**B**) of operating point.

of point A. If one requires a reasonable tolerance for variations in I_D between different units, point B should lie considerably to the right of point A, as discussed in Example 4.

5.2b. Other FET Devices as Small-Signal Amplifiers

The MOSFET with an n-type channel operates in exactly the same way as the JFET with n-type channel insofar as small-signal amplification is concerned. The JFET with p-type channel operates in exactly the opposite way, since all currents and voltages have opposite polarity. For the MOSFET with a p-type channel one must bear in mind that the device only operates in the enhancement mode. We shall illustrate these differences with two examples.

EXAMPLE 8: A JFET circuit uses a p-channel device having $V_P = 4$ V, $V_{GS0} = 2$ V, $K = 1.0$ mA/V^2, $V_{DD} = -20$ V, and $V_{DS0} = -8$ V. Find the current I_{D0}, the transconductance g_m, the load resistance R_d, and the voltage gain g_{v0}.

ANSWER:

$$I_{D0} = -K(V_{GS0} - V_P)^2 = -1.0 \times 10^{-3}(2 - 4)^2 = -4.0 \text{ mA}$$

$$g_m = -2K(V_{GS0} - V_P) = -2 \times 10^{-3}(2 - 4) = 4.0 \text{ millimhos}$$

$$R_d = \frac{V_{DD} - V_{DS0}}{I_{D0}} = \frac{-20 + 8}{-4 \times 10^{-3}} = 3000 \text{ } \Omega$$

$$g_v = -g_m R_d = -4.0 \times 10^{-3} \times 3000 = -12$$

EXAMPLE 9: Repeat the calculation for MOSFET circuit with a p-channel device having $V_T = -3$ V, $V_{GS0} = -5$ V, $K = 1.0$ mA/V^2, $V_{DD} = -20$ V, and $V_{DS0} = -8$ V.

ANSWER:

$$I_{D0} = -K(V_{GS0} - V_T)^2 = -1.0 \times 10^{-3}(-5 + 3)^2 = -4.0 \text{ mA}$$

$$g_m = -2K(V_{GS0} - V_T) = -2 \times 10^{-3}(-5 + 3) = 4.0 \text{ millimhos}$$

$$R_d = \frac{V_{DD} - V_{DS0}}{I_{D0}} = \frac{-20 + 8}{-4 \times 10^{-3}} = 3000 \text{ } \Omega$$

$$g_v = -g_m R_d = -4.0 \times 10^{-3} \times 3000 = -12$$

5.2c. Taylor Expansion Approach to Small-Signal Operation

We shall now look at the problem discussed in Section 5.2a from a somewhat generalized point of view. This will at the same time answer the question why in linear amplifiers the device should operate in the pinch-off region.

Let the device have a drain current

$$I_D = f(V_{GS}, V_{DS}) \tag{5.34}$$

Let V_{GS0} and V_{DS0} be the quiescent operating voltages giving rise to a quiescent current I_{D0}. Let $V_{GS} = V_{GS0} + \Delta V_{GS}$ and $V_{DS} = V_{DS0} + \Delta V_{DS}$, where ΔV_{GS} and ΔV_{DS} are small ac signals. Then the small-signal ac current is

$$\Delta I_D = \frac{\partial I_D}{\partial V_{GS}}\bigg|_0 \Delta V_{GS} + \frac{\partial I_D}{\partial V_{DS}}\bigg|_0 \Delta V_{DS} \tag{5.35}$$

where the symbol $|_0$ means that the partial derivative must be evaluated at the operating point. We recognize the first partial derivative as the trans-conductance g_m and the second one as the drain conductance g_d, so that

$$\Delta I_D = g_m \Delta V_{GS} + g_d \Delta V_{DS} \tag{5.36}$$

We may represent this relationship by the ac equivalent circuit of Fig. 5.8. If we connect a load resistance R_d to the output, we have

$$\Delta I_D = \frac{-\Delta V_{DS}}{R_d} \tag{5.37}$$

Substitution into Eq. (5.36) yields

$$g_m \Delta V_{GS} + \left(g_d + \frac{1}{R_d}\right)\Delta V_{DS} = 0; \qquad \Delta V_{DS} = -\frac{g_m}{g_d + 1/R_d}\Delta V_{GS} \tag{5.38}$$

so that the voltage gain may be written as

$$g_v = \frac{\Delta V_{DS}}{\Delta V_{GS}} = -\frac{g_m R_d}{1 + g_d R_d} \tag{5.39}$$

Therefore $|g_v|$ is as large as possible if g_m is as large as possible and g_d as small as possible. We shall see that this is the case when the device operates in the pinch-off region.

We first evaluate g_m and g_d for an n-channel JFET or MOSFET and obtain from Eq. (5.8)

$$g_m = \frac{\partial I_D}{\partial V_{GS}}\bigg|_0 = 2KV_{DS0}; \qquad g_d = \frac{\partial I_D}{\partial V_{DS}}\bigg|_0 = 2K(V_{GS0} - V_P - V_{DS0}) \tag{5.40}$$

for $0 \leq V_{DS0} \leq V_{GS0} - V_P$, so that g_m goes from zero to $2K(V_{GS0} - V_P)$ and g_d goes from $2K(V_{GS0} - V_P)$ to zero if V_{DS0} is gradually raised from zero to $V_{GS0} - V_P$. For the pinch-off region we have from Eq. (5.8a)

$$g_m = 2K(V_{GS0} - V_P); \qquad g_d = 0 \tag{5.40a}$$

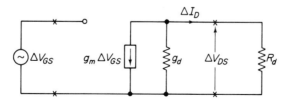

Fig. 5.8. Low-frequency small-signal equivalent circuit of field effect transistor.

We thus conclude that a linear FET amplifier attains the highest voltage gain when the device is operating in the pinch-off region.

For JFETs or MOSFETs with a p-type channel the same discussion can be held. In that case, for the non-pinch-off region [Eq. (5.12)] we have

$$g_m = \frac{\partial I_D}{\partial V_{GS}}\bigg|_0 = -2KV_{DS0}; \qquad g_d = \frac{\partial I_D}{\partial V_{DS}}\bigg|_0 = -2K(V_{GS0} - V_P - V_{DS0})$$

$$(5.41)$$

for $V_{GS} - V_P \leq V_{DS} \leq 0$, whereas for the pinch-off region, $V_{DS} \leq V_{GS} - V_P$ [Eq. (5.12)],

$$g_m = -2K(V_{GS0} - V_P); \qquad g_d = 0 \qquad (5.41a)$$

We note that g_m and g_d are positive. This is the merit of giving the expression for I_D a negative sign in these devices.

If one investigates the characteristics more carefully, one finds that I_D increases slowly with increasing $|V_{DS}|$ in the pinch-off region because the parameter K increases slowly with increasing $|V_{DS}|$ in that region. Thus for an n-type channel we have

$$g_d = \frac{\partial I_D}{\partial V_{DS}}\bigg|_0 = \frac{\partial K}{\partial V_{DS}}\bigg|_0 (V_{GS0} - V_P)^2 = I_{D0}\left[\frac{1}{K}\frac{\partial K}{\partial V_{DS0}}\bigg|_0\right] \qquad (5.42)$$

where $|_0$ again indicates evaluation at the operating point. Since the form between brackets is practically a constant, g_d varies linearly with I_{D0}. The effect thus becomes more pronounced at higher currents. It seems that the value of g_d is larger for MOSFETs than for JFETs.

In RC-coupled amplifiers the allowed values of R_d are relatively small. Since no large error is made in the evaluation of the gain as long as $g_d R_d \ll 1$, one can often ignore the effect. In tuned amplifiers where the load resistance R_d is replaced by a high-Q-tuned circuit, the effect can be quite large since there are no restrictions on the tuned circuit impedance in this case.

According to Eq. (5.39),

$$|g_v| = \frac{g_m R_d}{1 + g_d R_d} = g_m \frac{R_0 R_d}{R_0 + R_d} \qquad (5.42a)$$

where $R_0 = 1/g_d$, so that the effective resistance consists of the resistors R_0 and R_d in parallel.

EXAMPLE 10: If $g_m = 2 \times 10^{-3}$ mho, $R_0 = 20,000 \ \Omega$, and $R_d = 5000 \ \Omega$, find $|g_v|$ and compare it with $g_m R_d$.

ANSWER:

$$g_m \frac{R_0 R_d}{R_0 + R_d} = 2 \times 10^{-3}\frac{20 \times 5}{25}10^3 = 8$$

whereas $g_m R_d = 10$.

We finally turn to the distortion. If the device operates in the pinch-off we write

$$I_D = f(V_{GS}) \tag{5.43}$$

Hence if $V_{GS} = V_{GS0} + \Delta V_{GS}$, where ΔV_{GS} is an ac signal, we have for the ac drain current

$$\Delta I_D = \frac{\partial I_D}{\partial V_{GS}}\bigg|_0 \Delta V_{GS} + \frac{1}{2}\frac{\partial^2 I_D}{\partial V_{GS}^2}\bigg|_0 \Delta V_{GS}^2 \tag{5.44}$$

Higher-order terms can be neglected since all derivatives higher than the second order are zero because the characteristic is quadratic. Substituting for I_D, we obtain, if $\Delta V_{GS} = v_{gs} \cos \omega t$,

$$\Delta I_D = 2K(V_{GS} - V_P)v_{gs}\cos \omega t + \tfrac{1}{2}Kv_{gs}^2 + \tfrac{1}{2}Kv_{gs}^2 \cos 2\omega t \tag{5.44a}$$

in agreement with Eq. (5.27). The harmonic distortion can thus be evaluated with the Taylor expansion method.

5.3. BIASING CIRCUITS

Up to now we biased the gate of an FET by means of a dc supply voltage V_{GS0}. In junction FETs and in depletion mode MOSFETs it is common practice to supply V_{GS0} by a source resistor R_s bypassed by a capacitor C_s. Figure 5.9(a) shows the circuit. We want to choose R_s such that the operating point is $P(I_{D0}, V_{GS0})$. To that end, we require that the voltage drop across R_s be $-V_{GS0}$. That is,

$$-V_{GS0} = I_{D0}R_s \quad \text{or} \quad R_s = -\frac{V_{GS0}}{I_{D0}} \tag{5.45}$$

This is graphically illustrated in Fig. 5.9(b), which shows the (I_D, V_D) characteristic and the *bias line*,

$$-V_{GS} = I_D R_s \tag{5.45a}$$

connecting the operating point P to the origin O. The slope of this bias line gives R_s.

We must require $V_{DS0} \geq V_{GS0} - V_P$, so that the voltage between drain and ground must satisfy the condition

$$V_{D0} = V_{DS0} - V_{GS0} \geq -V_P \tag{5.46}$$

The capacitance C_s must be so chosen that it effectively bypasses the resistance R_s. We shall see in Section 6.2c that this is the case if

$$\omega C_s \geq g_m \tag{5.47}$$

for all frequencies of practical interest. This is somewhat stricter than the requirement

$$\omega C_s \geq \frac{1}{R_s} \tag{5.47a}$$

Of course, the whole problem can be avoided by omitting the resistor

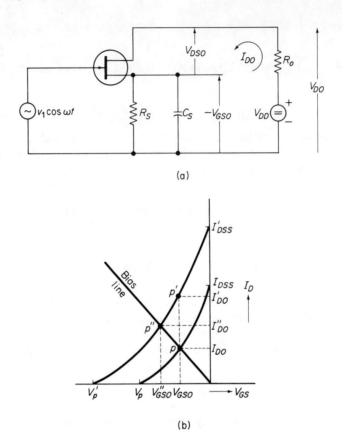

(a)

(b)

Fig. 5.9. (a) Amplifier circuit, in which gate bias is provided by a bypassed source resistor R_s. (b) Diagram showing the stabilizing effect of the source resistor R_s on the operating point.

R_s and the capacitor C_s altogether and using zero dc bias for the gate. However, we saw in one of the examples of Section 5.2a that this reduces the gain. If one wants a larger gain, one should use a bypassed bias resistor R_s.

The bypassed source resistor makes the operation of the circuit less sensitive to variations in K and V_p between different units. For example, Fig. 5.9(b) shows another unit having values I'_{DSS} and V'_p instead of I_{DSS} and V_p. If the two FETs are interchanged and the circuit was biased by a fixed EMF V_{GS0}, the operating point would change to $P'(V_{GS0}, I'_{D0})$, so that the quiescent current would change from I_{D0} to I'_{D0}. If, however, the circuit is biased by the resistor R_s, the operating point changes to $P''(V''_{GS0}, I''_{D0})$, so that the quiescent current changes only from I_{D0} to I''_{D0}. This makes it much easier to keep the device operating in the pinch-off region.

Except in direct-coupled amplifiers, which we shall discuss in a moment, it is common practice to connect the output of the one FET stage to the input of the next FET stage by capacitive coupling. Figure 5.10 shows such a circuit. It is then necessary to provide a dc path between gate and ground. This is done by a gate leak resistor which has a value of about 1 MΩ or more. The coupling capacitor C_b must be so chosen that it transmits the signal to the gate for all frequencies of practical interest. According to Section 4.3b, this is the case if

$$\omega C_b R_g \geq 1 \tag{5.48}$$

for all frequencies of practical interest.

The biasing circuit discussed at the beginning of this section can be used for n- and p-channel JFETs and for MOSFETs operating in the depletion mode. For MOSFETs operating in the enhancement mode the circuit cannot be used, since it gives a bias voltage of the wrong polarity.

This problem is easily solved if the device can operate in the enhancement mode only, that is, for n-channel MOSFETs with $V_T > 0$ and p-channel MOSFETs with $V_T < 0$. It consists of connecting the gate to the drain by means of a resistor R_g of the order of 1 MΩ or more, so that $V_{GS0} = V_{DS0}$ (Fig. 5.11).

As an example we shall consider a p-channel MOSFET with $V_T < 0$. The pinch-off condition here is

$$V_{DS0} \leq V_{GS0} - V_T$$

However, since $V_{GS0} = V_{DS0}$, this condition implies that $0 \leq -V_T$ and this is always true because V_T is negative.

Fig. 5.10. Circuit with gate leak resistor R_g to provide dc path from gate to ground and coupling capacitor C_b to provide dc separation from previous stage.

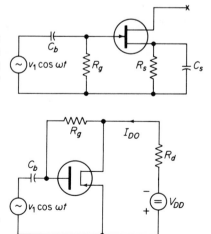

Fig. 5.11. Biasing circuit for enhancement mode MOSFET.

We now have, since $V_{DS0} = V_{GS0}$,

$$I_{D0} = -K(V_{GS0} - V_T)^2 = \frac{V_{DD} - V_{GS0}}{R_d} \tag{5.49}$$

It is often simplest to require a given value for I_{D0} and then find from Eq. (5.49) what values of V_{GS0} and R_d are needed.

We see that

$$R_d = \frac{V_{DD} - V_{GS0}}{I_{D0}} = -\frac{V_{DD} - V_{GS0}}{K(V_{GS0} - V_T)^2} \tag{5.49a}$$

Since the transconductance g_m of the device is

$$g_m = -2K(V_{GS0} - V_T) \tag{5.49b}$$

the absolute voltage gain of the circuit is

$$g_m R_d = 2\frac{(V_{DD} - V_{GS0})}{V_{GS0} - V_T} \tag{5.50}$$

A great advantage of the enhancement mode devices is that individual amplifier stages can be directly coupled, as long as the individual transistors do not differ very much in K-value or in V_T. This is shown in Fig. 5.12. All one needs here is a gate leak resistor R_g in the first stage. The circuit can therefore be put in integrated form; it only needs the provision of supplying an external gate leak R_g, the supply voltage V_{DD}, and the input and output coupling capacitors C_b.

EXAMPLE: A p-channel enhancement mode MOSFET has $K = 0.5 \text{ mA/V}^2$ and $V_T = -2 \text{ V}$. The supply voltage $V_{DD} = -20 \text{ V}$. Find (1) the value of V_{GS0} needed to make $I_{D0} = -2 \text{ mA}$, (2) the value of R_d needed to produce this current, and (3) the voltage gain of the amplifier.

ANSWER: (1) $-2 \times 10^{-3} = -\frac{1}{2} \times 10^{-3}(V_{GS0} + 2)^2$. $(V_{GS0} + 2)^2 = 4$; $(V_{GS0} + 2) = -2$ [we need the negative root, for $V_{GS0} < V_T$ or $(V_{GS0} - V_T) < 0$], or $V_{GS0} = -4 \text{ V}$. (2) $R_d = (-20 + 4)/(2 \times 10^{-3}) = 8000 \text{ }\Omega$. (3) $g_m R_d = [2(-20 + 4)]/(-4 + 2) = 16$.

It should be noted that the gate leak resistor R_g causes a loading of the

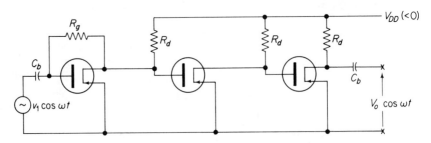

Fig. 5.12. Diagram of an integrated amplifier using enhancement mode MOSFETs.

input circuit. If the amplifier has a voltage gain g_v, then the resistance R_g presents an apparent resistance (Miller resistance)

$$R_{\text{in}} = \frac{R_g}{1 + |g_v|} \tag{5.51}$$

to the input. The proof is similar to the proof of the expression for the Miller effect input capacitance in Section 4.4. It should be investigated whether such an input resistance can be tolerated. If this is not the case, R_g should be made larger.

If an *n*-type MOSFET with $V_T < 0$ is operated in the enhancement mode, one must use a potentiometer-type bias circuit (Fig. 5.13). This is not as convenient as operating in the depletion mode. Moreover, the absolute value of the voltage gain $g_m R_d$ decreases if the gate voltage is made more positive, as was shown in Section 5.2a. Therefore the enhancement mode is not as useful as the depletion mode operation, unless reduced distortion or higher power are the overriding requirements.

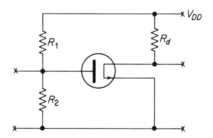

Fig. 5.13. Biasing circuit for depletion mode MOSFET operating in the enhancement mode.

5.4. POWER CONSIDERATIONS

All FETs have an individual power dissipation limit P_{max} that must not be exceeded. It should therefore be carefully investigated whether a particular FET circuit meets this requirement.

5.4a. Power Dissipation Under Quiescent Conditions

If I_{D0}, V_{DS0} is the quiescent operating point, then the power dissipated by the device is $(P_{\text{diss}})_0 = I_{D0} V_{DS0}$. This power must not exceed the limit P_{max}.

According to Fig. 5.6(b), the power supplied by the battery is $P_{\text{supp}} = V_{DD} I_{D0}$ and the power dissipated in the load is $P_{\text{dcL}} = I_{D0}^2 R_d$. Since $V_{DS0} = V_{DD} - I_{D0} R_d$,

$$(P_{\text{diss}})_0 = I_{D0} V_{DS0} = I_{DD} V_{DD} - I_{D0}^2 R_L = P_{\text{supp}} - P_{\text{dcL}} \tag{5.52}$$

This is nothing but the energy law for the circuit.

EXAMPLE 1: If $V_{DD} = 20$ V, $V_{GS0} = -2$ V, $V_P = -4$ V, $K = 1.0$ mA/V², and $R_d = 1000 \ \Omega$, find the values of P_{supp}, P_{dcL}, and $(P_{\text{diss}})_0$.

ANSWER:

$$I_{D0} = K(V_{GS0} - V_P)^2 = 4\,\text{mA}$$

$$P_{\text{supp}} = I_{D0}V_{DD} = 4 \times 10^{-3} \times 20\,\text{W} = 80\,\text{mW}$$

$$P_{\text{dcL}} = I_{D0}^2 R_d = (4 \times 10^{-3})^2 \times 10^3\,\text{W} = 16\,\text{mW}$$

$$(P_{\text{diss}})_0 = P_{\text{supp}} - P_{\text{dcL}} = 80 - 16 = 64\,\text{mW}$$

5.4b. Power Dissipation Under ac Conditions

If a small ac signal $v_1 \cos \omega t$ is applied to the input, the power P_{supp} supplied by the battery does not change, and the power P_{dcL} dissipated by the load does not change either. However, now an ac power

$$P_{\text{ac}} = \tfrac{1}{2}(g_m v_1)^2 R_d \tag{5.53}$$

is dissipated in the load, where g_m is the transconductance of the device. Thus for the power P_{diss} dissipated in the device we have

$$P_{\text{diss}} = P_{\text{supp}} - P_{\text{dcL}} - P_{\text{ac}} \tag{5.54}$$

so that less power is dissipated by the device. An FET used as an ac power amplifier therefore has less power dissipation under operating conditions than under quiescent conditions. Therefore, if the circuit was so designed that $(P_{\text{diss}})_0 < P_{\text{max}}$, then certainly $P_{\text{diss}} < P_{\text{max}}$.

We see that P_{ac} increases with the square of the ac amplitude. Therefore more ac power can be generated by going to a larger ac input signal. One cannot go too far in this direction, however, for the device generates considerable harmonic distortion. Since the amplitude of the second harmonic current in the output is $\tfrac{1}{2}Kv_1^2$, the second harmonic power generated is

$$P_{\text{ac2}} = \tfrac{1}{2}(\tfrac{1}{2}Kv_1^2)^2 R_d \tag{5.55}$$

which varies as the fourth power of v_1. Care must be taken that the second harmonic power stays within tolerable limits.

There are two other restrictions on the ac amplitude in order to avoid even more serious distortion. We shall discuss them for an n-channel JFET.

1. V_{GS} should never drop below V_P, for otherwise the drain current is cut off completely during part of the cycle.
2. V_{DS} should never drop below the maximum value of $V_{GS} - V_P$, for otherwise the drain current is limited during part of the cycle.

We shall now show that the second requirement puts a serious restriction on the load resistance R_d. This is illustrated in Fig. 5.14, which shows the (I_D, V_D) characteristic for quiescent operation $(V_{GS} = V_{GS0})$ and the (I_D, V_D) characteristic when the value of V_{GS} has its maximum value $V_{GS\,\text{max}}$. The small-signal load line must be so drawn that operating point B is to the right of point $A(I_{D0}, V_{GS0} - V_P)$, whereas the large-signal load line must be so drawn

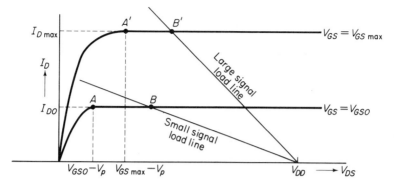

Fig. 5.14. (I_D, V_{DS}) characteristic and load line, showing the difference in the choice of load line between small-signal and large-signal operation.

that point B' is to the right of point $A'(I_{max}, V_{GS\,max} - V_P)$. Since the value of R_d is determined by the slope of the load line, we see that the large-signal value of the load resistance R_d is much smaller than the small-signal value of R_d. This must be borne in mind in the design of FET power amplifiers.

Because of the serious second harmonic distortion, FET power amplifiers are much less popular than transistor power amplifiers. In the next section we shall discuss a circuit that eliminates the distortion problem.

Finally, we shall illustrate the distortion problem with the help of an example.

EXAMPLE 2: An FET power amplifier has $V_P = -4$ V, $V_{GS0} = -1.5$ V, and an ac amplitude v_1 of 2 V, so that V_{GS} swings from -3.5 V to $+0.5$ V. Find d_2.

ANSWER:

$$d_2 = \frac{v_1}{4(V_{GS} - V_P)} = \frac{2}{4(-1.5 + 4.0)} = 20\%$$

This distortion is so large that it is clearly intolerable. Even if the amplitude v_1 is reduced to 1 V, the distortion is still quite objectionable. For that reason circuits that reduce the distortion are highly desirable.

5.4c. Push-Pull Operation

Figure 5.15(a) shows a circuit consisting of two identical JFETs, with an n-type channel with ac signals $v_1 \cos \omega t$ applied to the gates in opposite phase. The ac voltages of frequency ω at points A and B are 180 *degrees out of* phase, so that their contributions to the output voltage v_0 add, whereas the ac voltages of frequency 2ω at points A and B are *in phase*, so that their effects cancel each other. We now express this mathematically as follows:

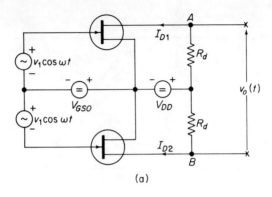

(a)

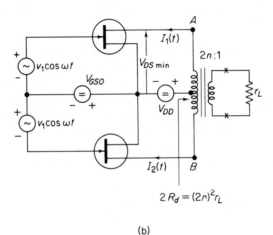

(b)

Fig. 5.15. (a) Push-pull operation, eliminating second harmonic distortion in the output. (b) Transformer-coupled push-pull circuit, giving higher efficiency than the circuit of (a).

$$I_{D1}(t) = K(V_{GS0} - V_P + v_1 \cos \omega t)^2; \quad I_{D2}(t) = K(V_{GS0} - V_P - v_1 \cos \omega t)^2 \tag{5.56}$$

$$v_0(t) = -I_{D1}R_d + I_{D2}R_d = -R_d[I_{D1}(t) - I_{D2}(t)] = -R_d 2g_m v_1 \cos \omega t \tag{5.57}$$

so that the second harmonic signal has completely disappeared. The circuit is called a *push-pull* circuit; it eliminates all even harmonics but leaves the odd ones.

It should be born in mind that the resistance R_d should be determined with the help of the large-signal load line method illustrated in Fig. 5.14. This can be expressed mathematically as follows. The largest voltage applied to the gate is $V_{GS0} + v_1$, so that the device is always operating at pinch off if the drain voltage

$$V_{DS} \geq V_{DS\,min} = V_{GS0} + v_1 - V_P \tag{5.58}$$

The maximum value of the load resistance R_d that can now be used is

$$R_d = \frac{V_{DD} - V_{DS\,min}}{I_{max}} \tag{5.59}$$

where

$$I_{max} = K(V_{GS0} + v_1 - V_P)^2 \tag{5.60}$$

is the maximum instantaneous drain current that flows through the device.
The transconductance g_m of each device is

$$g_m = 2K(V_{GS0} - V_P) \tag{5.61}$$

and the maximum ac power that can be delivered to the load is twice the power developed by a single stage:

$$P_{ac} = 2 \times \tfrac{1}{2}(g_m v_1)^2 R_d \tag{5.62}$$

The dc current flowing through each device is

$$I_D = K(V_{GS0} - V_P)^2 + \tfrac{1}{2}K v_1^2 \tag{5.63}$$

Here the first term is the quiescent current I_{D0} and the second term the rectified current I_r of the device. The total dc power supplied by the voltage supply is therefore

$$P_{supp} = 2I_D V_{DD} = 2(I_{D0} + I_r)V_{DD} \tag{5.64}$$

The ratio

$$\eta = \frac{P_{ac}}{P_{supp}} \tag{5.65}$$

is called the *efficiency* of the power amplifier. It is not very high since so much dc power is wasted in heating the load resistors.

We shall illustrate the above disussion with the help of two examples.

EXAMPLE 3: An FET power amplifier has $V_{DD} = 25$ V, $K = 0.8$ mA/V^2, $V_P = -4$ V, $V_{GS0} = -1.5$ V, and the ac gate amplitude $v_1 = 2.0$ V. Find the maximum value of R_d that can be used and determine the ac power dissipated in the load for push-pull operation.

ANSWER:

$$I_{max} = 0.8(-1.5 + 4.0 + 2.0)^2 = 16.2 \text{ mA}$$

$$V_{DS\,min} = -1.5 + 4.0 + 2.0 = 4.5 \text{ V}$$

$$R_d = \frac{25 - 4.5}{16.2 \times 10^{-3}} = 1265 \ \Omega$$

$$g_m = 2 \times 0.8 \times 10^{-3}(-1.5 + 4.0) = 4.0 \times 10^{-3} \text{ mho}$$

$$g_m v_1 = 8.0 \text{ mA}$$

$$P_{ac} = (g_m v_1)^2 R_d = (8.0 \times 10^{-3})^2 \times 1265 = 81 \text{ mW}$$

EXAMPLE 4: Find the efficiency of the power amplifier in Example 3.

ANSWER:

$$I_{D0} = 0.8(-1.5 + 4.0)^2 = 5.0 \text{ mA}$$

$$I_r = \tfrac{1}{2} \times 0.8 \times (2.0)^2 = 1.6 \text{ mA}$$

$$P_{\text{supp}} = 2 \times (5.0 + 1.6)10^{-3} \times 25 \text{ W} = 330 \text{ mW}$$

$$\eta = \frac{P_{\text{ac}}}{P_{\text{supp}}} = \frac{81}{330} = 24.6\%$$

An even better circuit is shown in Fig. 5.15(b). Here the two load resistors R_d are replaced by a center-tapped transformer of turns ratio 2n:1 terminated in a load resistor r_L. The ac impedance between points A and B is

$$2R'_d = (2n)^2 r_L \tag{5.66}$$

which must be compared with the resistance $2R_d$ of the previous case. If the transformer is sufficiently close to ideal, the harmonic current does not produce a voltage between points A and B. Therefore, for maximum ac power generation

$$V_{DS\,\min} = V_{GS0} + v_1 - V_P \tag{5.67}$$

if the device stays in pinch off at all times. Here we have taken into account that the ac current has an amplitude $g_m v_1$ and that the drain voltage swings from $V_{DS\,\min}$ to $2V_{DD} - V_{DS\,\min}$ so that the ac voltage between points A and B has a maximum amplitude $2(V_{DD} - V_{DS\,\min})$. Therefore the maximum value of $2R'_d$ is

$$2R'_d = \frac{2(V_{DD} - V_{DS\,\min})}{g_m v_1} \tag{5.68}$$

It should be noticed here that the maximum instantaneous current $I_{\max}$ must now be replaced by the ac current amplitude $g_m v_1$.

The ac power developed into r_L thus has a maximum value of

$$P_{\text{ac}} = \tfrac{1}{2}(g_m v_1)^2 2R'_d = g_m v_1(V_{DD} - V_{DS\,\min}) \tag{5.69}$$

The supplied dc power is again given by Eq. (5.64). The efficiency of the power amplification is much better than in the previous case, since no dc power is wasted in the load resistor.

It is necessary to choose V_{DD} such that the device can withstand maximum drain voltage $2V_{DD} - V_{DS\,\min}$. It is therefore advisable to choose V_{DD} somewhat lower than in the previous case.

We shall illustrate this with the help of examples.

EXAMPLE 5: The same power amplifier of Example 3 is used with a center-tapped output transformer and V_{DD} is therefore lowered to 15 V. Find $2R'_d$, P_{ac}, and the turns ratio $2n$ if $r_L = 10 \ \Omega$.

ANSWER:

$$V_{DS\ min} = 4.5\ \text{V}; \qquad g_m = 4.0 \times 10^{-3}\ \text{mho}; \qquad g_m v_1 = 8\ \text{mA}$$

$$2R'_d = \frac{2(15 - 4.5)}{8.0 \times 10^{-3}} = 2625\ \Omega$$

$$P_{ac} = 8.0 \times 10^{-3}(15 - 4.5) = 84\ \text{mW}$$

$$2n = \left(\frac{2R'_d}{r_L}\right)^{1/2} = \left(\frac{2625}{10}\right)^{1/2} = 16.2$$

EXAMPLE 6: Find the efficiency of the power amplifier in Example 5.

ANSWER: The dc current I_D is the same as in Example 4, i.e., 6.6 mA. Hence

$$P_{supp} = 2 \times 15 \times 6.6 = 198\ \text{mW}$$

$$\eta = \frac{84}{198} = 42.4\%$$

PROBLEMS

5.1. In JFETs with an n-type channel the current I_D is often expressed in terms of the current I_{DSS} for zero drain bias. (a) Show that at pinch off the transconductance at zero bias is

$$g_m = g_{m0} = -\frac{2I_{DSS}}{V_P}$$

(b) Show that the current I_D and the transconductance g_m at any other drain bias may be written as

$$I_D = I_{DSS}\left(1 - \frac{V_{GS}}{V_P}\right)^2; \qquad g_m = g_{m0}\left(1 - \frac{V_{GS}}{V_P}\right)$$

5.2. The circuit of Fig. 5.16 is used as a quadratic voltmeter. The MOSFETs are identical n-channel depletion mode devices that operate in the pinch-off mode under all conditions. The meter resistance is small in comparison with R_d and the meter responds to dc currents only. (a) Find the dc current flowing through the meter for $v_{gso} > 0$. (b) Find the range of input amplitude for which the response is accurately quadratic. (c) What happens if $v_{gso} > -V_T$? (d) If Q_1 and Q_2 are not fully identical, show how the bias circuit in the gate of Q_2 can be used for zero point adjustment.

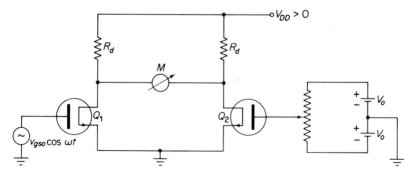

Fig. 5.16.

5.3. A JFET operating in the pinch-off mode is so driven that it gives $p\%$ distortion. Plot the normalized amplitude $-v_1/V_P$ versus the normalized quiescent gate voltage V_{GSO}/V_P.

5.4. The circuit of Fig. 5.9(a) is so designed that for $V_{DD} = 20$ V and for a device with $K = 1.0$ mA/V^2 and $V_P = -4.0$ V the circuit is operating at $V_{GSO} = -2.0$ V, $V_{DS} = 10$ V. (a) Design the circuit; that is, choose R_s and R_d so that this operating point is realized. (b) Evaluate the small-signal voltage gain of the circuit.

5.5. Having chosen the values of R_s and R_d as in Problem 5.4(a), a device is now used that has the same $V_P = -4.0$ V but has $K = 2.0$ mA/V^2 instead. (a) Investigate whether the device is now operating in the pinch-off mode. (b) If this is the case, evaluate the small-signal voltage gain of the circuit and compare with the result of Problem 5.4(b). *Hint for* (a): First evaluate V_{GSO}, then I_D, and then V_{DO}.

5.6. Having chosen the values of R_s and R_d as in Problem 5.4(a), a device is now used that has the same K-value of 1.0 mA/V^2 but has $V_P = -6.0$ V. (a) Investigate whether this device is operating in the pinch-off mode. (b) If this is the case, evaluate the small-signal voltage gain of the circuit and compare with the result of Problem 5.4(b).

5.7. In the circuit of Fig. 5.10 the value of R_g is 1.0 MΩ. Find the value of C_b so that the lower cutoff frequency of the circuit is 10 Hz.

5.8. The circuit of Fig. 5.11 has a voltage gain of -9 and $R_g = 1.0$ MΩ. (a) Find the input impedance of the circuit due to the Miller effect. (b) Find the value of C_b so that the lower cutoff frequency of the circuit is 10 Hz and compare with the result of Problem 5.7.

5.9. (a) Design the circuit of Fig. 5.12 so that the amplifier gain is 1000. $V_{DD} = -20$ V, $K = 1.0$ mA/V^2, and $V_T = -4$ V. (b) If the output voltage is 1 V, what is the distortion generated in the last stage? *Hint:* Solve for V_{GSO} from the expression for $g_m R_d$.

5.10. In the three-stage amplifier of Fig. 5.12 a resistor R_f is connected from the

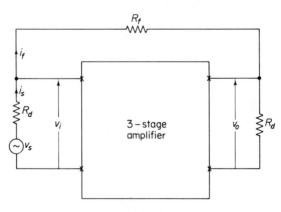

Fig. 5.17.

output to the input (Fig. 5.17). The input is fed from a stage identical to the other stages; that is, the source impedance at the input is R_d. You may assume that $R_f \gg R_d$. Without $R_f(R_f = \infty)$, $v_0/v_s = -1000$. Adjust the value of R_f so that $v_0/v_s = -100$. The values of the device parameters, and hence R_d, are the same as in Problem 5.9. This feedback circuit is used to reduce distortion and improve the frequency response. *Hint:* Bear in mind that the input impedance of the amplifier is infinite so that the input current i_s equals the current i_f through R_f.

5.11. The circuit of Fig. 5.13 is to be operated at $V_{GS0} = 2.0 \text{ V}$. $V_{DD} = 20 \text{ V}$, $V_T = -2.0 \text{ V}$, $V_{DS0} = 8 \text{ V}$, and $K = 0.5 \text{ mA/V}^2$. (a) Find the value of R_d and the voltage gain of the stage. (b) Adjust R_1 and R_2 so that when the output of the stage is capacitively coupled to the next stage, the gain drops only by 10%. (c) Find the amplitude v_1 for 1% distortion. (d) How much does the voltage gain improve if V_{GS0} is lowered to 0 V? What are the advantages and disadvantages of such a step?

5.12. In the circuit of Fig. 5.15(a) the JFETs are replaced by n-channel MOSFETs having $K = 1.0 \text{ mA/V}^2$ and $V_T = -4 \text{ V}$. (a) If $V_{GS0} = 0 \text{ V}$ and $V_{DD} = 32 \text{ V}$, choose v_1 so that each device just draws current for the full cycle. (b) Choose R_d so that the device is at the limit of pinch off when the maximum current is passed. (c) Find the total ac power dissipated in the two load resistors R_d under the above conditions. (d) What is the highest value to which the drain voltage swings? (e) What is the dc power supplied by the voltage V_{DD}, and what is the dc power dissipated in the load resistors?

5.13. In Problem 5.12 the load resistors are replaced by a center-tapped transformer [Fig. 5.15(b)]. (a) Find the value of $2R'_d$ in this case. (b) Find the maximum drain voltage that the device must be able to withstand. (c) Find the ac power delivered into the load. (d) Find the efficiency of the amplifier.

5.14. In the circuit of Fig. 5.18 the identical devices have $V_T = -5 \text{ V}$ and each

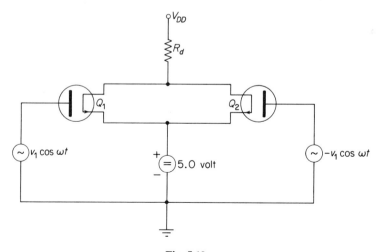

Fig. 5.18.

device is so biased that the current under quiescent conditions is just zero. $K = 1.0 \, \text{mA}/\text{V}^2$ for each device. The input voltage is applied in push-pull. (a) Find the current wave form passing through R_d. Note that each device draws current half the time. (b) Find the dc current and the second harmonic current passing through R_d, assuming that the device is in pinch off at all times. (c) If $V_{DD} = 30$ V and $v_1 = 5$ V, choose R_d so that the device is just at pinch off when it passes the maximum current.

5.15. In Problem 5.14, R_d is replaced by a low-resistance dc current meter. Show that the circuit now acts as a quadratic detector and evaluate the meter reading for $v_1 = 5$ V.

5.16. In Problem 5.14, R_d is replaced by a tuned circuit, tuned at the frequency 2ω (frequency doubler) and damped by a resistance R_d'. $v_1 = 5$ V, $V_{DD} = 30$ V. (a) Find the value of R_d' so that the device is just operating at pinch off when the peak current is passed. (b) Find the power of frequency 2ω dissipated in the load. (c) How much power is needed to drive the input? (d) How much power is supplied by V_{DD}? (e) What is the efficiency of this second harmonic generator? (f) What is the power dissipated in each device?

6

Miscellaneous
FET Circuits

6.1. CONSTANT CURRENT GENERATOR

In many electronic circuits it is convenient to have as a building block a constant current generator, i.e., a device or circuit that gives a constant current if the applied voltage is changed over a wide range.

We saw that an FET operating in the pinch-off mode has a very small drain conductance. Therefore such a device is ideally suited for approximating the constant current generator. Figure 6.1(a)–(c) shows three possibilities: Fig. 6.1(a) gives a constant current generator made with an n-channel JFET, Fig. 6.1(b) gives a constant current generator made with a p-channel JFET, and Fig. 6.1(c) gives a constant current generator made with an n-channel MOSFET that can operate in the depletion mode. We observe that p-channel MOSFETs cannot be used for this purpose, since they can operate in the enhancement mode only. In the first and third circuit the current $|I_{DSS}|$ is flowing *into* the drain and $V_{DS} > -V_P > 0$, whereas in the second circuit the current $|I_{DSS}|$ is flowing *out of* the drain and $V_{DS} < -V_P < 0$.

One of the drawbacks of this arrangement is that the constant current is fully determined by the device. It is often desirable to make the constant current adjustable. This can be done by putting a resistor in the source lead. Figure 6.1(d) shows such an arrangement for an n-channel JFET. We now

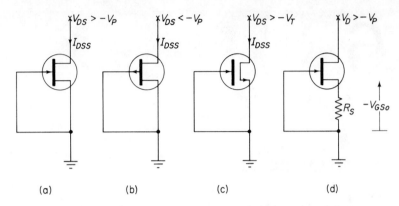

Fig. 6.1. (a) Constant current generator made of n-channel JFET. (b) Constant current generator made of p-channel JFET. (c) Constant current generator made of depletion mode n-channel MOSFET. (d) Constant current generator as of (a) with the current controlled by the source resistor R_s.

have

$$I_{D0} = -\frac{V_{GS0}}{R_S} = K(V_{GS0} - V_P)^2 = \frac{I_{DSS}}{V_P^2}(V_{GS0} - V_P)^2 \qquad (6.1)$$

from which R_s can be determined. We note that in this device V_{GS0} and V_P are both negative and $V_{GS0} > V_P$, so that $V_{GS0} - V_P$ is positive. Since I_{D0} is usually given, we have

$$V_{GS0} - V_P = -\left(\frac{I_{D0}}{I_{DSS}}\right)^{1/2} V_P; \qquad V_{GS0} = \left[1 - \left(\frac{I_{D0}}{I_{DSS}}\right)^{1/2}\right] V_P \qquad (6.1a)$$

so that

$$R_s = -\frac{V_{GS0}}{I_{D0}} = -V_P\frac{1 - (I_{D0}/I_{DSS})^{1/2}}{I_{D0}} \qquad (6.1b)$$

EXAMPLE: If $I_{DSS} = 9$ mA and $V_P = -3$ V, find the value of R_s needed to adjust the current to 1 mA.

ANSWER:

$$R_S = -V_P\frac{1 - (I_{D0}/I_{DSS})^{1/2}}{I_{D0}} = 3\frac{(1 - \frac{1}{3})}{10^{-3}} = 2000\,\Omega$$

In this case the channel is pinched off at the drain if $V_D > -V_P > 0$, just as in the circuits of Fig. 6.1(a) and (c).

6.2. SOURCE FOLLOWER CIRCUIT

If a large resistance R_s is put in series with the source and a voltage V_G is applied between gate and ground, then the voltage V_S between source and ground follows the gate voltage rather accurately. For that reason the circuit is called a *source follower* circuit. We shall now investigate several of these circuits.

6.2a. Source Follower Circuit as a dc Circuit

We shall first discuss the dc operation of a source follower with a depletion mode MOSFET having an n-type channel. Figure 6.2 shows the circuit. We require pinch-off operation, so that

$$V_{DD} - V_S \geq V_{GS} - V_T \qquad (6.2)$$

Now

$$V_{GS} = V_G - V_S; \qquad I_{DO} = \frac{V_S}{R_s} = K(V_{GS} - V_T)^2 \qquad (6.3)$$

Substituting for V_{GS}, we have

$$(V_G - V_S - V_T)^2 = \frac{V_S}{KR_s} \qquad (6.3a)$$

or

$$V_S^2 - 2V_S\left(V_G - V_T + \frac{1}{2KR_s}\right) + (V_G - V_T)^2 = 0$$

so that

$$V_S = V_G - V_T + \frac{1}{2KR_s} - \left[\left(V_G - V_T + \frac{1}{2KR_s}\right)^2 - (V_G - V_T)^2\right]^{1/2} \qquad (6.4)$$

We need the minus sign in front of the square root sign, for $V_G - V_S - V_T$ must be positive. Equation (6.4) can be rewritten as

$$V_S = V_G - V_T - \frac{[4(V_G - V_T)KR_s + 1]^{1/2} - 1}{2KR_s} \qquad (6.4a)$$

If the product KR_s is sufficiently large, the last term is small and V_S follows V_G reasonably accurately.

We see that $V_S = 0$ if $V_G = V_T$. How high can the gate voltage V_G be raised so that the device stays in the pinch-off condition? For pinch off $V_{DS} \geq V_{GS} - V_T$, but $V_{DS} = V_{DD} - V_S$ and $V_{GS} = V_G - V_S$, so that $V_{DS} \geq V_{GS}$

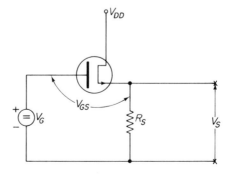

Fig. 6.2. dc source follower circuit.

$- V_T$ means that

$$V_{DD} - V_S \geq V_G - V_S - V_T \quad \text{or} \quad V_G \leq V_{DD} + V_T \qquad (6.5)$$

In the limiting case $V_G = V_{DD} + V_T$,

$$V_S = V_{DD} - \frac{(4V_{DD}KR_s + 1)^{1/2} - 1}{2KR_s}$$

EXAMPLE 1: Take $V_{DD} = 20$ V, $K = 1.0$ mA/V², and $R_s = 5000\ \Omega$. Find V_S for $V_G = V_{DD} + V_T$.

ANSWER: $2KR_s = 10$ V^{-1} and $4V_{DD}KR_s = 400$.

$$V_S \simeq V_{DD} - \frac{2(V_{DD}KR_s)^{1/2}}{2KR_s} = V_{DD}\left[1 - \frac{1}{(V_{DD}KR_s]^{1/2}}\right]$$
$$= 20(1 - 0.10) = 18 \text{ V}$$

This means that if V_G is raised from V_T to $20 + V_T$ V, V_S is raised by only 18 V, so that the circuit is not completely linear.

6.2b. Source Follower Circuit as an ac Circuit

We now look at the circuit as an ac circuit. To make the circuit more flexible, the source resistance R_s is split into a series combination of a small resistance R_{s1} and a large resistance R_{s2}. The resistance R_{s1} provides the dc gate bias via the gate leak resistor R_g; this sets the dc current level [Fig. 6.3(a)]. We shall now show that this circuit can be represented by the equivalent circuit of Fig. 6.3(b) if the coupling capacitor C_b is chosen so large that it presents an ac short circuit for all frequencies of practical interest.

The proof is simple. Since the gate voltage is $v_g - v_s$, and it gives rise to the current generator $g_m(v_g - v_s)$ between drain and source, the voltage

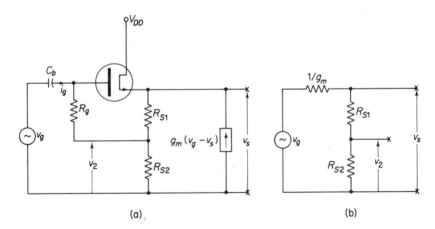

(a)   (b)

Fig. 6.3. (a) ac source follower circuit with gate bias arrangement. (b) Small-signal equivalent circuit of (a).

v_s is caused by the current generator $g_m(v_g - v_s)$. Hence we have, if $R_s = R_{s1} + R_{s2}$,

$$v_s = g_m(v_g - v_s)R_s; \qquad v_s(1 + g_m R_s) = g_m R_s v_g$$

or

$$v_s = v_g \frac{R_{s1} + R_{s2}}{1/g_m + (R_{s1} + R_{s2})} = v_g \frac{g_m(R_{s1} + R_{s2})}{1 + g_m(R_{s1} + R_{s2})} \tag{6.6}$$

which corresponds to the output signal of Fig. 6.3(b).

This is important, for $1/g_m$ can be a relatively low resistance by proper choice of the FET and proper design of the biasing circuit. The circuit can therefore transform from a very high impedance level to a much lower impedance level.

We shall now calculate the input impedance of the source follower. To that end, we turn to Fig. 6.3(a) and observe that the ac voltage v_2 developed across R_{s2} is

$$v_2 = \frac{R_{s2}}{R_{s1} + R_{s2}} v_s = v_g \frac{R_{s2}}{1/g_m + R_{s1} + R_{s2}} \tag{6.6a}$$

so that

$$v_g - v_2 = v_g \frac{1/g_m + R_{s1}}{1/g_m + R_{s1} + R_{s2}} = v_g \frac{1 + g_m R_{s1}}{1 + g_m(R_{s1} + R_{s2})} \tag{6.6b}$$

Consequently, the current i_g flowing into the input is

$$i_g = \frac{v_g - v_2}{R_g} = \frac{1 + g_m R_{s1}}{1 + g_m(R_{s1} + R_{s2})} \frac{v_g}{R_g}$$

so that the input resistance is

$$R_{in} = \frac{v_g}{i_g} = R_g \frac{1 + g_m(R_{s1} + R_{s2})}{1 + g_m R_{s1}} \tag{6.6c}$$

EXAMPLE 2: If $R_{s1} = 500\ \Omega$, $R_{s2} = 5000\ \Omega$, $g_m = 2000\ \mu\text{mho}$, and $R_g = 1.0$ MΩ, find v_s/v_g and R_{in}.

ANSWER:

$$\frac{v_s}{v_g} = \frac{2 \times 10^{-3} \times 5500}{1 + 2 \times 10^{-3} \times 5500} = \frac{11}{12}$$

$$R_{in} = 10^6 \frac{1 + 2 \times 10^{-3} \times 5500}{1 + 2 \times 10^{-3} \times 500} = 6 \times 10^6\ \Omega$$

We also note that for $(R_{s1} + R_{s2}) \gg 1/g_m$ all reference to the FET has disappeared. Therefore the circuit remains practically linear as long as the gate voltage swings between the limits $V_{DD} + V_T$ and V_T, i.e., for amplitudes v_g less than $\frac{1}{2}V_{DD}$.

6.2c. Choice of the Bypass Capacitor in the Gate Bias Circuit

We are now able to answer the question of how large the bypass capacitor C_s for the gate bias resistor R_s in Fig. 5.9 must be and we shall show that the requirement is

$$\omega C_s \geq g_m \tag{6.7}$$

for all frequencies of practical interest.

To that end, we redraw Fig. 6.3(a) as shown in Fig. 6.4(a). We see that the current generator $g_m(v_g - v_s)$ also flows through the collector lead. We now equate

$$g_m(v_g - v_s) = g'_m v_g \tag{6.8}$$

call g'_m the *apparent transconductance* of the circuit, and require that

$$\left| \frac{g'_m}{g_m} \right| \geq \frac{1}{2}\sqrt{2}$$

for all frequencies of practical interest. Substitution of Eq. (6.8) yields

$$\left| 1 - \frac{v_s}{v_g} \right| \geq \frac{1}{2}\sqrt{2} \quad \text{or} \quad \left| 1 - \frac{v_s}{v_g} \right|^2 \geq \frac{1}{2} \tag{6.9}$$

for all frequencies of interest.

In analogy with Fig. 6.3, we can now replace Fig. 6.4(a) by the equivalent circuit of Fig. 6.4(b). With the help of this equivalent circuit we can easily evaluate $| 1 - v_s/v_g |^2$. To simplify matters further, we ignore the effect of R_s. This is not allowed at low frequencies but is sufficiently accurate for our purpose. We then have

$$v_s \simeq \frac{1/(j\omega C_s)}{1/g_m + 1/(j\omega C_s)} v_g = \frac{g_m}{g_m + j\omega C_s} v_g$$

$$\left| 1 - \frac{v_s}{v_g} \right|^2 = \left| \frac{j\omega C_s}{g_m + j\omega C_s} \right|^2 = \frac{\omega^2 C_s^2}{g_m^2 + \omega^2 C_s^2} \geq \frac{1}{2}$$

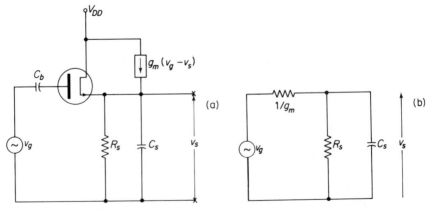

Fig. 6.4. (a) Calculation of the effect of the bypass capacitor of the source resistor R_s. (b) Small-signal equivalent circuit used in the calculation.

or

$$\omega^2 C_s^2 \geq g_m^2 \quad \text{or} \quad \omega C_s \geq g_m$$

as had to be proved.

EXAMPLE 3: If the device used in the circuit has $g_m = 4 \times 10^{-3}$ mho, find the value of C_s needed to have condition (6.7) satisfied down to 10 cycles.

ANSWER:

$$C_s = \frac{g_m}{2\pi f} \quad \text{at } f = 10 \text{ cycles}$$

$$= \frac{4 \times 10^{-3}}{20\pi} = 62 \ \mu\text{F}$$

6.2d. Source Follower with a Constant Current Generator Load

By using an FET constant current generator instead of the resistance R_s, we obtain an even more linear circuit if both devices are operated at pinch off. Figure 6.5 shows the circuit for *n*-channel MOSFETs.

The FET Q_1 has a current

$$I_{D1} = K_1 V_{T1}^2 \tag{6.10}$$

and FET Q_2 has a current

$$I_{D2} = K_2(V_G - V_S - V_{T2})^2 \tag{6.11}$$

since $V_{GS2} = V_G - V_S$, according to Fig. 6.5. Here V_{T1} and V_{T2} are the turn-on voltages of Q_1 and Q_2, respectively.

Since the two devices are in series, the currents are equal and hence

$$K_1 V_{T1}^2 = K_2(V_G - V_S - V_{T2})^2; \quad \left(\frac{K_1}{K_2}\right)^{1/2} V_{T1} = V_G - V_S - V_{T2} \tag{6.12}$$

or

$$V_S = V_G - V_{T2} - \left(\frac{K_1}{K_2}\right)^{1/2} V_{T1} \tag{6.13}$$

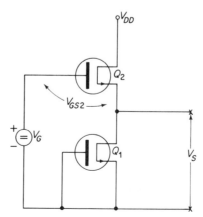

Fig. 6.5. Integrated source follower, using two depletion mode MOSFETs with *n*-type channel.

which is accurately linear. The circuit thus operates as a true source follower.

The limitations to the linearity of the circuit are set by the requirement that Q_1 and Q_2 must be in pinch-off condition.

1. Q_1 in pinch off means that $V_S \geq -V_{T1}$. Substitution into Eq. (6.13) yields

$$V_G \geq V_{T2} - V_{T1}\left[1 - \left(\frac{K_1}{K_2}\right)^{1/2}\right] \tag{6.14}$$

2. Q_2 in pinch off means that $V_{DD} - V_S \geq V_{GS2} - V_{T2} = V_G - V_S - V_{T2}$ or

$$V_G \leq V_{DD} + V_{T2} \tag{6.14a}$$

Hence linearity is fully ensured if

$$V_{T2} - V_{T1}\left[1 - \left(\frac{K_1}{K_2}\right)^{1/2}\right] \leq V_G \leq V_{DD} + V_{T2} \tag{6.14b}$$

For identical devices, $K_1 = K_2 = K$ and $V_{T1} = V_{T2} = V_T$. For ac operation the gate of Q_2 should then be dc-biased at $\frac{1}{2}V_{DD} + V_T$ and the maximum allowed amplitude of the ac signal is $\frac{1}{2}V_{DD}$.

EXAMPLE 4: If $V_{DD} = 20$ V and $V_T = -3$ V, find the dc gate bias V_{G0} that allows for the maximum ac gate signal; determine this signal v_g and the ac output amplitude v_s.

ANSWER:

$$V_{G0} = \tfrac{1}{2}V_{DD} + V_T = 10 - 3 = 7 \text{ V}$$
$$v_g = \tfrac{1}{2}V_{DD} = 10 \text{ V}; \qquad v_s = 10 \text{ V}$$

6.2e. Integrated Source Follower with Two Enhancement Mode MOSFETs

Figure 6.6(a) shows an integrated source follower made of two p-channel MOSFETs. The circuit is somewhat different from the previous case in that the gate of the load FET is connected to the drain rather than to the source, because it is an enhancement device. For the sake of simplicity we assume that Q_2 and Q_1 have the same turn-on voltage V_T. They may have different parameters K, however. We have

$$I_{D1} = -K_1(V_{GS1} - V_T)^2 = -K_1(V_S - V_T)^2 \tag{6.15}$$

since $V_{GS1} = V_S$, whereas

$$I_{D2} = -K_2(V_{GS2} - V_T)^2 = -K_2(V_G - V_S - V_T)^2 \tag{6.16}$$

because $V_{GS2} = V_G - V_S$. However, since the two devices are in series, the currents are equal, so that

$$K_1(V_S - V_T)^2 = K_2(V_G - V_S - V_T)^2 \tag{6.17}$$

or

$$\left(\frac{K_1}{K_2}\right)^{1/2}(V_S - V_T) = V_G - V_S - V_T$$

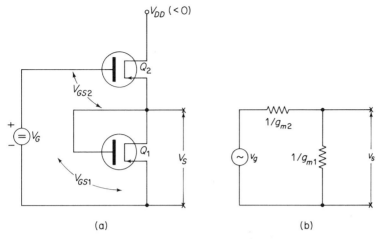

Fig. 6.6. Integrated source follower, using two enhancement mode MOSFETs with p-type channel.

Solving for V_S yields

$$V_S = \frac{V_G - V_T}{1 + (K_1/K_2)^{1/2}} + V_T \frac{(K_1/K_2)^{1/2}}{1 + (K_1/K_2)^{1/2}} \qquad (6.17a)$$

The output follows the input linearly. The best operation is obtained if $(K_1/K_2)^{1/2}$ is small, for in that case $V_S \simeq V_G - V_T$. This can be achieved by proper choice of $K_1/K_2 = w_1 L_2/(w_2 L_1)$, where w_1, L_1 and w_2, L_2 are the channel width and channel length of Q_1 and Q_2, respectively.

What is the range of linear operation? The output voltage V_S no longer follows V_G if Q_2 is either cut off, i.e., nonconducting, or if Q_2 is not in the pinch-off condition. Q_1 is always in pinch-off condition since the gate is returned to the drain.

1. Q_2 is nonconducting. The limit of nonconduction occurs if

$$V_{GS2} = V_G - V_S = V_T \quad \text{or} \quad V_S = V_G - V_T \qquad (6.18)$$

Substitution into (6.17a) yields

$$V_G - V_T = \frac{V_G - V_T}{1 + (K_1/K_2)^{1/2}} + V_T \frac{(K_1/K_2)^{1/2}}{1 + (K_1/K_2)^{1/2}} \quad \text{or} \quad V_G - V_T = V_T$$

or

$$V_G = 2V_T; \qquad V_S = V_T \qquad (6.18a)$$

2. Q_2 is not in pinch off. The limit of pinch off occurs if

$$V_{DS2} = V_{DD} - V_S = V_{GS2} - V_T = V_G - V_S - V_T \qquad (6.19)$$

or

$$V_G = V_{DD} + V_T \quad \text{or} \quad V_S = \frac{V_{DD} + V_T(K_1/K_2)^{1/2}}{1 + (K_1/K_2)^{1/2}} \qquad (6.19a)$$

as is found by substituting $V_G - V_T = V_{DD}$ into Eq. (6.17a). Thus linear operation occurs for

$$V_{DD} + V_T \leq V_G \leq 2V_T \tag{6.20}$$

and the corresponding range for V_S is

$$\frac{V_{DD} + V_T(K_1/K_2)^{1/2}}{1 + (K_1/K_2)^{1/2}} \leq V_S \leq V_T \tag{6.21}$$

If the circuit is used for ac purposes and v_g is the ac input signal, then the ac output signal v_s is

$$v_s = \frac{v_g}{1 + (K_1/K_2)^{1/2}} \tag{6.22}$$

so that a voltage gain of approximately unity is obtained if K_1/K_2 is small.

We can also look at this problem from an equivalent circuit point of view. Since Q_2 has a transconductance $g_{m2} = K_2(V_{GS2} - V_T)$ and Q_1 has an output conductance $g_{m1} = K_1(V_{GS1} - V_T)$, we have, in analogy with Fig. 6.4(b), the equivalent circuit of Fig. 6.6(b). As is seen from that circuit,

$$v_s = v_g \frac{1/g_{m1}}{1/g_{m2} + 1/g_{m1}} = v_g \frac{g_{m2}}{g_{m1} + g_{m2}} = \frac{v_g}{1 + g_{m1}/g_{m2}} \tag{6.23}$$

Since $K_1(V_{GS1} - V_T)^2 = K_2(V_{GS2} - V_T)^2$,

$$\frac{g_{m1}}{g_{m2}} = \frac{K_1(V_{GS1} - V_T)}{K_2(V_{GS2} - V_T)} = \left(\frac{K_1}{K_2}\right)^{1/2} \tag{6.23a}$$

so that Eq. (6.22) agrees with the equivalent circuit.

EXAMPLE 5: If $V_{DD} = -20$ V, $V_T = -3$ V, and $K_2 = 9K_1$, find the dc gate bias that allows for the maximum ac gate amplitude; find this amplitude v_g and determine the corresponding ac output amplitude v_s.

ANSWER: The most desirable dc gate bias is

$$V_{G0} = \tfrac{1}{2}(V_{DD} + 3V_T) = \tfrac{1}{2}(-20 - 9) = -14.5 \text{ V}$$

$$\text{ac gate amplitude } v_g = \tfrac{1}{2}(V_{DD} - V_T) = 8.5 \text{ V}$$

$$\text{ac output amplitude } v_s = \frac{v_g}{1 + (K_1/K_2)^{1/2}} = 6.4 \text{ V}$$

6.3. HIGH-FREQUENCY OPERATION OF FETs

At higher frequencies the equivalent circuit of an FET becomes more complicated because of the device capacitances: C_{gs} between gate and source, C_{ds} between drain and source, and C_{dg} between drain and gate. The first two limit the HF response of *RC*-coupled amplifiers, whereas the last provides feedback from output to input and gives rise to a Miller effect capacitance at the input, which further reduces the HF response (Fig. 6.7).

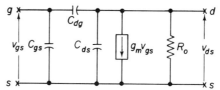

Fig. 6.7. High-frequency small-signal equivalent circuit of the FET, showing the device capacitances.

The effect of the capacitances C_{gs} and C_{ds} can be removed by tuning input and output to the center frequency of the desired passband, but the effect of the capacitance C_{dg} cannot be eliminated in that manner. Tuned HF amplifiers using FETs therefore often have the tendency to oscillate. For that reason it is important to make the capacitance C_{dg} as small as possible. While considerable improvement can be made by proper device design, it is nevertheless worthwhile to investigate other methods for preventing oscillations.

One way of improving the stability of the circuit consists of loading the input and/or output circuit quite heavily. This method is discussed in Section 6.3a. Another method consists of tuning the capacitance C_{dg} at the midband frequency of the desired passband. This is acceptable for a single passband receiver but is undesirable if one wants to tune over a range of frequencies.

6.3a. Common Gate Connection

One way of overcoming the feedback difficulty is to use the source as the input electrode and the drain as output electrode. This is called the *common gate* circuit. The reason for the improvement is twofold:

1. The input admittance of the circuit is g_m, the transconductance of the device, so that the input circuit is heavily loaded.
2. The source and the drain are far enough apart so that one can make the capacitance C_{ds} very small by proper device design. For example, in the MOSFET one can bypass both the substrate and the gate to ground, thus preventing coupling between output and input.

We now prove the statement about the input conductance g_m of the circuit. Direct-current bias voltages V_{GSO} and V_{DSO} are applied and a small EMF v_1 is connected between source and ground [Fig. 6.8(a)]. For an n-channel JFET the dc drain current I_D and the small-signal ac drain current i_d flow from drain to source as shown, but if g_m is the transconductance of the device and v_{gs} the ac signal between gate and source, then, since $v_{gs} = -v_1$,

$$i_d = g_m v_{gs} = -g_m v_1 \tag{6.24}$$

Consequently, the input conductance g_{in} of the circuit is

$$g_{in} = -\frac{i_d}{v_1} = g_m \tag{6.24a}$$

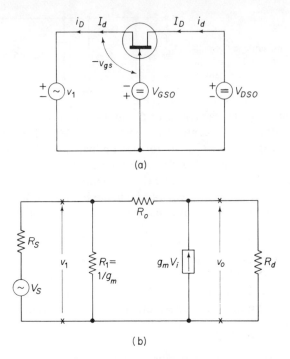

(a)

(b)

Fig. 6.8. (a) Common gate FET amplifier. (b) Small-signal equivalent circuit of (a).

as had to be proved. The equivalent circuit of the common gate connection is thus as shown in Fig. 6.8(b). Usually the effect of the resistance R_0 can be neglected.

We now add a signal source and a load resistance R_d to the circuit of Fig. 6.8(b) to make a complete amplifier stage and we match the source to the input. Then $R_s = R_1$, and

$$v_1 = \tfrac{1}{2}v_s; \qquad v_0 = g_m v_1 R_d = \tfrac{1}{2}g_m R_d v_s \qquad (6.25)$$

To demonstrate that this is indeed useful, we evaluate the power gain. If P_{av} is the power available at the source and P_{out} the power fed into the load R_d, then

$$P_{av} = \frac{1}{8}\frac{|v_s|^2}{R_s}; \qquad P_{out} = \frac{1}{2}\frac{|v_0|^2}{R_d} = \frac{1}{8}g_m^2 R_d |v_s|^2 \qquad (6.26)$$

so that the power gain G is

$$G = \frac{P_{out}}{P_{av}} = g_m^2 R_d R_s = g_m R_d \qquad (6.26a)$$

since $R_s = 1/g_m$. The circuit is thus indeed useful if $g_m R_d > 1$.

Other FETs can be operated in the same way.

6.3b. Cascode Circuit

The common gate circuit has as a drawback that its input impedance is so low. This can be overcome by feeding the common gate circuit from a common source circuit. The configuration thus obtained is called a *cascode* circuit. Usually the drain of the first device and the source of the second device are directly connected and the appropriate bias voltages are applied to G_1, G_2, and D_2. Figure 6.9(a) shows a cascode connection of two *n*-channel JFETs.

The equivalent circuit of the arrangement is shown in Fig. 6.9(b). As seen from this circuit, the output voltage of the first stage is $v_2 = -g_m R_1 v_{gs1} = -v_{gs1}$, so that the output current generator $-g_m v_2 = g_m v_{gs1}$ with the polarity as shown. The two devices together thus have the same transfer characteristic as a single device but with one great benefit: The circuit is stable at high frequencies even when input and output are tuned, because the direct capacitance $C_{d_2 g_1}$ can be kept quite small, whereas the intermediate stage has a very low impedance ($1/g_m$) so that the feedback effects via $C_{d_2 s_2}$ and $C_{d_1 g_1}$ are very small too.

Cascode circuits made on a single chip of silicon are commercially available. Only the leads S_1, G_1, G_2, and D_2 are taken out of the can containing the circuit. Because of these *four* leads, the circuit is called an *FET tetrode*. Another name commonly used is the *dual gate FET*.

While the circuit was discussed for *n*-channel JFETs, it should be emphasized that other types of FETs can be used for the same purpose.

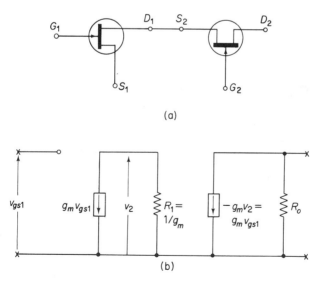

(a)

(b)

Fig. 6.9. (a) The dual gate FET or FET tetrode. (b) Small-signal equivalent circuit of (a).

6.4. DIFFERENTIAL AMPLIFIER

In many electronic problems it is necessary to find the difference between two voltages. A circuit performing this function is called a *differential amplifier* or *difference amplifier*. We shall now show how FETs can perform this function.

Figure 6.10(a) shows the circuit consisting of two identical MOSFETs Q_2 and Q_3 and a third MOSFET Q_1 connected as a current generator, which means that it is a device with $V_T < 0$. Since Q_1 acts as a constant current generator, it does not pass ac current; the ac currents through Q_2 and Q_3 are $g_m(v_2 - v_s)$ and $g_m(v_3 - v_s)$, respectively, where v_s is the ac drain voltage of Q_1. Applying Kirchhoff's law to point A of the circuit, we thus have

$$g_m(v_2 - v_s) + g_m(v_3 - v_s) = 0 \quad \text{or} \quad v_s = \tfrac{1}{2}(v_2 + v_3) \qquad (6.27)$$

so that

$$g_m(v_2 - v_s) = \tfrac{1}{2}g_m(v_2 - v_3); \qquad g_m(v_3 - v_s) = -\tfrac{1}{2}g_m(v_2 - v_3) \qquad (6.27a)$$

Consequently

$$v_{02} = -\tfrac{1}{2}g_m R_d(v_2 - v_3); \qquad v_{03} = \tfrac{1}{2}g_m R_d(v_2 - v_3) \qquad (6.28)$$

$$v_{03} - v_{02} = g_m R_d(v_2 - v_3) \qquad (6.28a)$$

so that the two outputs indeed respond to $v_2 - v_3$, as required. It should be noted that v_2 and v_3 must give the proper dc level so that Q_1 is operated in

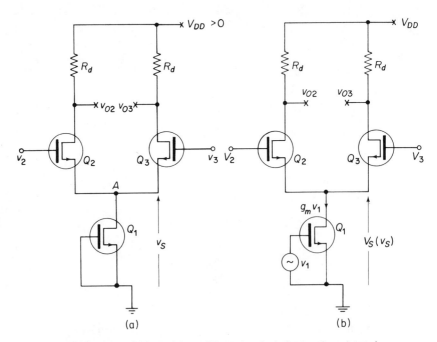

Fig. 6.10. (a) Differential amplifier using depletion-mode *n*-channel MOSFETs. (b) The circuit of (a) used as a volume control circuit.

pinch off. As long as Q_1 is operated in pinch off and the two devices are identical, $v_{03} - v_{02}$ does not respond to the dc level. Response to the dc level is called *common mode* response.

While the above derivation was held for ac signals, it should be obvious that it also holds for small dc signals taken with respect to the common dc level as the reference level.

One has two further possibilities: One can take the output either from the drain of Q_2 or from the drain of Q_3. In the first case the output voltage is 180 degrees out of phase with respect to $v_2 - v_3$, and in the second case it is in phase. These two possibilities can be important in feedback applications.

One can also apply an ac or dc signal to only *one* gate, e.g., to the gate of Q_2. The circuit then has only half the voltage gain of a single FET ($\frac{1}{2}g_m R_d$ instead of $g_m R_d$), but it has the advantage of much greater stability against drift. This is especially important in dc amplifiers.

Finally, it is possible to use the circuit as an HF amplifier with an HF input voltage v_2 and an HF output voltage v_{03}. The resistance R_d should then be replaced by a tuned circuit, tuned at the center frequency of the passband to be amplified. The advantage of this circuit is that it is perfectly stable against oscillations, since the coupling between the drain of Q_3 and the gate of Q_2 can be made very small, just as in the cascode circuit. The disadvantage in comparison with the cascode circuit is that it only has *half* the gain of the latter circuit at the same value of the output load impedance.

We now redraw Fig. 6.10(a) somewhat to understand a different application. A small ac voltage v_1 is applied to Q_1 and different dc gate bias voltages V_2 and V_3 are applied to Q_2 and Q_3. By changing V_2 relative to V_3, one can shift more or less dc and ac current through Q_2 or Q_3. The circuit can thus be used for *volume control* purposes.

To understand this in greater detail, we assume that for the given dc bias conditions the devices Q_1, Q_2, and Q_3 have transconductances g_{m1}, g_{m2}, and g_{m3}, respectively. The ac drain current of Q_1 thus "sees" a conductance $g_{m2} + g_{m3}$, formed by Q_2 and Q_3, so that the ac source voltage is given by

$$v_s = -\frac{g_{m1}v_1}{g_{m2} + g_{m3}} \qquad (6.29)$$

and hence the ac current flowing into the drain of Q_3 is

$$-g_{m3}v_s = \frac{g_{m3}}{g_{m2} + g_{m3}}g_{m1}v_1 \qquad (6.29a)$$

We now start with V_2 so much less than V_3 that g_{m2} is just zero; g_{m3} then has its maximum value g_{max} and the ac output current is $g_{m1}v_1$. If V_2 is now raised, g_{m2} increases and g_{m3} decreases, so that the ac output current decreases. Finally, V_2 is so large that g_{m3} is just zero and g_{m2} has its maximum value g_{max}; the ac output current is then zero. Therefore the ac current in Q_3

can vary continuously from $g_{m1}v_1$ to zero, which is useful for volume control \
purposes.

The circuit has several other applications, e.g., as an amplitude modulator or as a superheterodyne mixer, but it is beyond the scope of this discussion to go into greater detail.

6.5. INTEGRATED CIRCUIT INVOLVING TWO ENHANCEMENT MODE MOSFETs

Let two MOSFETs with a p-type channel have the same value of V_T but different values of K, which is achieved by making the two devices on the same silicon chip with widely different w/L ratios. One of the MOSFETs is used as an amplifier and the other one is used as a load resistor by connecting its gate to the drain (Fig. 6.11). The advantages of this circuit arrangement are the following:

1. No resistors occur, which simplifies integrated manufacturing.
2. The circuit works as a linear amplifier for a rather wide range of operating conditions.
3. The circuit works extremely well in logic circuits.

We shall now discuss the circuit in greater detail.

6.5a. Circuit as an Amplifier

Since the two devices are in series, they have the same current. Hence

$$I_D = -K_1(V_{GS1} - V_T)^2 = I_{D2} = -K_2(V_{GS2} - V_T)^2 \qquad (6.30)$$

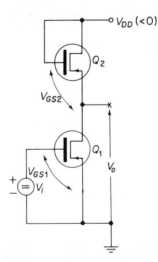

Fig. 6.11. Linear integrated MOSFET amplifier with p-type channels, using a MOSFET as a load resistor.

or

$$(V_{GS1} - V_T)\left(\frac{K_1}{K_2}\right)^{1/2} = (V_{GS2} - V_T) \tag{6.30a}$$

as long as both devices are operating at pinch off.

Since the gate of Q_2 is connected to the drain and Q_2 is an enhancement device, it is always operating at pinch off. However, Q_1 is operating at pinch off for a certain range of values of V_i that will be determined in a moment.

We must now bear in mind that

$$V_{GS1} = V_i; \qquad V_{GS2} = V_{DD} - V_0 \tag{6.31}$$

Substitution into (6.30a) yields

$$(V_i - V_T)\left(\frac{K_1}{K_2}\right)^{1/2} = V_{DD} - V_0 - V_T$$

$$V_0 = V_{DD} - V_T - (V_i - V_T)\left(\frac{K_1}{K_2}\right)^{1/2} \tag{6.32}$$

We thus see that V_0 varies linearly with V_i, so that the circuit acts as a *linear* amplifier when Q_1 is operating under pinch-off condition and not turned off. We also see that the circuit has a voltage gain

$$g_v = -\left(\frac{K_1}{K_2}\right)^{1/2} = -\left(\frac{w_1}{L_1}\frac{L_2}{w_2}\right)^{1/2} \tag{6.32a}$$

which can be made quite large by proper choice of the geometries. Finally, the output voltage varies as $-V_i$, indicating a 180-degree phase difference.

For what range of V_i is the circuit linear? The limits of linear operation are the following:

1. If $V_i = V_T$, Q_1 is just off. Substitution into Eq. (6.32) yields

$$V_0 = V_{DD} - V_T \tag{6.33}$$

2. If $V_0 = V_i - V_T$, Q_1 is just operating at pinch off. Substitution into Eq. (6.32) gives

$$V_i - V_T = V_{DD} - V_T - \left(\frac{K_1}{K_2}\right)^{1/2}(V_i - V_T)$$

or

$$V_i - V_T = \frac{V_{DD} - V_T}{1 + (K_1/K_2)^{1/2}} \tag{6.34}$$

The device is thus linear for

$$V_T + \frac{V_{DD} - V_T}{1 + (K_1/K_2)^{1/2}} \le V_i \le V_T \tag{6.35}$$

The most suitable dc operating point for Q_1 is

$$V_{GS1} = V_T + \frac{1}{2}\frac{V_{DD} - V_T}{1 + (K_1/K_2)^{1/2}} \tag{6.35a}$$

and the largest possible ac amplitude v_i is

$$v_i = -\frac{1}{2}\frac{(V_{DD} - V_T)}{1 + (K_1/K_2)^{1/2}} \tag{6.35b}$$

(note that $V_{DD} - V_T$ is negative, whereas v_i should be a positive number). The corresponding range for V_0 is

$$\frac{V_{DD} - V_T}{1 + (K_1/K_2)^{1/2}} \geq V_0 \geq V_{DD} - V_T \tag{6.36}$$

and the output amplitude is

$$v_0 = +\frac{1}{2}(V_{DD} - V_T)\frac{(K_1/K_2)^{1/2}}{1 + (K_1/K_2)^{1/2}} \tag{6.36a}$$

indicating that v_0 is 180 degrees out of phase with respect to v_i.

EXAMPLE 1: If $K_1/K_2 = 25$, $V_{DD} = -27$ V, and $V_T = -3$ V, find the dc gate bias V_{GS1} for maximum ac gate amplitude, the ac gate amplitude v_i, and the ac output amplitude v_0.

ANSWER:

$$V_{GS1} = -3.0 + \frac{1}{2}\frac{(-27 + 3)}{6} = -3.0 - 2.0 = -5.0 \text{ V}$$

$$v_i = -\frac{1}{2}\frac{(-27 + 3)}{6} = 2.0 \text{ V}$$

$$v_0 = +\frac{1}{2}(-27 + 3)\frac{5}{6} = -10.0 \text{ V}$$

so that $v_0/v_i = -5$, as expected.

While the circuit has interesting features as an amplifier, it has even more interesting features as a logic circuit. We shall discuss this in the next section.

6.5b. Circuit as a Logic Circuit

Since all voltages in the circuit are negative and logic up to now has been defined for positive voltages (so-called *positive* logic), we have to redefine logic for negative voltages (so-called *negative* logic). A 0 is now defined as a negative voltage close to 0 V, whereas a 1 is defined as a voltage that is far negative; say $V > -2$ V is a 0 and $V < -10$ V is a 1.

We shall see now that in our circuit a 0 input gives a 1 output and a 1 input gives a 0 output so that the circuit acts as an *inverter*. Let the devices have the same turn-on voltage $V_T < 0$; then $V_i > V_T$ is called a 0 input and $V_i \ll V_T$ is called a 1 input.

Now if the circuit has a 0 input, then the output voltage is a 1 for $V_i > V_T$, and hence, according to Eq. (6.33),

$$V_0 = V_{DD} - V_T \tag{6.37}$$

so that the device is off. If, however,

$$V_i - V_T \ll \frac{V_{DD} - V_T}{1 + (K_1/K_2)^{1/2}} \tag{6.38}$$

(1 input), then the device operates in the ohmic region and V_0 is quite close to zero, say $V_0 > V_T$, so that the output is a 0, as previously stated.

We shall now calculate V_0 for that case. Since Q_1 is operating in the non-pinch-off region, Eq. (6.30) must be replaced by

$$I_{D1} = -K_1[2V_i - V_T)V_0 - V_0^2] = I_{D2} = -K_2(V_{DD} - V_T - V_0)^2 \tag{6.39}$$

We now observe that $|V_0|$ is quite small, say $|V_0| \ll |V_i - V_T|$ and $|V_0| \ll |V_{DD} - V_T|$. Equation (6.39) may therefore be written as

$$2K_1(V_i - V_T)V_0 = K_2(V_{DD} - V_T)^2 \tag{6.39a}$$

or

$$V_0 = \frac{K_2(V_{DD} - V_T)^2}{2K_1(V_i - V_T)} = I_{D0}R_{on} \tag{6.40}$$

where

$$R_{on} = -\frac{1}{2K_1(V_i - V_T)}; \quad I_{D0} = -K_2(V_{DD} - V_T)^2 \tag{6.40a}$$

This has the following simple interpretation: If V_0 is close to zero, Q_2 acts as a constant current generator carrying a current $I_{D0} = -K(V_{DD} - V_T)^2$ and Q_1 acts as a conductance

$$g_{d1} = \frac{I_{D1}}{V_0} = -2K_1(V_i - V_T) = \frac{1}{R_{on}} \tag{6.40b}$$

and the output is simply the product of I_{D0} and R_{on}, as expected for the ohmic regime of the device.

EXAMPLE 2: If $V_T = -3$ V, $V_{DD} = -21$ V, $K_1 = 1.25$ mA/V^2, and $K_2 = 0.05$ mA/V^2, find the output voltage V_0 for (a) $V_i = -18$ V and (b) $V_i = 0$ V.

ANSWER:

(a)

$$I_{D0} = -K_2(V_{DD} - V_T)^2 = -0.05 \times 10^{-3} \times 18^2 = -16.2 \text{ mA}$$

$$R_{on} = -\frac{1}{2K_1(V_i - V_T)} = \frac{1}{2 \times 1.25 \times 10^{-13} \times 15} = 26.7 \ \Omega$$

$$V_0 = I_{D0}R_{on} = -16.2 \times 10^{-3} \times 26.7 = -0.43 \text{ V}$$

(b)

$$V_i = 0; \quad V_0 = V_{DD} - V_T = -21 + 3 = -18 \text{ V}$$

6.6. FETs AS OFF-ON SWITCHES

FETs used in logic circuits are either *off* (= carrying zero current) or *on* (carrying current), and in the on condition the drain voltage is so close to zero that the output acts as a resistor of value R_{on}. We have already seen an example in Section 6.5b. We shall now discuss the devices in greater detail.

6.6a. *n*-Channel JFET

Let the *n*-channel JFET have a pinch-off voltage $V_p < 0$ and let the gate circuit have a protecting resistor R_g to limit the gate current [Fig. 6.12(a)].

The device is off if $V_i < V_P < 0$ and in that case $V_0 = V_{DD}$. To bring the device to the on condition, a sufficiently large positive voltage V_i is applied, driving the gate into the gate current regime. The gate voltage V_{GS0} is then approximately equal to the turn-on voltage of a silicon diode, or about 0.70 V. The condition for V_i is therefore $V_i \gg V_{GS0}$. If the device were operating in the pinch-off region ($V_0 \geq V_{GS0} - V_P$), the drain current would be

$$I_D = K(V_{GS0} - V_P)^2 \qquad (6.41)$$

However, we want the device to operate in the ohmic regime, i.e., with V_0 close to zero. In that case

$$I_D = \frac{V_{DD} - V_0}{R_d} \simeq \frac{V_{DD}}{R_d} = I_{D0} \qquad (6.42)$$

R_d must therefore be so chosen that

$$\frac{V_{DD}}{R_d} \ll K(V_{GS0} - V_P)^2; \qquad R_d \gg \frac{V_{DD}}{K(V_{GS0} - V_P)^2} \qquad (6.43)$$

Suppose that this has been done; then the output of the device acts as a

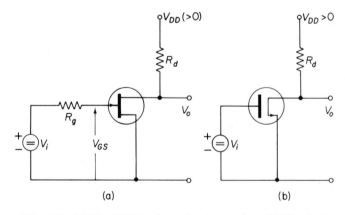

(a)   (b)

Fig. 6.12. (a) The JFET in large-signal operation. (b) The deple-tion mode *n*-channel MOSFET in large-signal operation.

resistance

$$R_{\text{on}} = \frac{1}{g_{d0}} = \frac{1}{2K(V_{GS0} - V_P)} \tag{6.44}$$

so that

$$V_0 = V_{DD} \frac{R_{\text{on}}}{R_d + R_{\text{on}}} \simeq V_{DD} \frac{R_{\text{on}}}{R_d} \simeq I_{D0} R_{\text{on}} \tag{6.45}$$

EXAMPLE: If $V_{GS0} = 0.70$ V, $V_P = -4.0$ V, $V_{DD} = 20$ V, and $K = 1.0$ mA/V^2, find R_{on}, choose R_d, and find V_0.

ANSWER:

$$R_{\text{on}} = \frac{1}{2 \times 10^{-3} \times 4.70} = 106\ \Omega$$

$$\frac{V_{DD}}{K(V_{GS0} - V_P)^2} = \frac{20}{10^{-3} \times (4.7)^2} = 900\ \Omega$$

Therefore choose $R_d = 10,000\ \Omega$. Then

$$I_{D0} \simeq \frac{V_{DD}}{R_d} = \frac{20}{10^4} = 2\ \text{mA}$$

$$V_0 = I_{D0} R_{\text{on}} = 2 \times 10^{-3} \times 106 = 0.21\ \text{V}$$

6.6b. *p*-Channel JFET

This is exactly as in the previous case (Section 6.6a) except that all polarities are the opposite, most $>$ signs become $<$ signs and vice versa, and $V_P > 0$.

The device is off if $V_i > V_P$. The device is on if a sufficiently large negative voltage V_i is applied that drives the device into gate current. In this case $V_{GS} = V_{GS0} \simeq -0.70$ V and the condition for V_i is $V_i \ll V_{GS0}$. To make the device operate in the ohmic mode, R_d must be so chosen that the appropriately modified Eq. (6.43) is satisfied. The equation for R_{on} is

$$R_{\text{on}} = \frac{1}{g_{d0}} = -\frac{1}{2K(V_{GS0} - V_P)} \tag{6.46}$$

so that

$$V_0 = V_{DD} \frac{R_{\text{on}}}{R_d + R_{\text{on}}} \simeq V_{DD} \frac{R_{\text{on}}}{R_d} \tag{6.47}$$

6.6c. *n*-Channel MOSFET with $V_T < 0$

This case is quite similar to the case of Section 6.6a, with two exceptions:

1. There is no restriction on V_i, so that it is always possible to drive the device into the ohmic mode of operation by making V_i sufficiently large.
2. There is no gate current, so that no protective resistor R_g is needed. This leads to Fig. 6.12(b).

One should not make R_d too small, for in that case the drain current in the ohmic regime

$$I_D = \frac{V_{DD} - V_0}{R_d} \simeq \frac{V_{DD}}{R_d} = I_{D0} \qquad (6.48)$$

becomes too large. We must now require for ohmic operation that

$$K(V_i - V_T)^2 \gg \frac{V_{DD}}{R_d} \qquad (6.49)$$

from which the value of V_i for a given R_d follows. Except for this, the discussion is the same as the case of Section 6.6a.

6.6d. *n*-Channel MOSFET with $V_T > 0$

This is quite similar to the case of Section 6.6c; the only difference is that $V_T > 0$. The device is off for $V_i < V_T$ and on for $V_i \gg V_T$. The conditions for operation in the ohmic regime are again given by Eq. (6.48) and (6.49).

6.6e. *p*-Channel MOSFET

Since the *p*-channel MOSFET has $V_T < 0$ and all polarities are opposite to those of the case of Section 6.6d, we have the following: The device is off for $V_i > V_T$ and on for $V_i \ll V_T$. The conditions for operating in the ohmic regime are again given by Eq. (6.48), but Eq. (6.49) must be replaced by

$$-K(V_i - V_T)^2 \ll \frac{V_{DD}}{R_d} \qquad (6.50)$$

because the current is negative and V_{DD} is negative.

6.6f. Cascaded Logic FET Circuits

Often one wants to use an FET logic circuit to drive a similar logic circuit. This means that the following requirements must be met: If the one stage is turned off, then the next stage must be driven into the ohmic regime, and if the one stage is driven into the ohmic regime, the next stage must be turned off. All the circuits of Sections 6.6a–e meet the first requirement, but only the circuits of Sections 6.6d–e meet the second. Therefore only enhancement MOSFETs can be used in logic chains. For this reason most MOSFET integrated logic circuits use enhancement mode MOSFETs.

We shall illustrate this with an example. The circuits of Section 6.6d have $V_T > 0$. It should now be possible to design the preceding similar circuit in such a way that if it is driven into the ohmic regime then $V_0 < V_T$. The following circuit is then turned off. This is not possible for the circuits of Section 6.6c, since $V_T < 0$ and $V_0 > 0$.

6.7. FET CASCADED AMPLIFIERS

Here we shall apply the theory of cascaded amplifiers, as developed in Chapter 4, to FET amplifier stages.

6.7a. Interstage Network

First we shall discuss the case of untuned amplifiers; to that end we turn to the interstage network of Fig. 4.10 and identify the network elements for the FET.

The two parameters g_m and R_0 were defined in Chapter 5 by the relations

$$g_m = \frac{\partial I_D}{\partial V_{GS}}; \qquad \frac{1}{R_0} = \frac{\partial I_D}{\partial V_{DS}} \tag{6.51}$$

The resistance R_i is infinite for all practical purposes. The resistance R_b corresponds to the gate leak resistance R_g, used to connect the gate to ground for dc; its value is of the order of 1–2 MΩ. The device capacitances are shown in Fig. 6.7; consequently $C_0 = C_{ds}$, $C_i = C_{gs}$, and the Miller effect capacitance is $C_f = C_{dg}(1 + |g_{v0}|)$, where g_{v0} is the midband gain of the cascaded stages. Hence the total interstage capacitance is

$$C_{\text{par}} = C_{ds} + C_w + C_{gs} + C_{dg}(1 + |g_{v0}|) \tag{6.52}$$

where C_w is the wiring capacitance.

The gain-bandwidth product is defined as

$$|g_{v0}|f_1 = \frac{g_m}{2\pi C_{\text{par}}} \tag{6.53}$$

Its value is of the order of 20–200 MHz, depending on the individual FETs involved. Because the transconductance increases and the capacitances decrease with decreasing channel length L, the gain-bandwidth product increases with decreasing L. Because of the Miller effect capacitance, $|g_{v0}|f_1$ depends on the midband gain $|g_{v0}|$, increasing with decreasing $|g_{v0}|$.

For most wideband applications $R_g \gg R_L$ and $R_0 \gg R_L$, so that the midband gain $g_{v0} \simeq -g_m R_L$. As was shown in Chapter 5, $|g_{v0}|$ is usually between 5 and 20, limited by the restriction in R_L because of the dc voltage drop in R_L; therefore the upper cutoff frequency is of the order of 1–40 MHz.

EXAMPLE 1: FETs having $g_m = 10^{-2}$ mho, $C_{gs} = 5.0$ pF, $C_{ds} = 2.0$ pF, and $C_{dg} = 2.0$ pF are used in a cascaded amplifier having a required midband gain per stage of 10. Find the interstage circuit capacitance, the load resistance, and the upper cutoff frequency f_1. The wiring capacitance is 3 pF.

ANSWER:

$$|g_{v0}| = g_m R_L = 10; \qquad R_L = \frac{10}{g_m} = 10^3 \ \Omega$$

$$C_{\text{par}} = (2.0 + 3.0 + 5.0 + 2.0 \times 11) \text{ pF} = 32 \text{ pF}$$

$$f_1 = \frac{g_m}{2\pi C_{\text{par}} |g_{v0}|} = \frac{10^{-2}}{2\pi \times 32 \times 10^{-12} \times 10} = 5.1 \text{ MHz}$$

Example 1 shows that the Miller effect capacitance can have a large effect.

Having discussed the untuned amplifier, the case of the tuned amplifier, depicted in Fig. 4.14(a), can be treated in the same way. If one uses dual gate FETs to improve the stability of the circuit, the feedback capacitance between output and input is usually negligible. If each device half has capacitances C_{gs} and C_{dg}, the input capacitance of the dual gate FET is

$$C_i = C_{gs} + 2C_{dg} \tag{6.54}$$

since the first half of the dual gate FET stage has a voltage gain of -1, so that the Miller effect capacitance is $2C_{dg}$.

In wideband amplifiers all resistances in Fig. 4.14(a) are large in comparison with R_c.

EXAMPLE 2: In a dual gate FET each device half has $g_m = 10^{-2}$ mho, $C_{gs} = 5$ pF, $C_{ds} = 2$ pF, and $C_{dg} = 2$ pF. These FETs are being used to build a cascaded amplifier tuned at 100 MHz. If the interstage network is so designed that the midband gain is 10 and the tuned circuit capacitance C is 2 pF, find the load resistance R_c, the tuning inductance L, and the bandwidth B.

ANSWER:

$$|g_{v0}| = g_m R_c = 10; \qquad R_c = \frac{10}{g_m} = 1000\ \Omega$$

$$C_{par} = C_{ds} + C + C_{gs} + 2C_{dg} = (2 + 2 + 5 + 4)\ \text{pF} = 13\ \text{pF}$$

$$\omega_0^2 L C_{par} = 1; \qquad L = \frac{1}{\omega_0^2 L C_{par}} = \frac{1}{4\pi^2 \times 10^{16} \times 1.3 \times 10^{-11}} = 0.19\ \mu\text{H}$$

$$|g_{v0}|B = \frac{g_m}{2\pi C_{par}} \quad \text{or} \quad B = \frac{g_m}{2\pi C_{par}|g_{v0}|}$$

$$= \frac{10^{-2}}{2\pi \times 13 \times 10^{-12} \times 10} = 12.2\ \text{MHz}$$

Sometimes one does not want a wide-band response but needs instead a narrow-band response for better selectivity. This means that one needs high-Q tuned circuits for the interstage networks. In this case R_c may be much larger than R_0, so that the resistance R_0 determines the bandwidth. In such situations it is even more important that the FETs are of the dual gate variety. The drain of the previous stage should then not be connected to the nongrounded side of the inductance L, but rather it should be connected only partway up the inductance L; such an arrangement is called *tapping*.

The inductance L now acts as a step-up transformer of turns ratio n. Figure 6.13(a) shows the circuit. Assuming an ideal transformer situation, this circuit can be represented by the equivalent circuit of Fig. 6.13(b). If the tuned circuit has a Q-factor Q_0 at the center frequency f_0, then

$$R_c = \frac{Q_0}{\omega_0 C} = \omega_0 L Q_0 \tag{6.55}$$

where C is the circuit capacitance and $\omega_0^2 L C = 1$.

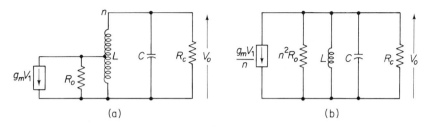

Fig. 6.13. (a) High-impedance tuned output circuit with the drain tapped to the core to decrease the dumping effect of the ac output resistance R_0. (b) Equivalent circuit of (a) used for calculating the ac response.

EXAMPLE 3: The circuit of Fig. 6.13(a) must be tuned at $f_0 = 10$ MHz, and the required bandwidth is 200 kHz. The dual gate FET has $g_m = 6 \times 10^{-3}$ mho and $R_0 = 10^4$ Ω, and the transformer has a turns ratio of 3. If $C = 50$ pF, find the value of L needed to tune the circuit; determine also the value of R_c needed to achieve the required bandwidth, the Q-factor of the circuit, and the midband gain. The input capacitance of the next stage is incorporated into the tuning capacitance C, and the effect of the capacitance C_{ds} of the previous stage can be neglected.

ANSWER:

$$\omega_0^2 LC = 1; \qquad L = \frac{1}{4\pi^2 \times 10^{14} \times 50 \times 10^{-12}} = 5.0 \ \mu\text{H}$$

$$2\pi BCR_{\text{par}} = 1; \qquad R_{\text{par}} = \frac{1}{2\pi BC} = \frac{1}{2\pi \times 2 \times 10^5 \times 50 \times 10^{-12}}$$

$$= 16,000 \ \Omega \qquad n^2 R_0 = 90,000 \ \Omega$$

$$\frac{1}{R_c} = \frac{1}{R_{\text{par}}} - \frac{1}{n^2 R_0} = (0.625 - 0.111) \times 10^{-4} = 0.514 \times 10^{-4} \ \text{mho}$$

$$R_c = 19,500 \ \Omega; \qquad Q_0 = \omega_0 CR_c = 2\pi \times 10^7 \times 5 \times 10^{-11} \times 1.95 \times 10^4 = 61$$

$$\text{midband gain} \ |g_{v0}| = \frac{g_m}{n} R_{\text{par}} = \frac{6 \times 10^{-3}}{3} \times 16 \times 10^3 = 32$$

6.7b. Shunt Peaking and Series Peaking

Here we shall give a short discussion of the improved response that is possible with the help of the shunt peaking and series peaking circuits mentioned in Chapter 4.

First we shall turn to the *shunt-peaked* circuit of Fig. 6.14(a). It can be replaced by the equivalent circuit of Fig. 6.14(b). As is seen by inspection,

$$v_0 = -g_m v_1 \frac{(R_L + j\omega L)/(j\omega C_{\text{par}})}{R_L + j\omega L + 1/(j\omega C_{\text{par}})} = -g_m v_1 \frac{1 + j\omega L/R_L}{1 - \omega^2 LC_{\text{par}} + j\omega C_{\text{par}} R_L} \tag{6.56}$$

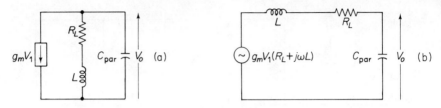

Fig. 6.14. (a) Shunt-peaked interstage circuit to improve the frequency response. (b) Alternate equivalent circuit used for the calculation of the response.

Introducing the dimensionless parameter m with the help of the definition

$$L = R_L^2 C_{par} m \qquad (6.57)$$

the square of the voltage gain may be written as

$$|g_v(f)|^2 = \left|\frac{v_0}{v_0}\right|^2 = |g_{v0}|^2 \frac{1 + m^2\omega^2 C_{par}^2 R_{par}^2}{1 + (1 - 2m)\omega^2 C_{par}^2 R_{par}^2 + m^2\omega^4 C_{par}^4 R_{par}^4} \qquad (6.58)$$

If $m^2 = 1 - 2m$ or $m = \sqrt{2} - 1 \doteq 0.414$, $|g_v/g_{v0}|^2$ is very close to unity as long as

$$m^2\omega^4 C_{par}^4 R_{par}^4 < 1 + (1 - 2m)\omega^2 C_{par}^2 R_{par}^2 \qquad (6.58a)$$

This is called the case of the *optimally flat response*.

The frequency response has a maximum at $\omega = 0$ for $m < \sqrt{2} - 1$. For $m > \sqrt{2} - 1$ the response has a minimum at $\omega = 0$ and a maximum at some elevated frequency; this maximum becomes more pronounced when m becomes larger.

Solving

$$\left|\frac{g_v(f_1)}{g_{v0}}\right| = \frac{1}{2}\sqrt{2} \qquad (6.59)$$

gives the following: The upper cutoff frequency f_1 satisfies

$$2\pi f_1 C_{par} R_{par} = 1.72 \qquad (6.59a)$$

for $m = \sqrt{2} - 1$. The upper cutoff frequency has thus increased by 72% in comparison with the nonpeaked case ($m = 0$).

If one evaluates the step function response, one finds that the response is speeded up by the inductance L. For $m < \frac{1}{4}$ there is no overshoot; for $m > \frac{1}{4}$ there is some overshoot that increases with increasing m; for $m = \sqrt{2} - 1$ the overshoot is 3%. This indicates that m should not be chosen too large.

EXAMPLE 4: If $C_{par} = 30$ pF and $R_L = 1000\ \Omega$, find the value of the inductance L needed to give the optimally flat response.

ANSWER: $L = mR_L^2 C_{par} = 0.414 \times 10^6 \times 3 \times 10^{-11} = 12.4\ \mu$H.

Next we shall discuss the *series-peaked* circuit. Here we make a small simplification in that we neglect the effect of $C_0 = C_{ds}$ and consider the

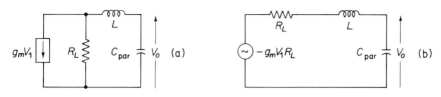

Fig. 6.15. (a) Series-peaked interstage circuit to improve the frequency response. (b) Alternate equivalent circuit used for calculating the response.

peaking inductance L to be in series with C_{par}. We then obtain the circuit of Fig. 6.15(a). It can be represented by the equivalent circuit of Fig. 6.15(b). As is seen by inspection,

$$v_0 = -g_m R_L v_1 \frac{1/(j\omega C_{par})}{R_L + j\omega L + 1/(j\omega C_{par})} = -\frac{g_m R_L v_1}{1 - \omega^2 L C_{par} + j\omega C_{par} R_L}$$

$$(6.60)$$

Again introducing the dimensionless parameter m given by Eq. (6.57), the square of the voltage gain may be written as

$$|g_v(f)|^2 = |g_{v0}|^2 \frac{1}{1 + (1 - 2m)\omega^2 C_{par}^2 R_{par}^2 + m^2 \omega^4 C_{par}^4 R_{par}^4} \qquad (6.61)$$

For $m = \frac{1}{2}$ the middle term in the denominator disappears and

$$\left|\frac{g_v(f)}{g_{v0}}\right|^2 = \frac{1}{1 + \frac{1}{4}\omega^4 C_{par}^4 R_{par}^4} \qquad (6.61a)$$

This is again called the case of *optimally flat response*.

The upper cutoff frequency f_1 is defined by

$$\left|\frac{g_v(f_1)}{g_{v0}}\right| = \frac{1}{2}\sqrt{2} \quad \text{or} \quad 2\pi f_1 C_{par} R_L = \sqrt{2} \qquad (6.61b)$$

The upper cutoff frequency is thus 41.4% larger than in the nonpeaked case. This is not so good as in the series-peaked case. For $m > \frac{1}{2}$ the response shows a maximum at an elevated frequency that becomes more pronounced if m becomes larger. For that reason the parameter m should not be chosen too large.

If one evaluates the pulse response, one finds that the inductance L speeds up the response. As in the previous case, there is no overshoot for $m < \frac{1}{4}$, whereas overshoot occurs for $m > \frac{1}{4}$ that increases with increasing m. Consequently, m should not be made too large.

PROBLEMS

6.1. In the circuit of Fig. 6.1(d) the FET has $I_{DSS} = 8$ mA, $V_P = -4$ V. Find the value of R_s needed to make a constant current generator of 0.5 mA.

6.2. In radiation detectors one is often faced by the problem that a small ac signal is developed across the large internal resistance R of the detector. For example, the detector can be represented by a signal of $10^{-3} \cos \omega t$ V in series with 10^{10} Ω. With the help of a source follower this signal is converted into a signal across a low impedance. Figure 6.16 shows the circuit. Find the ac voltage developed across R_s if $V_T = -3.00$ V, $K = 1.00$ mA/V^2, and $R_s = 1000$ Ω. *Hint:* First find the dc operating point, then determine the transconductance at that point, and finally use the equivalent circuit of Fig. 6.3(b), replacing $R_{s1} + R_{s2}$ by R_s.

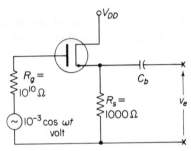

Fig. 6.16.

6.3. An alternative way of solving the difficulty discussed in Problem 6.2 is to use an ac FET amplifier. One then often runs into the difficulty that the detector has a capacitance C in parallel with R and that this limits the frequency response of the detector. To overcome this, one uses a two-stage amplifier and feeds back to the input by means of a small capacitance C_f (see Fig. 6.17). Show that the total input capacitance is now $C_{par} = C + (1 - \mu)C_f$, where μ is the voltage gain of the two-stage amplifier, and adjust C_f so that $C_{par} = 0$.

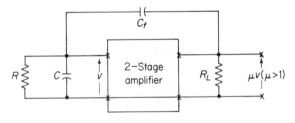

Fig. 6.17.

6.4. A dual gate FET is used as an HF amplifier with tuned output. The output resistance R_0 of the device is 20,000 Ω and the tuned circuit impedance R_c of the tuned output circuit is also 20,000 Ω. If the device has a transconductance of 4×10^{-3} mho, find the ac voltage gain of the tuned amplifier stage.

6.5. In the circuit of Fig. 6.10(a) the ac signal to be amplified is applied to the gate of Q_2 and the output signal is taken from the gate of Q_3. $V_{DD} = 20$ V. The FETs have $I_{DSS} = 4$ mA and $V_T = -2.0$ V. The FETs Q_2 and Q_3 are identical. $V_T = -2.0$ V. (a) If the gates of Q_2 and Q_3 are returned to a reference voltage $V_G = 4.0$ V, show that Q_1 is operating at pinch off. (Find the value of $V_s = V_{DS1}$.) (b) Find the transconductance of Q_2 and Q_3. (c) What are the minimum values of V_{D2} and V_{D3}, so that Q_2 and Q_3 are operated at pinch off? Choose the actual values of V_{D2} and V_{D3} about 1 V higher, making it a round number. (d) Now determine R_d and the voltage gain of the amplifier if the output is taken from Q_3.

6.6. In the circuit of Fig. 6.10(b) the resistance R_d in the drain of Q_2 is eliminated and the output is taken from Q_3. Q_1, Q_2, and Q_3 have $I_{DSS} = 4.0$ mA, $V_T = -2.0$ V, and $V_{DD} = 20$ V. The voltages V_2 and V_3 have a fixed reference level of 5.0 V, V_3 is fixed, and V_2 is variable; that is, $V_2 = 5.0 + \Delta V_2$ and $V_3 = 5.0$ V. By choice of ΔV_2 one can now either make $I_{D2} = 0$ or $I_{D3} = 0$. (a) For what value of ΔV_2 is $I_{D2} = 0$, and what is the value of V_s in this case? (b) For what value of ΔV_2 is $I_{D3} = 0$, and what is the value of V_s in this case? (c) What value of R_d must be used so that Q_3 is always operating at pinch off? Choose V_{D3} about 1 V higher than the minimum value required, making it a round number. (d) What are the maximum and the minimum voltage gains of the circuit?

6.7. An integrated MOS amplifier has $V_{DD} = -27$ V, $V_T = -3$ V, and $K_1/K_2 = 9$. (a) Find the dc gate bias for maximum ac gate amplitude, the maximum ac gate amplitude v_i, and the maximum ac drain amplitude v_0. (b) Calculate the maximum and minimum current if $K_2 = 0.1$ mA/V². (c) The circuit is used as a logic circuit. Find V_0 when the device Q_1 is off. (d) Find V_{GS1} when $V_0 = -1$ V and calculate the current flowing through the circuit under that condition.

6.8. The currents flowing in the logic circuit of Problem 6.7 are too large. For that reason V_{DD} is lowered to -8 V, $K_1 = 0.9$ mA/V², and $K_2 = 0.1$ mA/V². The logic circuit is driven directly from a previous identical stage. (a) Find V_{GS1} when the input is a logical 1. (b) Calculate $V_0 = V_{on}$ for that case and investigate whether this is low enough so that the next stage will be turned off. (c) Calculate the device current when the input is at a logical one.

6.9. A p-channel JFET has $V_{DD} = -20$ V, $V_P = +4.0$ V, $V_{GS0} = -0.70$ V, and $K = 1.0$ mA/V². Find R_{on}, choose R_d, and find V_{on}. Show that $R_d = 5000\ \Omega$ is sufficient.

6.10. The JFET in the circuit of Fig. 6.18 has $r_d = 30,000\ \Omega$, $I_{DSS} = 10$ mA, and $V_P = -4$ V. (a) Choose R so that $V_0 = 15$ V and investigate whether the FET is indeed operating at pinch off. (b) What is the output ripple?

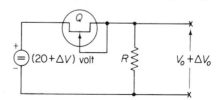

Fig. 6.18.

7

Transistor Operation and Simple Circuits

7.1. CURRENT FLOW IN SILICON TRANSISTORS

A transistor consists of two diodes connected back to back with one region, the base, in common. When the device is used as an amplifier, the base-emitter junction is forward-biased and the base-collector junction is back-biased. The current flow is then mainly due to minority carrier *injection into* the base by the emitter and minority carrier *extraction from* the base by the collector [Fig. 7.1(a)]. There can also be a slight contribution to the emitter current by the injection of majority carriers from the base into the emitter, but usually this effect is quite small.

In well-designed transistors practically all the injected minority carriers are collected by the collector; only relatively few disappear by recombination in the base. In this recombination process majority carriers also disappear; they must be replenished by majority carriers entering the base region through the base contact.

The three electrodes E, C, and B of the transistor thus carry the following currents [Fig. 7.1(b)]:

1. The emitter current I_E, caused by minority carrier injection into the base by the emitter.
2. The collector current I_C, caused by minority carrier extraction from the base by the collector.

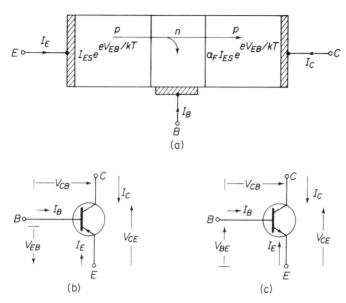

Fig. 7.1. (a) Schematic cross-section of a *p-n-p* transistor used to illustrate the current flow. (b) Symbol for a *p-n-p* transistor with current flow. (c) Symbol for an *n-p-n* transistor with current flow.

3. The base current I_B, caused by recombination of minority and majority carriers in the base.

We shall count these three currents positive when flowing *into* the device. We shall now discuss these currents in greater detail.

7.1a. *p-n-p* Silicon Transistor under Normal Bias

First we shall discuss the *p-n-p* silicon transistor. For normal operation the emitter-base junction is forward-biased and the emitter injects holes into the base. The collector is back-biased and extracts holes from the base. The injected hole current I_E, flowing into the emitter contact from the outside, may therefore be written (in analogy with the diode case) as

$$I_E = I_{ES} \exp\left(\frac{eV_{EB}}{kT}\right) = I_{ES} \exp\left(-\frac{eV_{BE}}{kT}\right) \tag{7.1}$$

where V_{EB} is the emitter-base voltage and $V_{BE} = -V_{EB}$ is the base-emitter voltage, whereas I_{ES} is a kind of saturation current.

Here we have neglected the very small saturation currents of the emitter-base and the collector-base diodes. This is usually allowed, for in normal operation I_E is of the order of 1 mA, whereas these saturation currents are of the order of picoamperes or less. The effect of these saturation currents is that for a floating (= open) emitter a very small current I_{CBO}, called the collector-saturated current, flows from base to collector.

As in the diode case, the turn-on voltage of a silicon *p-n-p* transistor, defined as the emitter-base voltage for an emitter current of 1 mA, is about 0.70 V.

The part α_F of the injected hole current I_E is collected by the collector, where α_F is practically independent of operating conditions for a wide current range. Since the collector current I_C is taken positive when flowing *into* the collector, we may write

$$I_C = -\alpha_F I_E = -\alpha_F I_{ES} \exp\left(\frac{eV_{EB}}{kT}\right) \tag{7.2}$$

The case in which the emitter is the input and the collector is the output is called the *common base connection* [Fig. 7.2(a)]. Therefore the factor α_F can be called the *common base current amplification factor*; in well-designed transistors, α_F can be very close to unity, say $\alpha_F = 0.99$.

In the discussion given up to now, it was assumed that all current was carried by holes; the deviation of α_F from unity is then due to hole-electron recombination in the base. In actual transistors there are two small additional effects:

    1. Injection of electrons from the base into the emitter region. This contribution to I_E has the same dependence on V_{EB} as the contribu-

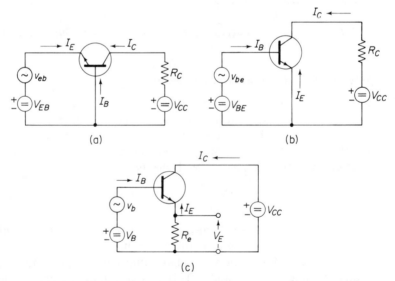

Fig. 7.2. (a) Common base operation of a transistor. (b) Common emitter operation of a transistor. (c) Common collector, or emitter follower, operation of a transistor.

tion due to hole injection from the emitter, but it gives no contribution to the collector current. This effect is important only at high currents in most units.

2. Recombination of electrons and holes in the emitter-base space-charge region. This depends more slowly on V_{EB} than the injection of holes by the emitter; therefore this effect is most pronounced at relatively small currents.

Applying Kirchhoff's law to the transistor, we may calculate the base current I_B. Calling I_B positive when flowing *into* the base contact, we have

$$I_E + I_C + I_B = 0 \tag{7.3}$$

so that

$$I_B = -(1 - \alpha_F)I_E = -(1 - \alpha_F)I_{ES} \exp\left(\frac{eV_{EB}}{kT}\right) \tag{7.4}$$

We can now express I_E and I_C in terms of I_B as follows:

$$I_E = -\frac{I_B}{(1 - \alpha_F)}; \qquad I_C = \frac{\alpha_F}{1 - \alpha_F} I_B \tag{7.5}$$

Since $1 - \alpha_F$ is a small quantity, I_E and I_C are much larger than I_B. It thus makes sense to use the device with the base as input; the small base current I_B then controls the much larger emitter and collector currents. This gives rise to two possibilities:

1. The base is the input and the collector is the output electrode. This is called the *common emitter* connection [Fig. 7.2(b)]. We may then write

$$I_C = \beta_F I_B; \qquad \beta_F = \frac{\alpha_F}{1 - \alpha_F} \tag{7.6}$$

Since α_F is close to unity (0.99 or higher), β_F can be a large number, of the order of 100 or more. The factor β_F is called the *common emitter current amplification factor*; in transistor manuals it is designated by the symbol h_{FE}.

2. The base is the input and the emitter is the output electrode. This is called the *common collector* or *emitter follower* connection [Fig. 7.2(c)]. Again introducing the parameter β_F, we may write

$$I_E = -(1 + \beta_F)I_B \tag{7.7}$$

The factor $1 + \beta_F$ is called the *common collector* current amplification factor.

The three transistor circuits—common base, common emitter, and common collector—are shown in Fig. 7.2(a)–(c). Figure 7.2(a) shows the common base circuit; here $V_{EB} = -V_{BE}$ is negative, Figure 7.2(b) shows the common emitter circuit, and Fig. 7.2(c) shows the common collector circuit. In the first two circuits the load resistance R_c is in the collector lead; in the third circuit the load resistance R_e is in the emitter lead.

In well-designed transistors β_F is not very strongly dependent on current, so that I_C and I_E vary practically linearly with I_B. Since $\beta_F \gg 1$, the device then operates as a linear current amplifier. We shall see that the device is far from linear when used as a voltage amplifier.

7.1b. *n-p-n* Transistor for Normal Bias Conditions

The *n-p-n* transistor is similar to the *p-n-p* transistor, but the currents flow in opposite directions and the bias voltages have opposite polarity. We thus have

$$I_E = -I_{ES} \exp\left(-\frac{eV_{EB}}{kT}\right) = -I_{ES} \exp\left(\frac{eV_{BE}}{kT}\right) \tag{7.8}$$

$$I_C = -\alpha_F I_E = \alpha_F I_{ES} \exp\left(\frac{eV_{BE}}{kT}\right) \tag{7.9}$$

$$I_B = -(1 - \alpha_F)I_E = (1 - \alpha_F)I_{ES} \exp\left(\frac{eV_{BE}}{kT}\right) \tag{7.10}$$

If we again introduce $\beta_F = \alpha_F/(1 - \alpha_F)$, Eq. (7.6) and (7.7) remain valid. That is, when the devices are used as a current amplifier, both types operate the same way.

In the *n-p-n* transistor the directions of positive current flow have been chosen in the same way as for the *p-n-p* transistor. Since all currents flow in the opposite direction, the arrow in the emitter lead, symbolizing the direction of the current for minority carrier injection, must be reversed [Fig. 7.1(c)].

7.1c. Currents in Transistors with Both Junctions Forward-Biased

In some pulse applications the emitter-base junction and the collector-base junction are *both* forward-biased. We shall now investigate the current flow for this situation in two steps, first for a *p-n-p* transistor.

In the first step we reverse the bias voltages; i.e., the collector-base junction is now forward-biased and the emitter-base junction is back-biased. This is called *reverse* operation. In analogy with Eq. (7.1) and (7.2) we then have

$$I_C = I_{CS} \exp\left(\frac{eV_{CB}}{kT}\right); \qquad I_E = -\alpha_R I_{CS} \exp\left(\frac{eV_{CB}}{kT}\right) \tag{7.11}$$

Here V_{CB} is the (forward) collector-base voltage, I_{CS} is a kind of saturation current, and α_R is the so-called *reverse* current amplification factor. If all current is carried by holes, α_R represents the part of the hole current injected by the collector that is collected by the emitter.

If both junctions are forward-biased, i.e., if V_{EB} and V_{CB} are both positive, the current is found by combining Eq. (7.1), (7.2), and (7.11). Hence

$$I_E = I_{ES} \exp\left(\frac{eV_{EB}}{kT}\right) - \alpha_R I_{CS} \exp\left(\frac{eV_{CB}}{kT}\right) \tag{7.12}$$

$$I_C = -\alpha_F I_{ES} \exp\left(\frac{eV_{EB}}{kT}\right) + I_{CS} \exp\left(\frac{eV_{CB}}{kT}\right) \tag{7.13}$$

Since $I_B + I_E + I_C = 0$, the base current I_B is

$$I_B = -(1 - \alpha_F)I_{ES} \exp\left(\frac{eV_{EB}}{kT}\right) - (1 - \alpha_R)I_{CS} \exp\left(\frac{eV_{CB}}{kT}\right) \tag{7.14}$$

Equations (7.12)–(7.14) are called the *Ebers-Moll equations* for the *p-n-p* transistor.

Similar equations hold for the silicon *n-p-n* transistor, except that the directions of all currents and the polarities of all voltages are reversed. We thus have

$$I_E = -I_{ES} \exp\left(\frac{eV_{BE}}{kT}\right) + \alpha_R I_{CS} \exp\left(\frac{eV_{BC}}{kT}\right) \tag{7.15}$$

$$I_C = \alpha_F I_{ES} \exp\left(\frac{eV_{BE}}{kT}\right) - I_{CS} \exp\left(\frac{eV_{BC}}{kT}\right) \tag{7.16}$$

$$I_B = (1 - \alpha_F)I_{ES} \exp\left(\frac{eV_{BE}}{kT}\right) + (1 - \alpha_R)I_{CS} \exp\left(\frac{eV_{BC}}{kT}\right) \tag{7.17}$$

Something should be said about the parameters α_F, α_R, I_{ES}, and I_{CS}. Many *p-n-p* transistors are of the p^+-n-p^--p^+ variety. That is, they are made on a p^+-wafer with an epitaxial, weakly *p*-type layer (p^--layer) grown on it; afterwards the *n*-type base region and the p^+-type emitter region are diffused in. In such a structure, I_{CS} is often considerably larger than I_{ES}. Part of this is due to the fact that the collector junction has a much larger junction area than the emitter junction (Fig. 7.3). In addition, the injection of electrons by the base into the collector can be larger than the extraction of holes from the base by the collector.

The transistor is generally so designed that α_F is close to unity. It can be shown on general grounds that (see Appendix A)

$$\alpha_R I_{CS} = \alpha_F I_{ES} \tag{7.18}$$

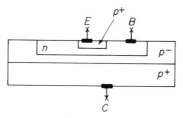

Fig. 7.3. Cross-section of a typical transistor, showing that the base-collector junction has a much larger junction area than the emitter-base junction.

so that α_R is relatively small if I_{CS} is considerably larger than I_{ES}; values as low as 0.1 are quite possible.

We can now solve the following problem.

EXAMPLE 1: In a p-n-p transistor with a forward-biased emitter-base junction the collector is left floating. Find the potential difference V_{CE} between the floating collector and the emitter, in particular for $\alpha_R = 0.1$.

ANSWER: As seen from Fig. 7.1(b), we have

$$V_{CB} = V_{CE} + V_{EB}$$

Substituting into Eq. (7.13) and putting $I_C = 0$ yields

$$0 = -\alpha_F I_{ES} \exp\left(\frac{eV_{EB}}{kT}\right) + I_{CS} \exp\left(\frac{eV_{EB}}{kT}\right) \exp\left(\frac{eV_{CE}}{kT}\right)$$

so that

$$\exp\left(\frac{eV_{CE}}{kT}\right) = \frac{\alpha_F I_{ES}}{I_{CS}} = \alpha_R; \qquad V_{CE} = \frac{kT}{e} \ln \alpha_R$$

because of Eq. (7.18). Hence V_{CE} is slightly negative; for $\alpha_R = 0.10$, $V_{CE} \simeq -0.060$ V.

EXAMPLE 2: Repeat the calculation of Example 1 for an n-p-n transistor if $\alpha_R = 0.10$.

ANSWER: V_{CE} has the opposite value as in the previous case; that is,

$$V_{CE} = -\frac{kT}{e} \ln \alpha_R$$

so that $V_{CE} \simeq 0.060$ V if $\alpha_R = 0.10$.

In the normal operation of transistors some restrictions on the bias conditions must be met:

1. Under some conditions the base-emitter junction of a transistor can be back-biased. Care must be taken that the junction does not show avalanche breakdown in this case.
2. The collector-base junction is normally back-biased. Care must be taken that this bias is not made so large that avalanche breakdown occurs.
3. If the collector-base junction is more strongly back-biased, the collector space-charge region expands into the base region, thereby reducing the width of the base region. Care must be taken that the bias cannot be chosen such that the base region disappears (punch through), since this results in a very large current that can destroy the transistor.
4. The power dissipated in the transistor under quiescent conditions is $I_C V_{CE}$. Care must be taken that this power does not exceed the

limit set for the transistor. This is especially important for power applications.

7.1d. Silicon Transistors Connected as Diodes

In transistor integrated circuits one often needs diodes to perform certain functions. It would be expensive to diffuse these diodes in separately, but it is rather inexpensive to diffuse extra transistors in and connect them as diodes. When doing so, it must be born in mind that integrated transistors are practically always of the *n-p-n* variety.*

The following connections are possible:

1. *Collector connected to the base.* In this case $V_{BC} = 0$, and as long as V_{BE} is larger than a few tenths of a volt, $|I_E| \gg I_{CS}$. Therefore, according to Eq. (7.15),

$$|I_E| = I_B = I_{ES} \exp\left(\frac{eV_{BE}}{kT}\right) \tag{7.19}$$

2. *Emitter connected to the base.* This is similar to the previous case, only the emitter and collector are interchanged. Hence $V_{BE} = 0$, and, according to Eq. (7.16),

$$|I_C| = I_B = I_{CS} \exp\left(\frac{eV_{BC}}{kT}\right) \tag{7.20}$$

3. *Emitter connected to collector.* In this case $V_{BE} = V_{BC}$ and hence from Eq. (7.17) we have

$$I_B = [(1 - \alpha_F)I_{ES} + (1 - \alpha_R)I_{CS}] \exp\left(\frac{eV_{BE}}{kT}\right) \tag{7.21}$$

This is smaller than in cases 1 and 2 unless α_R is small.

4. *Collector floating ($I_C = 0$).* According to Eq. (7.16),

$$I_{CS} \exp\left(\frac{eV_{BC}}{kT}\right) = \alpha_F I_{ES} \exp\left(\frac{eV_{BE}}{kT}\right) \tag{7.22}$$

Substituting into Eq. (7.15),

$$|I_E| = I_B = (1 - \alpha_R \alpha_F)I_{ES} \exp\left(\frac{eV_{BE}}{kT}\right) \tag{7.23}$$

This is smaller than in cases 1 and 2 unless α_R is small.

5. *Emitter floating ($I_E = 0$).* In analogy with the previous case

$$|I_C| = I_B = (1 - \alpha_R \alpha_F)I_{CS} \exp\left(\frac{eV_{BC}}{kT}\right) \tag{7.24}$$

This is smaller than in cases 1 and 2 unless α_R is small.

---

*The reason is that phosphorus diffusion, used for making an *n*-type emitter, gives a higher surface impurity concentration than boron diffusion, used for a *p*-type emitter; therefore phosphorus *n*-type emitters make better emitters.

Connection 1 gives the shortest response time. This is best seen from Fig. 7.3. Since the emitter has the smallest junction area and the emitter is forward-biased, practically all the injected electrons are in the region right under the emitter. As shown in Appendix A, the electrons are collected in a time $\tau_d = w^2/2D_n$, the so-called carrier diffusion time, where w is the width of the base region and D_n the diffusion constant of the electrons in the base. In cases 3–5, the collector also injects electrons into the base and it now takes a time τ_n, the electron lifetime in the base region, before they are removed. Usually τ_n is much larger than τ_d.

7.1e. Simple Transistor Amplifier

Figure 7.4 shows a simple *n-p-n* transistor amplifier. A dc EMF V_{BB} and an ac EMF v_b are applied to the base via a base resistor R_b. As a consequence dc currents I_B and I_C and ac currents i_b and i_c flow into the base and the collector leads, respectively, so that a dc voltage V_{CE} and an ac voltage v_{ce} are developed between the collector and ground.

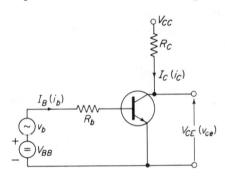

Fig. 7.4. Simple transistor amplifier circuit as a current amplifier.

First we shall investigate the dc operation. Since the base has a turn-on voltage $V_{BE0} \simeq 0.70$ V,

$$I_B = \frac{V_{BB} - V_{BE0}}{R_b} \tag{7.25}$$

If the collector is satisfactorily biased,

$$I_C = \beta_F I_B \tag{7.26}$$

so that

$$V_{CE} = V_{CC} - I_C R_C = V_{CC} - \beta_F \frac{R_c}{R_b}(V_{BB} - V_{BE0}) \tag{7.27}$$

We must now investigate what is meant by "the collector is satisfactorily biased." For normal operation the collector-base junction should be back-biased, say $V_{CB} \geq 0.10$ V. We saw for a forward-biased transistor in which the device draws a reasonable amount of current that $V_{BE} \simeq 0.70$ V. Hence $V_{CE} = V_{CB} + V_{BE} > 0.80$ V. Actually, this condition can be slightly relaxed, since only a very small current is injected by the collector if the collector is forward-biased by less than 0.30 V. That is, it is actually sufficient if $V_{CE} > 0.50$ V.

We shall now discuss examples.

EXAMPLE 3: If $V_{CC} = 10.0$ V, $V_{BE0} = 0.70$ V, $R_b = 50,000\ \Omega$, $R_C = 5000\ \Omega$, $\beta_F = 100$, and $I_C = 1.0$ mA, find I_B, V_{BB}, and V_{CE}.

ANSWER:

$$I_B = \frac{I_C}{\beta_F} = 10\ \mu A$$

$$V_{BB} - V_{BE0} = V_{BB} - 0.70 = I_B R_b = 0.50\ V \qquad V_{BB} = 1.20\ V$$

$$V_{CE} = 10 - \frac{100 \times 5000}{50,000} \times 0.50 = 5.0\ V$$

EXAMPLE 4: If all values of V_{CC}, V_{BB}, R_b, and R_c remain the same as in Example 3, how large can β_F be before V_{CE} drops to 0.50 V?

ANSWER:

$$I_B = \frac{V_{BB} - V_{BE0}}{R_b} = 10\ \mu A; \qquad I_C = \beta_F I_B$$

$$V_{CE} = V_{CC} - \beta_F I_B R_c = 0.50\ V$$

$$\beta_F = \frac{V_{CC} - 0.50}{I_B R_c} = \frac{10.0 - 0.5}{10^{-5} \times 5000} = \frac{9.5}{0.05} = 190$$

We see that the requirement $V_{CE} > 0.50$ V limits the allowed variation in β_F, or the acceptable value of R_c.

We shall now turn to the ac response. If we neglect the resistance r_π of the base-emitter diode, to be introduced in Section 7.2a, we have

$$i_b = \frac{v_b}{R_b}; \qquad i_c = \beta_F i_b = \beta_F \frac{v_b}{R_b} \tag{7.28}$$

$$v_{ce} = -i_c R_c = -\beta_F \frac{R_c}{R_b} v_b \tag{7.29}$$

so that the voltage gain can be defined as

$$g_v = \frac{v_{ce}}{v_b} = -\beta_F \frac{R_c}{R_b} \tag{7.30}$$

The circuit is useful if $|g_v| > 1$. We shall now illustrate this with an example.

EXAMPLE 5: Find the voltage gain in Example 3.

ANSWER:

$$|g_v| = \beta_F \frac{R_c}{R_b} = 100 \times \frac{5000}{50,000} = 10$$

so that the circuit is useful.

The advantage of this type of circuit is that it has very little distortion. We shall discuss this in greater detail in Section 7.2b. The disadvantage of the circuit is that the value of V_{CE} is very sensitive to changes in the value of β_F. Since for a given transistor type the value of β_F can easily vary by a factor of 2–3 between units, the resistance R_c must be so chosen that V_{CE} never drops below 0.50 V when transistors are interchanged. This limits the voltage gain that can be achieved with the circuit.

7.2. SMALL-SIGNAL OPERATION OF THE TRANSISTOR

Figure 7.5(a) shows a slightly modified bias circuit for the transistor, in which the base is returned to the collector supply voltage V_{CC} via a large resistor R_b that determines the base current. A small ac signal v_{be} is applied to the base via a blocking capacitor C_b that has a negligible impedance for all frequencies of practical interest, giving rise to an ac output voltage v_{ce}. The small-signal problem thus amounts to evaluating the response of the transistor to the small signals v_{be} and v_{ce}, i.e., to calculating the ac base and collector currents i_b and i_c, respectively, in Fig. 7.5(b).

7.2a. Small-Signal Equivalent Circuit of the Transistor

We shall first carry out the above evaluation for an *n-p-n* transistor by evaluating the response of the transistor to the ac base voltage v_{be}. Since

$$I_B = (1 - \alpha_F)I_{ES} \exp\left(\frac{eV_{BE}}{kT}\right); \qquad I_C = \alpha_F I_{ES} \exp\left(\frac{eV_{BE}}{kT}\right) \qquad (7.31)$$

we have

$$i_b = \frac{\partial I_B}{\partial V_{BE}} v_{be} = g_{be} v_{be}; \qquad i_c = \frac{\partial I_C}{\partial V_{BE}} v_{be} = g_m v_{be} \qquad (7.32)$$

where

$$g_{be} = \frac{\partial I_B}{\partial V_{BE}} = \frac{e}{kT}(1 - \alpha_F)I_{ES} \exp\left(\frac{eV_{BE}}{kT}\right) = \frac{eI_B}{kT} \qquad (7.32a)$$

$$g_m = \frac{\partial I_C}{\partial V_{BE}} = \frac{e}{kT}\alpha_F I_{ES} \exp\left(\frac{eV_{BE}}{kT}\right) = \frac{eI_C}{kT} \qquad (7.32b)$$

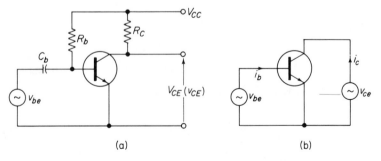

(a)   (b)

Fig. 7.5. (a) Bias circuit of transistor operated as a voltage amplifier. (b) Small-signal diagram of the transistor.

The parameter g_m is called the *transconductance* of the transistor; it represents the signal transfer characteristic of the device. The conductance g_{be} is the small-signal base-emitter conductance of the transistor; it is usually denoted by $1/r_\pi$:

$$r_\pi = \frac{1}{g_{be}} = \frac{kT}{eI_B} \tag{7.32c}$$

Figure 7.6(a) shows the position of r_π and of the current generator $g_m v_{be}$ in

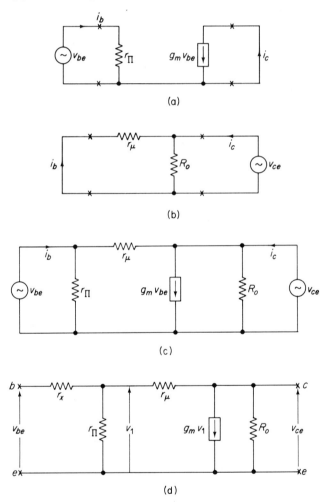

(a)

(b)

(c)

(d)

Fig. 7.6. (a) Small-signal equivalent circuit of a transistor, showing the response to the ac base voltage. (b) Small-signal equivalent circuit of a transistor, showing the response to the ac collector voltage. (c) Equivalent π-circuit, obtained by combining (a) and (b). (d) Hybrid π-circuit in which a series base resistance r_x is added.

the equivalent circuit. The polarity of the current generator is as shown, since I_C and i_c flow *into* the collector.

EXAMPLE 1: Find r_π and g_m for a transistor having $\beta_F = 100$ and operating at $I_C = 1$ mA at $T = 300°$K.

ANSWER:

$$g_m = \frac{eI_C}{kT} = \frac{10^{-3}}{25.8 \times 10^{-3}} = 38.7 \times 10^{-3} \text{ mho}$$

$$g_{be} = \frac{eI_B}{kT} = \frac{g_m}{\beta_F} = 3.87 \times 10^{-4} \text{ mho}$$

$$r_\pi = \frac{1}{g_{be}} = 2580 \ \Omega$$

Next we shall evaluate the effect of the EMF v_{ce} for the *n-p-n* transistor. It comes about because $\alpha_F I_{ES}$ decreases and $(1 - \alpha_F)I_{ES}$ increases with increasing base width w. Now the base width w decreases with increasing collector voltage V_{CE}, because the junction space-charge region extends deeper into the base region. Therefore I_C increases and I_B decreases with increasing collector voltage V_{CE}. An ac collector voltage thus modulates I_B and I_C, thereby producing ac currents i_b and i_c, respectively. We thus have

$$i_c = \frac{\partial I_C}{\partial V_{CE}} v_{ce} = g_{ce}v_{ce}; \qquad i_b = \frac{\partial I_B}{\partial V_{CE}} v_{ce} = -g_{cb}v_{ce} \qquad (7.33)$$

The parameter g_{ce} corresponds to the output conductance of the transistor; in the equivalent circuit it is designated by $1/R_0$. The parameter g_{cb} is called the feedback conductance of the transistor; in the equivalent circuit it is designated by $1/r_\mu$ [Fig. 7.6(b)]:

$$g_{ce} = \frac{1}{R_0}; \qquad g_{cb} = \frac{1}{r_\mu} \qquad (7.33a)$$

We have chosen a minus sign in Eq. (7.33), since i_b flows *into* the base, whereas the current $g_{cb}v_{ce}$ flows *out of* the base.

Thus from Eq. (7.9) and (7.10) we have

$$g_{ce} = \frac{\partial I_C}{\partial V_{CE}}\Big|_{V_{BE}=\text{const}} = \frac{\partial(\alpha_F I_{ES})}{\partial V_{CE}} \exp\left(\frac{eV_{BE}}{kT}\right)$$

$$= \frac{eI_C}{kT} \frac{kT}{e} \frac{1}{\alpha_F I_{ES}} \frac{\partial(\alpha_F I_{ES})}{\partial V_{CE}} = g_m\mu \qquad (7.33b)$$

$$g_{cb} = -\frac{\partial I_B}{\partial V_{CE}}\Big|_{V_{BE}=\text{const}} = -\frac{\partial[(1 - \alpha_F)I_{ES}]}{\partial V_{CE}} \exp\left(\frac{eV_{BE}}{kT}\right)$$

$$= -\frac{eI_B}{kT} \frac{kT}{e} \frac{1}{(1 - \alpha_F)I_{ES}} \frac{\partial}{\partial V_{CE}}[(1 - \alpha_F)I_{ES}] = g_{be}\mu' \qquad (7.33c)$$

where

$$\mu = \frac{kT}{e} \frac{1}{\alpha_F I_{ES}} \frac{\partial}{\partial V_{CE}}(\alpha_F I_{ES}); \qquad \mu' = -\frac{kT}{e} \frac{1}{(1 - \alpha_F)I_{ES}} \frac{\partial}{\partial V_{CE}}[(1 - \alpha_F)I_{ES}]$$

$$(7.33d)$$

Here g_{be} and g_m are given by Eqs. (7.32a) and (7.32b). A more detailed evaluation shows that $\mu' = \mu$ and that μ is a relatively small quantity, usually of the order of 10^{-3}–10^{-4} (see Appendix A).

EXAMPLE 2: Find R_0 and r_μ for a transistor with $\beta_F = 100$, operating at $T = 300°$K, and $I_C = 1$ mA if $\mu = 10^{-3}$.

ANSWER:

$$R_0 = \frac{1}{g_m\mu} = \frac{1}{3.87 \times 10^{-2} \times 10^{-3}} = 2.58 \times 10^4 \ \Omega$$

$$r_\mu = \frac{1}{g_{be}\mu} = \frac{1}{3.87 \times 10^{-4} \times 10^{-3}} = 2.58 \times 10^6 \ \Omega$$

This example shows, except when the amplifier has a large gain, that the effect of r_μ can often be neglected and that the effect of R_0 is also often relatively small.

We now return to the equivalent circuit of Fig. 7.6(c), which is the combination of Fig. 7.6(a) and (b). The reader can easily verify that if v_{ce} is removed, the circuit represents Eq. (7.32), whereas it represents Eq. (7.33) if v_{be} is removed. This is called the *equivalent π-circuit*.

We shall now put in a refinement by adding the series resistance r_x of the base region. The rationale of this modification is that the base hole current I_B flows through the narrow base region to the points where recombination takes place. This region represents a resistive path to the hole current flow in the base, and this is represented by r_x. The resistance r_x is of the order of 10 to a few hundred Ω; its effect can be neglected unless one operates at high frequencies or wants high accuracy. The refinement gives rise to Fig. 7.6(d). Except for the resistances r_x and r_μ, the circuit is identical to the one shown in Fig. 4.4(b). The circuit is called the *hybrid π-circuit*.

It is interesting to note that transistors in integrated circuits have a higher base resistance than the best discrete transistors. This is associated with the manufacturing process.

We observe that

$$g_m r_\pi = \frac{eI_C}{kT}\frac{kT}{eI_B} = \frac{I_C}{I_B} = \beta_F \tag{7.34}$$

This relationship links the equivalent π-circuit of Fig. 7.6(d) with the equivalent circuit of Fig. 4.4(a); it is important for the cascade connection of transistor amplifier stages.

Often r_x, r_μ, and R_0 can be neglected if one does not require great accuracy in the calculation.

We shall now turn to the *p-n-p* transistor and apply the same procedure. Since in this case

$$I_B = -(1 - \alpha_F)I_{ES}\exp\left(-\frac{eV_{BE}}{kT}\right); \qquad I_C = -\alpha_F I_{ES}\exp\left(-\frac{eV_{BE}}{kT}\right) \tag{7.35}$$

we obtain

$$g_{be} = \frac{\partial I_B}{\partial V_{BE}} = \frac{e}{kT}(1 - \alpha_F)I_{ES} \exp\left(-\frac{eV_{BE}}{kT}\right) = -\frac{eI_B}{kT} \qquad (7.35a)$$

$$g_m = \frac{\partial I_C}{\partial V_{BE}} = \frac{e}{kT}\alpha_F I_{ES} \exp\left(-\frac{eV_{BE}}{kT}\right) = -\frac{eI_C}{kT} \qquad (7.35b)$$

While I_B and I_C are now negative, g_{be} and g_m are positive numbers.

To calculate the effect of the collector EMF v_{ce}, one must again bear in mind that the *p-n-p* and the *n-p-n* transistors are each other's opposite. That is, V_{CE} is now negative, the base width w *increases* with increasing V_{CE}, and hence $\alpha_F I_{ES}$ decreases and $(1 - \alpha_F)I_{ES}$ increases with increasing V_{CE}. The net result is the same as for the *n-p-n* transistor: I_C increases and I_B decreases with increasing V_{CE}. We thus have

$$g_{ce} = \frac{\partial I_C}{\partial V_{CE}}\bigg|_{V_{BE}=\text{const}} = g_m\mu; \qquad \mu = -\frac{kT}{e}\frac{1}{\alpha_F I_{ES}}\frac{\partial}{\partial V_{CE}}(\alpha_F I_{ES}) \qquad (7.35c)$$

$$g_{cb} = -\frac{\partial I_B}{\partial V_{CE}}\bigg|_{V_{BE}=\text{const}} = g_{be}\mu';$$

$$\mu' = \frac{kT}{e}\frac{1}{(1 - \alpha_F)I_{ES}}\frac{\partial}{\partial V_{CE}}[(1 - \alpha_F)I_{ES}] \qquad (7.35d)$$

since we want μ and μ' to be positive parameters. Again, it may be shown that $\mu' = \mu$. Note the differences in sign in comparison with the *n-p-n* case. The equivalent circuit is the same as for the *n-p-n* transistor.

We are now able to understand why the collector must be so biased that it injects a negligible current into the base. According to Eq. (7.16), for arbitrary collector bias we have

$$I_C = \alpha_F I_{ES} \exp\left(\frac{eV_{BE}}{kT}\right) - I_{CS} \exp\left(\frac{eV_{BC}}{kT}\right) \qquad (7.36)$$

where the last term represents the current due to electrons injected by the collector. According to Fig. 7.1(c),

$$V_{CE} = V_{CB} + V_{BE} \quad \text{or} \quad V_{CE} = -V_{BC} + V_{BE}$$

so that $\partial/\partial V_{CE} = -\partial/\partial V_{BC}$, since V_{BE} is kept constant in the differentiation process. Hence

$$g_{ce} = \frac{\partial I_C}{\partial V_{CE}} = \frac{\partial}{\partial V_{CE}}(\alpha_F I_{ES}) \exp\left(\frac{eV_{BE}}{kT}\right) + \frac{eI_{CS}}{kT} \exp\left(\frac{eV_{BC}}{kT}\right) \qquad (7.37)$$

The first term is of the order of a fraction of 10^{-4} mho, whereas the second term is 0.387×10^{-4} mho at $T = 300°$K if the absolute value I_{CS} exp (eV_{BC}/kT) of the injected electron current is 1 μA; this is comparable to the first term.

Therefore, unless the electron current injected by the collector is less than a few microamperes, it seriously increases the collector conductance

g_{ce}, and hence it decreases R_0. To prevent this, the current injected into the base by the collector should be kept small.

7.2b. Distortion

We saw in Section 7.1 that the relationship between the base current I_B and the collector current I_C was practically linear, since in

$$I_C = \beta_F I_B \tag{7.38}$$

the current amplification factor β_F is practically independent of operating conditions. That is, the circuit of Fig. 7.4 acts as a linear *current* amplifier. However, in the circuit of Fig. 7.5(a), the EMF v_{be} is applied directly to the base. We shall see that the circuit then gives a considerable distortion if v_{be} is of the order of several millivolts.

According to Eq. (7.10), for the base current of an *n-p-n* transistor

$$I_B = (1 - \alpha_F)I_{ES} \exp\left(\frac{eV_{BE}}{kT}\right) \tag{7.39}$$

Therefore if $V_{BE} = V_{BE1} + v_{be} = V_{BE1} + v_{be1} \cos \omega t$ is the total base emitter voltage, where V_{BE1} is the quiescent base voltage, the base current may be written as $I_B = I_{B1} + i_b$, where I_{B1} is the quiescent base current. But, according to Eq. (7.39),

$$I_{B1} + i_b = (1 - \alpha_F)I_{ES} \exp\left(\frac{eV_{BE1}}{kT}\right) \exp\left(\frac{ev_{be}}{kT}\right) \tag{7.39a}$$

Since I_{B1} is the current for $v_{be} = 0$, we have

$$I_{B1} = (1 - \alpha_F)I_{ES} \exp\left(\frac{eV_{BE1}}{kT}\right) \tag{7.39b}$$

so that the current i_b may be written as

$$
\begin{aligned}
i_b &= I_{B1}\left[\exp\left(\frac{ev_{be1}}{kT}\right) - 1\right]\\
&= I_{B1}\left[\left(\frac{ev_{be1}}{kT}\right)\cos \omega t + \frac{1}{2}\left(\frac{ev_{be1}}{kT}\right)^2 \cos^2 \omega t + \frac{1}{6}\left(\frac{ev_{be1}}{kT}\right)^3 \cos^3 \omega t + \cdots\right]\\
&= \frac{1}{4}I_{B1}\left(\frac{ev_{be1}}{kT}\right)^2 + I_{B1}\left(\frac{ev_{be1}}{kT}\right)\cos \omega t + \frac{1}{4}I_{B1}\left(\frac{ev_{be1}}{kT}\right)^2 \cos 2\omega t + \cdots
\end{aligned}
\tag{7.40}
$$

since $2\cos^2 \omega t = 1 + \cos 2\omega t$ and $\exp x = 1 + x + \frac{1}{2}x^2 + \frac{1}{6}x^3 + \cdots$.

We thus see that the ac signal produces a dc current proportional to the square of the amplitude v_{be} (rectification) and also produces a second harmonic signal besides the wanted signal

$$\frac{eI_{B1}}{kT} v_{be1} \cos \omega t = g_{be} v_{be1} \cos \omega t \tag{7.40a}$$

of frequency ω. The second harmonic distortion is therefore

$$d_2 = 100\frac{\text{amplitude of frequency } 2\omega}{\text{amplitude of frequency}} = 100\frac{ev_{be1}}{4kT}\% = 25\frac{ev_{be1}}{kT}\% \qquad (7.41)$$

To keep the second harmonic distortion tolerable, the ac signal must be kept quite small, as can be seen from the following example.

EXAMPLE 3: Find the value of v_{be1} that gives 1% distortion if $T = 300°$K.

ANSWER: Since $kT/e = 25.8 \times 10^{-3}$ V at $T = 300°$K, d_2 is 1% if $v_{be1} = 1.03$ mV.

One usually feeds the base from a signal $v_{b0} \cos \omega t$ in series with a resistance R_b, as in Fig. 7.7. We shall now show that this reduces the distortion. To prove this, we write

$$i_b = a_1 \cos \omega t + a_2 \cos 2\omega t + \cdots \qquad (7.42)$$

and evaluate a_1 and a_2. From Fig. 7.7

$$v_b = i_b R_b + v_{be} \qquad (7.43)$$

However, $i_b = I_B[\exp(ev_{be}/kT) - 1]$ or $v_{be} = (kT/e)\ln[1 + (i_b/I_B)]$, or since $\ln(1 + x) = x - \frac{1}{2}x^2 + \cdots$,

$$v_{be} = \frac{kT}{e}\left[\frac{i_b}{I_B} - \frac{1}{2}\left(\frac{i_b}{I_B}\right)^2 + \cdots\right] \qquad (7.44)$$

Substituting into Eq. (7.43) and neglecting all terms higher than second order, we have, since $r_\pi = kT/eI_B$,

$$v_b = i_b R_b + \frac{kT}{eI_B}i_b + \frac{1}{2}\left(\frac{kT}{eI_B}\right)^2 \frac{e}{kT}i_b^2$$

$$= (R_b + r_\pi)i_b + \frac{1}{2}\frac{e}{kT}r_\pi^2 i_b^2 \qquad (7.45)$$

Substituting Eq. (7.42), putting $v_b = v_{b0}\cos\omega t$, and neglecting all terms higher than second order, we obtain

$$v_{b0}\cos\omega t = (R_b + r_\pi)(a_1\cos\omega t + a_2\cos 2\omega t) - \frac{1}{2}\frac{e}{kT}r_\pi^2 a_1^2\cos^2\omega t$$

$$= (R_b + r_\pi)a_1\cos\omega t + \left[(R_b + r_\pi)a_2 - \frac{1}{4}\frac{e}{kT}r_\pi^2 a_1^2\right]\cos 2\omega t \qquad (7.46)$$

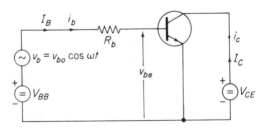

Fig. 7.7. Circuit used to calculate the distortion in transistor amplifiers with a large base resistor R_b.

since $\cos^2 \omega t = \frac{1}{2}(1 + \cos 2\omega t)$ and the dc term is negligible. Consequently

$$(R_b + r_\pi)a_1 = v_{b0}; \qquad a_1 = \frac{v_{b0}}{R_b + r_\pi} \tag{7.47}$$

$$(R_b + r_\pi)a_2 - \frac{1}{4}\frac{e}{kT}r_\pi^2 a_1^2 = 0; \qquad a_2 = \frac{1}{4}\frac{e}{kT}v_{b0}^2\frac{r_\pi^2}{(R_b + r_\pi)^3} \tag{7.48}$$

Therefore the distortion in i_b, and hence in i_c, is

$$d_2 = 100\frac{a_2}{a_1}\% = 25\frac{ev_{b0}}{kT}\left(\frac{r_\pi}{R_b + r_\pi}\right)^2\% = 25\frac{ev_{be1}}{kT}\left(\frac{r_\pi}{R_b + r_\pi}\right)\% \tag{7.49}$$

where $v_{be1} = v_{b0}r_\pi/(R_b + r_\pi)$ is the ac amplitude of frequency ω at the base. Since

$$25\frac{ev_{be1}}{kT}\% \tag{7.49a}$$

represents the distortion when the base is fed from a low-impedance source, the factor $(R_b + r_\pi)/r_\pi$ is the *distortion-reduction factor*.

EXAMPLE 4: Find the value of R_b for which the distortion is reduced by a factor of 2.

ANSWER:

$$\frac{R_b + r_\pi}{r_\pi} = 2 \quad \text{or} \quad R_b = r_\pi$$

7.3. TRANSISTOR BIAS CIRCUITS

Figure 7.8(a) shows the simplest bias circuit possible. It consists of having the base returned to V_{CC} via a large resistance R_b. The dc operating voltage V_{BE0} of the transistor is about 0.70 V. We see that

$$I_B = \frac{V_{CC} - V_{BE0}}{R_b} \tag{7.50}$$

$$I_C = \beta_F I_B \tag{7.51}$$

$$V_{CE} = V_{CC} - I_C R_C = V_{CC} - \beta_F \frac{R_C}{R_b}(V_{CC} - V_{BE0}) \tag{7.52}$$

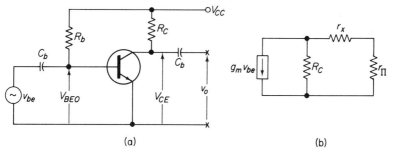

(a)   (b)

Fig. 7.8. (a) Bias circuit used to keep the base current constant. (b) Interstage equivalent circuit of a cascade amplifier.

We must now require that the collector-base junction is not forward-biased. This condition is well satisfied if

$$\beta_F \frac{R_C}{R_b} \leq 1 \tag{7.52a}$$

The trouble with this circuit is that this condition must remain satisfied if transistors are interchanged. However, β_F varies considerably for transistors of the same type; therefore R_c must be chosen so that condition (7.52a) remains satisfied under the worst possible conditions.

EXAMPLE 1: In the circuit of Fig. 7.8(a), $V_{CC} = 12$ V and $\beta_F = 100$. Adjust R_b such that $I_C = 1$ mA and choose R_c so that $V_{CE} = 9$ V.

ANSWER:

$$I_C = 1 \text{ mA}; \quad I_B = \frac{I_C}{\beta_F} = 10 \ \mu\text{A}$$

$$R_b = \frac{V_{CC} - V_{BE0}}{I_B} = (12 - 0.70) \times 10^5 = 1.13 \times 10^6 \ \Omega$$

$$R_c = \frac{V_{CC} - V_{CE}}{I_C} = \frac{12 - 9}{10^{-3}} = 3000 \ \Omega$$

EXAMPLE 2: What value of β_F can be tolerated in the circuit of Example 1 if V_{CE} must not drop below 0.50 V?

ANSWER:

$$I_B = 10 \ \mu\text{A}$$

$$(I_C)_{\max} = \frac{(V_{CC}) - V_{CE \min}}{R_c} = \frac{12 - 0.50}{3 \times 10^3} = 3.83 \text{ mA}$$

$$(\beta_F)_{\max} = \frac{(I_C)_{\max}}{I_B} = \frac{3.83 \times 10^{-3}}{10^{-5}} = 383$$

Since the transistor has a transconductance $g_m = eI_C/kT$, the small-signal voltage gain of the circuit is

$$g_v = \frac{v_0}{v_{be}} = -\frac{g_m v_{be} R_c}{v_{be}} = -g_m R_c \tag{7.53}$$

EXAMPLE 3: Find the voltage gain for Example 1.

ANSWER:

$$g_m = \frac{10^{-3}}{25.8 \times 10^{-3}} = 38.6 \times 10^{-3} \text{ mho}$$

$$g_m R_c = 38.6 \times 10^{-3} \times 3 \times 10^3 = 116$$

If the amplifier is part of a cascade connection, the resistance $(r_x + r_\pi)$ of Fig. 7.6(d) is connected in parallel with the resistance R_c insofar as ac is concerned. We thus have [Fig. 7.8(b)]

$$g_v = -g_m \frac{R_c(r_\pi + r_x)}{R_c + r_\pi + r_x} \tag{7.54}$$

Usually the effect of r_x can be neglected.

EXAMPLE 4: Find the voltage gain if the circuit of Example 1 is part of a cascade connection. The effect of the resistance r_x may be neglected.

ANSWER:

$$r_\pi = \frac{kT}{eI_B} = 25.8 \times 10^{-3} \times 10^5 = 2580 \ \Omega$$

$$g_v = -38.6 \times 10^3 \times \frac{3000 \times 2580}{5580} = -54$$

This example shows that the input resistance r_π of the circuit gives rise to a sizable reduction in voltage gain when stages are connected in cascade.

Circuits have been designed that show a much smaller dependence of the operating conditions on the value of β_F. An example is the circuit of Fig. 7.9. Here the base is fed from a potentiometer arrangement and a bypassed resistor R_e is inserted in the emitter lead. Thus we have

$$\frac{V_{CC} - V_2}{R_1} = \frac{V_2}{R_2} + (1 - \alpha_F)I_E \tag{7.55}$$

or, solving for V_2,

$$V_2 = V_{CC}\frac{R_2}{R_1 + R_2} - (1 - \alpha_F)I_E\frac{R_1 R_2}{R_1 + R_2} \tag{7.56}$$

Moreover,

$$V_E = I_E R_e \tag{7.57}$$

so that

$$V_{BE0} = V_2 - V_E = V_{CC}\frac{R_2}{R_1 + R_2} - I_E\left[(1 - \alpha_F)\frac{R_1 R_2}{R_1 + R_2} + R_e\right] \tag{7.58}$$

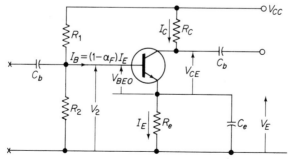

Fig. 7.9. Bias circuit used to compensate for the effect of differences in β_F between individual transistors.

Solving for I_E yields

$$I_E = \frac{V_{CC}R_2/R_1 - V_{BE0}(1 + R_2/R_1)}{(1 - \alpha_F)R_2 + R_e(1 + R_2/R_1)} \tag{7.59}$$

For V_{CE} we finally have

$$V_{CE} = V_{CC} - I_C R_c - I_E R_e \simeq V_{CC} - I_E(R_c + R_e) \tag{7.60}$$

since $I_C \simeq I_E$.

If $(1 - \alpha_F)R_2 \ll R_e(1 + R_2/R_1)$, we see that I_E, and hence I_C, is practically independent of operating conditions. R_c must be so chosen that $V_{CE} > 0.50$ V. The disadvantage of the circuit is that a resistance $R_1 R_2/(R_1 + R_2)$ is placed in parallel to the input resistance r_π of the transistor. This reduces the gain per stage in cascaded stages to a certain extent. We shall now illustrate this with the help of a few examples.

EXAMPLE 5: If $R_1 = 20,000 \ \Omega$, $R_2 = 10,000 \ \Omega$, $1 - \alpha_F = 0.010$, $V_{CC} = 9.0$ V, and $V_{BE0} = 0.70$ V, find the value of R_e so that $I_E = 1.0$ mA. Also find the value of R_c so that $V_{CE} = 2.00$ V.

ANSWER:

$$(1 - \alpha_F)R_2 + R_e\left(1 + \frac{R_2}{R_1}\right) = \frac{V_{CC}R_2/R_1 - V_{BE0}(1 + R_2/R_1)}{I_E}$$

$$100 + \tfrac{3}{2}R_e = (4.50 - 1.05) \times 10^3 = 3.45 \times 10^3; \qquad R_e = 2230 \ \Omega$$

$$R_c + R_E = \frac{V_{CC} - V_{CE}}{I_E} = 7000 \ \Omega; \ R_c = 4770 \ \Omega$$

EXAMPLE 6: How much does I_E change if $1 - \alpha_F$ changes from 0.010 to 0.000?

ANSWER: For $1 - \alpha_F = 0$ we have

$$I_E = \frac{V_{CC}R_2/R_1 - V_{BE0}(1 + R_2/R_1)}{R_e(1 + R_2/R_1)} = \frac{3.45}{3350} = 1.03 \text{ mA}$$

or I_E changes only by 3%.

EXAMPLE 7: What is the voltage gain of the amplifier stage?

ANSWER:

$$g_m = \frac{eI_E}{kT} = 38.6 \times 10^{-3} \text{ mho}$$

$$g_v = -g_m R_c = -38.6 \times 10^{-3} \times 4.77 \ 10^{+3} = -184$$

As in the case of the FET circuit, the condition for C_e is that

$$\omega C_e \geq g_m \tag{7.61}$$

for all frequencies of practical interest.

EXAMPLE 8: Find the value of C_e so that condition (7.61) is satisfied down to 10 cycles.

ANSWER:

$$C_e = \frac{g_m}{2\pi f} = \frac{38.6 \times 10^{-3}}{2\pi \times 10} \simeq 600 \ \mu\text{F}$$

Neither of the circuits discussed so far is very well suited for integrated circuits. The circuit of Fig. 7.8(a) requires far too high a value of R_b to put it in integrated form. The circuit of Fig. 7.9 could be so designed that none of the resistors would be larger than 10,000–20,000 Ω, but in that case one still has to provide an external capacitor C_e to bypass the integrated resistor R_e.

We shall see in the next chapter how integrated bias circuits can be designed.

7.4. POWER DISSIPATION IN TRANSISTORS

The power dissipated in a transistor having a collector-emitter bias voltage V_{CE} and carrying a collector current I_C is

$$P_{\text{diss}} = I_C V_{CE} \tag{7.62}$$

Because of the dissipated power, the temperature of the transistor rises and as a consequence the characteristic changes. Fortunately the circuits of Section 7.3 have the merit that the currents flowing in the circuits are rather insensitive to temperature changes. Part of this is caused by the fact that we neglected the leakage current I_{CBO}, flowing in the transistor at zero emitter current. This current increases very rapidly with increasing temperature [as $\exp(-eE_g/kT)$, where E_g is the band-gap width, which is 1.1 eV in silicon]. As a consequence this current becomes quite significant for device temperatures above 200°C. To prevent this, the temperature of the transistor should stay within more reasonable limits.

The rise ΔT in temperature of the device is proportional to the power dissipation P_{diss}, and for that reason it makes sense to call

$$\theta = \frac{\Delta T}{P_{\text{diss}}} \tag{7.63}$$

the *thermal resistance* of the transistor; it is usually expressed in degrees Celsius per milliwatt. To keep ΔT small for a given power dissipation, one should make θ small, which can, e.g., be done by providing the transistor with a *heat sink*. This is especially important for power transistors. For transistors used as small-signal amplifiers the heating effect is usually negligible.

EXAMPLE 1: A transistor has $\theta = 0.5°\text{C/mW}$. If the ambient environment has a temperature of 25°C, how much power dissipation can be tolerated if the temperature must stay below 125°C?

ANSWER: $\Delta T = 100°\text{C}$; $P_{\text{diss}} = \Delta T/\theta = 100/0.5 \ \text{mW} = 200 \ \text{mW}$.

EXAMPLE 2: A power transistor has a maximum power dissipation of 20 W. If the ambient environment has a temperature of 25°C and the device temperature must not rise above 150°C, find the required value of the thermal resistance θ.

ANSWER: $\Delta T = 125°C$, $\theta = \Delta T/P_{\text{diss}} = 125/20 = 6.25°C/W$. This can be achieved by an appropriately designed heat sink.

7.5. SMALL-SIGNAL DEVICE PARAMETERS

Up to now we have represented the small-signal properties of a common emitter transistor circuit by the equivalent networks of Fig. 7.6(c) and (d). In most device manuals another set of parameters, the so-called *h*-parameters, is given. We shall now define these parameters.

In this definition the independent variables are the small-signal base current i_b and the small-signal collector voltage v_{ce}, whereas the dependent variables are the open circuit base voltage v_{be} and the small-signal collector current i_c. Since we are dealing with small-signal parameters, the relationships between the two sets of variables are linear. We may thus write

$$v_{be} = h_{ie}i_b + h_{re}v_{ce} \tag{7.64}$$

$$i_c = h_{fe}i_b + h_{oe}v_{ce} \tag{7.65}$$

It is easily seen that these *h*-parameters have the following meaning:

1.

$$h_{ie} = \frac{v_{be}}{i_b}\bigg|_{v_{ce}=0} = \text{input impedance for short-circuited output} \tag{7.66}$$

If we look at Fig. 7.6(d), we see that

$$h_{ie} = r_x + r_\pi \tag{7.66a}$$

2.

$$h_{fe} = \frac{i_c}{i_b}\bigg|_{v_{ce}=0} = \text{short-circuit forward current gain} \tag{7.67}$$

In the approximation we have used so far,

$$h_{fe} = \beta_F \tag{7.67a}$$

3.

$$h_{re} = \frac{v_{be}}{v_{ce}}\bigg|_{i_b=0} = \text{reverse voltage gain for open input} \tag{7.68}$$

Inspection of Fig. 7.6(d) shows that

$$h_{re} = \frac{r_\pi}{r_\mu + r_\pi} \simeq \frac{r_\pi}{r_\mu} \tag{7.68a}$$

Since $1/r_\mu = g_{be}\mu$ and $1/r_\pi = g_{be}$, we see that

$$h_{re} = \mu \qquad (7.68b)$$

where μ is defined by Eq. (7.33d).

4.

$$h_{0e} = \frac{i_c}{v_{ce}}\Big|_{i_b=0} = \text{output admittance for open input} \qquad (7.69)$$

If in Fig. 7.6(d) we replace v_{be} by a high-impedance load to make $i_b = 0$, we have, as is seen by inspection,

$$v_{be} = \frac{r_\pi}{r_\mu + r_\pi} v_{ce} \simeq \frac{r_\pi}{r_\mu} v_{ce}$$

$$i_c = h_{0e} v_{ce} = \frac{v_{ce}}{R_0} + g_m v_{be} = \left(\frac{1}{R_0} + \frac{r_\pi}{r_\mu} g_m\right) v_{ce}$$

or

$$h_{0e} = \frac{1}{R_0} + \frac{r_\pi}{r_\mu} g_m \qquad (7.69a)$$

We now bear in mind that $1/r_\mu = g_{be}\mu$, $1/r_\pi = g_{be}$, and $1/R_0 = g_m\mu$. Therefore

$$h_{0e} = 2g_m\mu = \frac{2}{R_0} \qquad (7.69b)$$

The h-parameters are thus easily expressed in terms of the network parameters of the equivalent π-network and vice versa.

Besides the parameter h_{fe}, the device manuals also give the parameter

$$h_{FE} = \frac{I_C}{I_B} = \beta_F \qquad (7.70)$$

which is the dc current amplification factor. The reason for this distinction is that the parameter β_F often depends somewhat on I_B. We then have

$$I_C = \beta_F I_B \qquad (7.71)$$

$$i_c = \frac{\partial}{\partial I_B}(\beta_F I_B)i_b = h_{fe} i_b \qquad (7.71a)$$

so that

$$h_{fe} = \frac{\partial}{\partial I_B}(\beta_F I_B) = \beta_F + I_B \frac{\partial \beta_F}{\partial I_B} = \beta_f \qquad (7.71b)$$

We thus see that $h_{fe} > h_{FE}$ if β_F increases with increasing I_B, whereas $h_{fe} < h_{FE}$ if β_F decreases with increasing I_B. This is illustrated in Fig. 7.10.

The reason the h-parameters are found in device manuals is that these parameters are *much more easily measured* than the parameters of the equivalent π-network. One can, of course, object that at least the parameters r_π and g_m are easily calculated. Whereas the formula $g_m = eI_C/kT$ is usually well

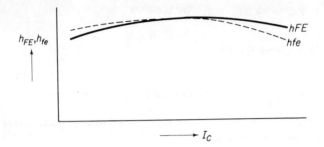

Fig. 7.10. dc current amplification factor h_{FE} and ac current amplification factor h_{fe} as a function of the collector current.

satisfied, except at very high currents, this is not always the case for the formula $1/r_\pi = eI_B/kT$, since this expression presupposes that there is no recombination in the emitter-base space-charge region. Finally, the resistance r_x is difficult to measure accurately.

PROBLEMS

7.1. In an *n-p-n* transistor with forward-biased collector junction the emitter is left floating; that is, $I_E = 0$. Show that $V_{EC} = -(kT/e)\ln\alpha_F$.

7.2. An *n-p-n* transistor having $V_{BE0} = 0.70$ V at $I_E = 1.0$ mA is used as a diode by connecting base and collector together. The diode is used in the circuit of Fig. 7.11. Find the resistance R so that $I_E = 1.0$ mA for $V_{BB} = 6.0$ V.

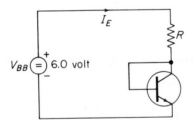

Fig. 7.11.

7.3. In a symmetrical transistor, with $\alpha_F = \alpha_R = 0.95$ and $I_{ES} = I_{CS}$, the emitter and collector are connected together to make a diode with the base contact as the other electrode. When the transistor is connected for normal bias, $I_E = 1.0$ mA at $V_{BE} = V_{BE0} = 0.70$ V. Find the value of V_{BE} for a diode current of 1.0 mA.

7.4. In an *n-p-n* transistor $\alpha_F = 0.995$ and $\alpha_R = 0.50$. The device is made into a diode by letting the collector float. When the transistor is connected as a normal forward-biased transistor, $V_{BE} = 0.700$ V at $I_E = 1.00$ mA. Find the value of V_{BE} for a diode current of 1.0 mA.

7.5. In a silicon transistor amplifier of the type shown in Fig. 7.4, $V_{CC} = 8.0$ V and $V_{BB} = 1.40$ V, whereas $V_{BE} = 0.70$ V at $I_E = 1.0$ mA. (a) Find the value of R_b needed to adjust the collector current I_C to 1.0 mA if $\beta_F = 100$. (b) Find the value of R_C so that $V_{CE} = 4.0$ V. (c) If a small ac signal v_b is applied in series with V_{BB}, find the voltage gain $|v_{ce}/v_b|$.

7.6. A transistor is operated so that $I_B = 25\ \mu A$ and $I_C = 2.5$ mA. For $T = 300°K$, find (a) the values of r_π and g_m and (b) the values of R_0 and r_μ if the measured value of h_{re} is 0.50×10^{-3}.

7.7. In the transistor circuit of Fig. 7.5(a), $\beta_F = 100$, $V_{CC} = 4.5$ V, and $V_{BE0} = 0.70$ V at $I_E = 1.0$ mA. (a) Find the value of R_b needed to adjust I_C to 1.0 mA. (b) Find the value of R_c so that $V_{CE} = 2.0$ V. (c) Find the small-signal voltage gain $|v_{ce}/v_b|$ of the stage.

7.8. In the transistor circuit of Fig. 7.5(a), R_b is chosen as in Problem 7.7 so that $I_B = 10\ \mu A$ under all cases. The transistors to be used can have values of β_F ranging between 50 and 300. (a) Find the value of R_c so that V_{CE} does not drop below 0.50 V under the worst-case condition. (b) Find the possible variation in the small-signal voltage gain $|v_{ce}/v_b|$ for that value of R_c.

7.9. Extend the discussion on distortion for a low-impedance signal source connected to the base and evaluate the base voltage amplitude for which the third harmonic distortion is 1.0%. *Hint:* Make use of the fact that $\cos^3 x = \frac{1}{4} \cos 3x + \frac{3}{4} \cos x$.

7.10. A transistor with $\beta_F = 100$ is operating in the circuit of Fig. 7.8(a) with $I_C = 1.0$ mA. The output of the amplifier is capacitively fed into a similar stage. $V_{CC} = 10$ V and $R_c = 2000\ \Omega$. Find the interstage voltage gain.

7.11. A transistor with $\beta_F = 100$ is operating in the circuit of Fig. 7.9 with $R_1 = 20,000\ \Omega$, $R_1 = 10,000\ \Omega$, and $R_c = 5000\ \Omega$. (a) Determine R_e so that $I_C \simeq I_E = 1.0$ mA. $V_{CC} = 10$ V, $V_{BE} = 0.70$ V. (b) Find the voltage gain per stage. (c) Find the distortion if the base is fed from a small-impedance source and v_{ce} has 1.0-V amplitude. (d) Find the distortion for $v_{ce} = 1.0$ V if the base is fed from an identical previous stage.

7.12. Figure 7.12 shows a two-stage amplifier that must deliver an output signal of 1.0-V amplitude to a load resistance $R_c = 5000\ \Omega$. The transistors are operated at a collector current of 1.0 mA and $\beta_F = 100$. Choose the value of R so that the distortion is 1.0%. You may assume that $R_b' + R$ has been so chosen that $I_B = 10\ \mu A$.

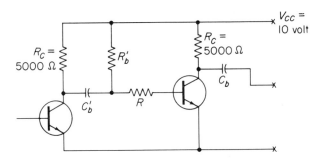

Fig. 7.12.

7.13. A transistor operating at 20-μA base current has $h_{ie} = 1540\ \Omega$. Find r_x.

7.14. A transistor has $h_{0e} = 10^{-4}$ mho. Find R_0.

7.15. A transistor has $h_{re} = 10^{-3}$, $g_m = 38.8 \times 10^{-3}$ mho, and $r_\pi = 2580\ \Omega$. Find r_μ and R_0.

8

Various

Transistor Circuits

8.1. VARIOUS INTEGRATED TRANSISTOR CIRCUITS

We saw in Section 7.1 that the output conductance $1/R_0$ of a transistor circuit is usually quite small. Just as in the FET case this means that a transistor with a forward-biased base can be used as a *constant current generator*.

The bias of the base circuit must be so chosen that it can be put in integrated form. One simple way of doing this is shown in Fig. 8.1(a). It consists of using two *identical* transistors Q_1 and Q_2; Q_1 is hooked up as a diode and Q_2 is hooked up as a normal transistor. The current flowing through Q_1 is

$$I_{C1} = \frac{V_{CC} - 0.70}{R_d} \qquad (8.1)$$

and hence the current I_{C2} flowing through R_L is equal to it, since the transistors are identical. We neglect here the small currents I_{B1} and I_{B2}.

To have the circuit work properly, the collector-base junction of Q_2 must be forward-biased. Since $V_{BE1} = V_{BE2} = 0.70$ V and $V_{CB1} = 0$ V, this is the case if $R_L < R_d$.

Figure 8.1(b) shows how this arrangement can be used to provide a constant current to a circuit known as a differential amplifier. Q_3 and Q_4 constitute a constant current generator, the current of which is controlled by the resistor R_d. If Q_1 and Q_2 are identical transistors, half the current is passed

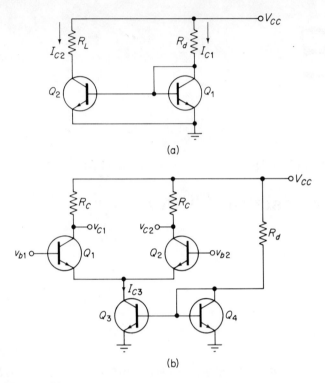

Fig. 8.1. (a) Constant current generator circuit using transistors. (b) Constant current generator circuit driving a differential amplifier.

through Q_1 and the other half through Q_2, so that

$$I_{E1} = I_{E2} = \frac{1}{2} \frac{(V_{CC} - 0.70)}{R_d} \tag{8.1a}$$

Since V_{CC} is of the order of 5–10 V and the currents I_{E1} and I_{E2} are of the order of 1 mA, R_d is a resistance of the order of several thousand ohms. This is easily put in integrated form.

EXAMPLE: If $I_{E1} = I_{E2} = 1$ mA and $V_{CC} = 7.5$ V, find R_d.

ANSWER: $R_d = \dfrac{V_{CC} - 0.70}{2I_{E1}} = \dfrac{6.8}{2 \times 10^{-3}} = 3400 \ \Omega$

A second circuit often used is the *constant voltage generator* (Fig. 8.2) It consists of a diode control circuit that keeps the base voltage V_{BB} of transistor Q constant. If a load resistor R_L is inserted in the emitter lead, the transistor can supply a large load current through this resistor at a constant voltage approximately 0.70 V below V_{BB}. One can change the load current by changing R_L.

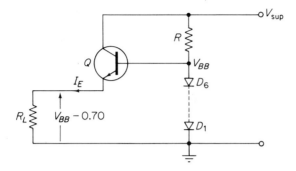

Fig. 8.2. Constant voltage generator using a diode voltage supply driving the base of a transistor.

The circuit is rather insensitive to changes in the load current. If the transistor Q has a current amplification factor β_F, a change ΔI_E in emitter current gives rise to a change in base current:

$$\Delta I_B = \frac{\Delta I_E}{\beta_F + 1} \tag{8.2}$$

This change in base current can easily be taken up by the diode control circuit without an appreciable change in V_{BB}. Therefore the voltage drop across R_L is practically independent of the load current. If necessary, it is not difficult to take the effect of ΔI_B on V_{BB} into account.

It is not necessary to use a load resistance R_L. One may also return the emitter of Q to the collectors of the differential amplifier of Fig. 8.1(b); the circuit then keeps the collector voltage of this amplifier practically constant at the value $V_{BB} - 0.70$ V. The advantage of this circuit over a diode control circuit is that one can supply a much larger current to the amplifier circuit. If I_E is the current needed for the amplifier circuit, the base current I_B required is only $I_E/(\beta_F + 1)$.

In dc amplifiers the dc output voltage V_{C1} of a transistor amplifier stage is usually higher than the input voltage V_{B1}, and therefore it is necessary to lower this voltage before feeding it into another dc amplifier stage.

Figure 8.3(a) shows such a circuit; it changes V_{C1} to $V_{C1} - 0.70$ V. By using more transistor circuits, each transistor lowers the output voltage by 0.70 V without altering the part of V_{C1} caused by V_{B1}. By using m transistors, the output voltage is $V_{C1} - 0.70m$ V.

Figure 8.3(b) shows a circuit using two diodes accomplishing the same purpose. Each diode lowers the output voltage by 0.70 V. Therefore in this case $V_{B2} = V_{C1} - 2 \times 0.70$ V. For m diodes in series we have

$$V_{B2} = V_{C1} - 0.70m \text{ V}$$

Both circuits can easily be put in integrated form since the resistors

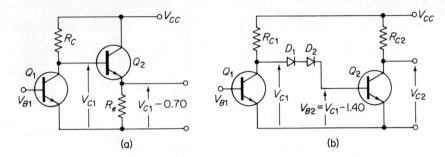

Fig. 8.3. (a) dc circuit for lowering the output voltage by 0.70 V with the help of an emitter follower. (b) dc circuit for lowering the output voltage by 1.40 V with the help of two series diodes.

required are of the order of a few thousand ohms and the diodes are easily made by connecting some transistors as diodes.

8.2. LARGE-SIGNAL TRANSISTOR CIRCUIT

The discussion in Chapter 7 mostly involved small-signal applications of transistors. Probably 80–90% of all transistors are used under large-signal conditions and therefore it is very desirable that this type of application be properly understood.

8.2a. Large-Signal Transistor Circuit as a Linear Circuit

Figure 8.4(a) shows a large-signal transistor circuit. We assume that V_{CC} is relatively large and that V_{BB} is gradually raised from 0 V. As long as

$$V_{BB} < V_{BE0} \simeq 0.70 \text{ V} \tag{8.3}$$

the device hardly draws any current and $V_{CE} \simeq V_{CC}$.

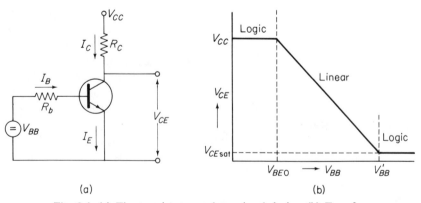

Fig. 8.4. (a) The transistor as a large-signal device. (b) Transfer characteristic of (a).

For $V_{BB} > 0.70$ V a base current

$$I_B = \frac{V_{BB} - V_{BE0}}{R_b} \qquad (8.4)$$

is flowing, where $V_{BE0} \simeq 0.70$ V. The resulting collector current is

$$I_C = \beta_F I_B = \beta_F \frac{(V_{BB} - V_{BE0})}{R_b} \qquad (8.5)$$

so that

$$V_{CE} = V_{CC} - I_C R_c = V_{CC} - \frac{\beta_F R_c}{R_b}(V_{BB} - V_{BE0}) \qquad (8.6)$$

The output voltage V_{CE} thus drops linearly with increasing V_{BB} until V_{CE} becomes so small that the collector-base junction is forward-biased. The *linear* mode of operation of the circuit is thus the mode where V_{CE} varies linearly with V_{BB}.

What happens if V_{BB} is further increased? The voltage V_{CE} then remains practically independent of current and the transistor is said to be *saturated*. The value of V_{CE} under that condition is denoted by $V_{CE \, sat}$ and is of the order of 0.10 V. The current equations are then

$$I_B = \frac{V_{BB} - V_{BE0}}{R_b}; \qquad I_C = \frac{V_{CC} - V_{CE \, sat}}{R_c} \qquad (8.7)$$

and $I_C < \beta_F I_B$. The relationship $I_C = \beta_F I_B$ can therefore not be used in saturation. What remains true, however, is that

$$I_E = I_B + I_C \qquad (8.8)$$

if the polarities are as shown in Fig. 8.4(a).

What are the limits of the linear mode of operation? One limit is $V_{BB} = V_{BE0}$; the transistor is then turned off. The other limit is found when the transistor is just saturated, that is, when V_{CE} in Eq. (8.6) has just reached the value V_{CEsat}. At that point it is still true that $I_C = \beta_F I_B$, so that Eq. (8.6) remains valid. Therefore, inverting Eq. (8.6),

$$V_{BB} = V'_{BB} = V_{BE0} + (V_{CC} - V_{CE \, sat})\frac{R_b}{\beta_F R_c} \qquad (8.9)$$

so that the linear range of operation extends from

$$V_{BE0} \le V_{BB} \le V_{BE0} + \frac{R_b}{\beta_F R_C}(V_{CC} - V_{CE \, sat}) \qquad (8.9a)$$

The corresponding range of V_{CE} is therefore

$$V_{CC} \ge V_{CE} \ge V_{CE \, sat} \qquad (8.9b)$$

The "width" of the linear range of operation is given by Eq. (8.9a). This also corresponds to the difference between the *off* and the *saturated* condition.

Figure 8.4(b) shows the transfer characteristic, showing the two logic ranges $V_{BB} < V_{BE0}$ and $V_{BB} > V'_{BB}$ and the linear range in between.

8.2b. Circuit as a Logic Circuit

Let $V_{BB} < V_{BE0}$ correspond to the logical 0 condition at the input; then the output voltage $V_{CE} \simeq V_{CC}$ is "high," corresponding to the logical 1 condition at the output. Let

$$V_{BB} > V_{BE0} + \frac{(V_{CC} - V_{CE\ sat})R_b}{\beta_F R_c} \tag{8.10}$$

correspond to a logical 1 condition at the input; then the output voltage $V_{CE} = V_{CE\ sat}$, corresponding to a logical 0 at the output.

The transistor is thus a logical *inverter*, converting a logical 1 into a logical 0 and vice versa. The only requirements to be made are that $V_{CE\ sat} < V_{BE0}$, which is always satisfied since $V_{BE0} \simeq 0.70$ V and $V_{CE\ sat} \simeq 0.10$ V, and that V_{BB} in Eq. (8.10) must be equal to or smaller than V_{CC}. The latter requires $R_b/(\beta_F R_c) < 1$.

8.2c. Further Investigation of the Saturated Condition

We shall evaluate the value of $V_{CE\ sat}$ with the help of the Ebers-Moll equations (7.15)–(7.17). We shall introduce for the *n-p-n-* transistor

$$I_F = I_{ES} \exp\left(\frac{eV_{BE}}{kT}\right); \qquad I_R = I_{CS} \exp\left(\frac{eV_{BC}}{kT}\right) \tag{8.11}$$

and rewrite the Ebers-Moll equations as

$$I_C = \alpha_F I_F - I_R; \qquad I_B = (1 - \alpha_F)I_F + (1 - \alpha_R)I_R \tag{8.12}$$

Solving for I_F and I_R yields

$$I_F = \frac{(1 - \alpha_R)I_C + I_B}{1 - \alpha_F \alpha_R}; \qquad I_R = \frac{-(1 - \alpha_F)I_C + \alpha_F I_B}{1 - \alpha_F \alpha_R} \tag{8.13}$$

Now, according to Eq. (8.11),

$$V_{BE} = \frac{kT}{e} \ln\left(\frac{I_F}{I_{ES}}\right); \qquad V_{BC} = \frac{kT}{e} \ln\left(\frac{I_R}{I_{CS}}\right) \tag{8.14}$$

but as is seen from Fig. 8.5,

$$V_{BE} = V_{BC} + V_{CE} \quad \text{or} \quad V_{BC} = V_{BE} - V_{CE} \tag{8.15}$$

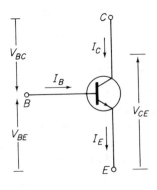

Fig. 8.5. Voltages and currents in an *n-p-n* transistor.

Since $\alpha_R I_{CS} = \alpha_F I_{ES}$, we have

$$
\begin{aligned}
V_{BC} &= \frac{kT}{e} \ln \frac{\alpha_R I_R}{\alpha_R I_{CS}} = \frac{kT}{e} \ln \left(\frac{\alpha_R I_R}{\alpha_F I_{ES}} \right) \\
&= \frac{kT}{e} \ln \left(\frac{I_F}{I_{ES}} \right) + \frac{kT}{e} \ln \left(\frac{\alpha_R I_R}{\alpha_F I_F} \right) = V_{BE} + \frac{kT}{e} \ln \left(\frac{\alpha_R I_R}{\alpha_F I_F} \right)
\end{aligned}
\tag{8.16}
$$

Applying Eq. (8.15), we see that

$$
V_{CE} = -\frac{kT}{e} \ln \left(\frac{\alpha_R I_R}{\alpha_F I_F} \right) = \frac{kT}{e} \ln \left(\frac{\alpha_F I_F}{\alpha_R I_R} \right)
\tag{8.17}
$$

Now I_F is larger than I_R, say by one order of magnitude, and α_F is larger than α_R, say also by an order of magnitude. Therefore at $T \simeq 300°\mathrm{K}$

$$
V_{CE} \simeq \frac{kT}{e} \ln 100 = 25.8 \times 10^{-3} \times 4.60 \simeq 0.12 \text{ V}
$$

V_{CE} is only slightly dependent on operating conditions, since it changes only by 60 mV if I_F/I_R changes by a factor of 10. Therefore it makes sense to introduce the saturated collector voltage $V_{CE \text{ sat}}$; we see that $V_{CE \text{ sat}} \simeq 0.10$ V.

It is often convenient to express V_{CE} directly in terms of I_C and I_B with the help of Eq. (8.13). This yields

$$
\begin{aligned}
V_{CE} &= \frac{kT}{e} \ln \left\{ \frac{\alpha_F[(1 - \alpha_R)I_C + I_B]}{\alpha_R[-(1 - \alpha_F)I_C + \alpha_F I_B]} \right\} \\
&= \frac{kT}{e} \ln \left[\frac{(1 - \alpha_R)I_C + I_B}{\alpha_R(-I_C/\beta_F + I_B)} \right]
\end{aligned}
\tag{8.17a}
$$

This indicates that for $V_{CE} \simeq 0.10$–0.20 V there must be a slight overdrive; that is, I_B must be somewhat larger than I_C/β_F.

8.3. EMITTER FOLLOWER

8.3a. Emitter Follower as a Large-Signal Circuit

Figure 8.6(a) shows an emitter follower circuit; here a signal V_{BB} is applied to the base via a resistor R_b and the output is taken from a resistor R_e in the emitter lead. We shall see that the output voltage follows the input voltage linearly; hence the name of the circuit.

If $V_{BB} < V_{BE0} \simeq 0.70$ V, the transistor is off and the output voltage V_E is zero. If now V_{BB} is increased beyond $V_{BE0} \simeq 0.70$ V, a current I_B flows so that

$$
V_{BB} - V_E - V_{BE0} = R_b I_B
\tag{8.18}
$$

but

$$
V_E = I_E R_e = (\beta_F + 1) R_e I_B
\tag{8.19}
$$

as long as the transistor is not saturated. Substitution into Eq. (8.18) yields

$$
I_B[R_b + (\beta_F + 1)R_e] = V_{BB} - V_{BE0}
\tag{8.20}
$$

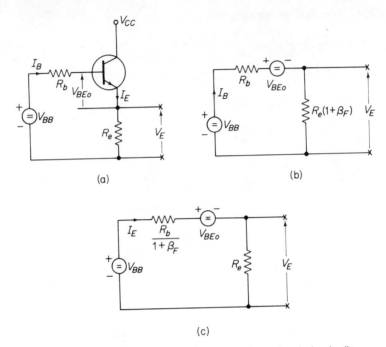

Fig. 8.6. (a) Emitter follower circuit as a large-signal circuit. (b) Large-signal equivalent circuit, seen from the base side. (c) Large-signal equivalent circuit, seen from the emitter side.

or since $I_B = I_E/(\beta_F + 1)$,

$$I_E\left[\frac{R_b}{\beta_F + 1} + R_e\right] = V_{BB} - V_{BE0} \tag{8.21}$$

Equations (8.19), (8.20), and (8.21) can be represented by the equivalent circuits of Fig. 8.6(b) and (c). Thus we see from Fig. 8.6(b) that the input circuit "sees" an input resistance $R_e(1 + \beta_F)$ in series with the EMF V_{BE0} (turn-on voltage) and from Fig. 8.6(c) that the output load resistance R_e "sees" an EMF $V_{BB} - V_{BE0}$ in series with an *internal resistance* $R_b/(1 + \beta_F)$. This indicates that the circuit transforms the emitter resistance R_e "up" and the base resistance R_b "down"; i.e., it transforms from a higher to a lower impedance level without loss of signal. We also see that

$$V_E = (V_{BB} - V_{BE0})\frac{R_e}{R_e + R_b/(1 + \beta_F)} \tag{8.22}$$

so that the output voltage V_E follows the EMF V_{BB} indeed linearly as long as the transistor is on ($V_{BB} > V_{BE0}$) and is not saturated ($V_E < V_{CC} - V_{CE\,sat}$). The upper limit of the output voltage is thus set by collector saturation

$$V_E = V_{CC} - V_{CE\,sat} \tag{8.23}$$

so that the corresponding value of V_{BB} is

$$V_{BB} = V_{BE0} + (V_{CC} - V_{CE\text{ sat}})\frac{R_e + R_b/(1 + \beta_F)}{R_e} \qquad (8.24)$$

The range of linear operation for V_{BB} is thus

$$V_{BE0} \le V_{BB} \le V_{BE0} + (V_{CC} - V_{CE\text{ sat}})\frac{R_e + R_b/(1 + \beta_F)}{R_e} \qquad (8.25)$$

corresponding to a range of output voltage

$$0 \le V_E \le V_{CC} - V_{CE\text{ sat}} \qquad (8.25a)$$

The circuit can also be used as a logic circuit. A 0 input ($V_{BB} < V_{BE0}$) corresponds to a 0 output, and a 1 input ($V_{BB} \simeq V_{CC}$) corresponds to a 1 output. *The circuit therefore transmits the logic information without altering it.*

8.3b. Small-Signal Theory of the Emitter Follower

In the previous discussion we replaced the base-emitter diode by an EMF $V_{BE0} \simeq 0.70$ V. This is an approximation, since for small-signal operation the diode has an input resistance $r_x + r_\pi$. The effect of r_x can usually be neglected but the effect of r_π should be taken into account. To evaluate this effect, we shall apply a small EMF v_b directly to the base, calculate the ac base current i_b, and define the input impedance as v_b/i_b [Fig. 8.7(a)]. As seen from the equivalent circuit,

$$v_b - v_e = i_b r_\pi; \qquad v_e = (\beta_F + 1)i_b R_e \qquad (8.26)$$

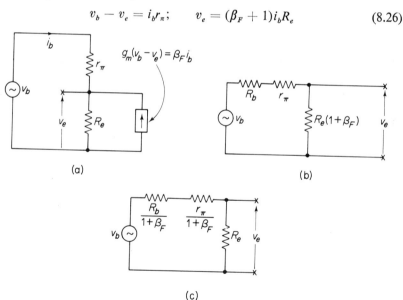

(a)　　　　　　　　　　(b)

(c)

Fig. 8.7. (a) Emitter follower as a small-signal circuit. (b) Small-signal equivalent circuit, seen from the base side. (c) Small-signal equivalent circuit, seen from the emitter side.

or, substituting for v_e,

$$v_b = i_b[r_\pi + (\beta_F + 1)R_e] \qquad (8.27)$$

so that the input impedance of the circuit is

$$Z_{in} = r_\pi + (\beta_F + 1)R_e \qquad (8.27a)$$

whereas the large-signal theory gave $(\beta_F + 1)R_e$. Often $r_\pi \ll (\beta_F + 1)R_e$; the difference between the two methods of approach is then quite small.

We can now correct the equivalent circuit of Fig. 8.6(b) and (c) by making it apply to the ac case. The results are shown in Fig. 8.7(b) and (c); Fig. 8.7(b) is the equivalent circuit seen from the base side and Fig. 8.7(c) is the equivalent circuit seen from the emitter side. Both are identical, since they give the same value of v_e.

As long as $r_\pi \ll R_b + (1 + \beta_F)R_e$, the difference between Fig. 8.6(b) and Fig. 8.7(b), and also the difference between Fig. 8.6(c) and Fig. 8.7(c), is quite small. Nevertheless it is usually worthwhile to investigate how large the correction is.

Since

$$v_e = v_b \frac{R_e}{(R_b/(\beta_F + 1) + r_\pi/(\beta_F + 1) + R_e} \qquad (8.28)$$

and only r_π depends appreciably on operating conditions, the output voltage hardly depends on operating conditions as long as $r_\pi \ll R_b + (1 + \beta_F)R_e$. Therefore the output signal will be practically undistorted, even if v_b is a relatively large signal. Since V_E can swing from zero to $V_{CC} - V_{CE\ sat}$, the maximum undistorted signal has an output amplitude $\frac{1}{2}(V_{CC} - V_{CB\ sat})$. The corresponding input signal has an amplitude

$$v_b = \frac{1}{2}(V_{CC} - V_{CE\ sat})\frac{R_b/(\beta_F + 1) + R_e}{R_e} \qquad (8.28a)$$

and the required dc bias is

$$V_{BB0} = V_{BE0} + \frac{1}{2}(V_{CC} - V_{CE\ sat})\frac{R_b/(\beta_F + 1) + R_e}{R_e} \qquad (8.28b)$$

8.3c. Darlington Circuit

Since the emitter current I_E in an emitter follower is $1 + \beta_F$ times the base current, the current amplification of an emitter follower is $1 + \beta_F$. If that is not sufficient, one can put two transistors Q_1 and Q_2 in series, feeding the emitter of Q_1 into the base of Q_2 and taking the output from a load resistance R_e in the emitter lead of Q_2. Such a circuit is shown in Fig. 8.8(a); it is called the *Darlington circuit*.

In this case, if the emitter and base currents of Q_1 and Q_2 are denoted by I_{E1}, I_{B1} and I_{E2}, I_{B2}, respectively, we have

$$I_{E1} = (\beta_{F1} + 1)I_{B1} \tag{8.29}$$

$$I_{B2} = I_{E1} \tag{8.30}$$

$$I_{E2} = (\beta_{F2} + 1)I_{B2} = (\beta_{F2} + 1)(\beta_{F1} + 1)I_{B1} \tag{8.31}$$

so that the total current amplification factor is $(\beta_{F2} + 1)(\beta_{F1} + 1)$.

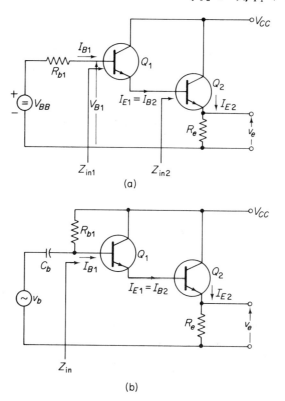

Fig. 8.8. Darlington circuit. (a) Driven from a high impedance source. (b) Driven with capacitive coupling.

In principle one can put more transistors in series and step up the current gain even more. The difficulty, however, is that at very low currents $\beta_F + 1$ decreases with decreasing current and that finally the base current of the first transistor becomes comparable to the collector saturated current I_{CB0}.

We now look at the dc voltage V_{B1} between the base of Q_1 and ground. If V_{BE1} and V_{BE2} are the base-emitter voltages of Q_1 and Q_2, respectively, then the voltage V_{B1} between base 1 and ground is

$$V_{B1} = V_E + V_{BE2} + V_{BE1} \simeq V_E + 1.4 \text{ V} \tag{8.32}$$

Next we shall investigate the input impedance of the circuit. The resistance R_e in the emitter lead of Q_2 gives rise to an input impedance

$$Z_{in2} = R_e(1 + \beta_{F2})$$

for Q_2, and this is also the apparent resistance between emitter 1 and ground. Therefore the input impedance seen at the base of Q_1 is

$$Z_{in} = R_e(1 + \beta_{F2})(1 + \beta_{F1}) \qquad (8.33)$$

We have neglected here the small effects of the base-emitter resistances $r_{\pi1}$ and $r_{\pi2}$; this is usually allowed, unless R_e is relatively small. By putting more transistors in series, one can boost the input impedance of the circuit still further. One of the great merits of the Darlington circuit is therefore that it uses low-impedance devices to build a circuit with a high input impedance.

EXAMPLE 1: If $R_e = 1000\ \Omega$, $I_{E2} = 1$ mA, $\beta_{F1} = \beta_{F2} = 100$, and $V_{BE0} = 0.70$ V at $I_E = 1$ mA, find the current amplification, the input impedance, and the currents.

ANSWER: The current amplification is $(1 + \beta_{F1})(1 + \beta_{F2}) \simeq 10^4$.

$$Z_{in\ 2} = R_e(1 + \beta_{F2}) \simeq 10^5\ \Omega$$

$$I_{E1} = I_{B2} = \frac{I_{E2}}{1 + \beta_{F2}} \simeq 10\ \mu A; \qquad r_{\pi2} = \frac{kT}{eI_{B2}} \simeq 2.6 \times 10^3\ \Omega \ll Z_{in\ 2}$$

$$Z_{in\ 1} = R_e(1 + \beta_{F1})(1 + \beta_{F2}) \simeq 10^7\ \Omega$$

$$I_{B1} = \frac{I_{E1}}{1 + \beta_{F1}} \simeq 0.1\ \mu A; \qquad r_{\pi1} = \frac{kT}{eI_{B1}} \simeq 2.6 \times 10^5\ \Omega \ll Z_{in\ 1}$$

EXAMPLE 2: The circuit of Fig. 8.8(a) is modified by returning R_{b1} to V_{CC} and applying the ac signal to the base of Q_1 via a capacitor $C_b = 100$ pF (Fig. 8.8(b). Find (a) the value of R_{b1} if $V_{CC} = 6.0$ V and $V_{BE2} = 0.70$ V, (b) the input impedance of the circuit, and (c) the lower cutoff frequency of the circuit. The other circuit parameters are as in Example 1.

ANSWER: (a) $V_E = 10^{-3} \times 1000 = 1$ V; $V_{BE2} = 0.70$ V. Since $I_{E1} \simeq I_{ES1}$ $\exp(eV_{BE1}/kT)$ and $I_{E2} \simeq I_{ES2} \exp(eV_{BE2}/kT)$ and $I_{ES1} \simeq I_{ES2}$,

$$V_{BE1} = V_{BE2} - \frac{kT}{e} \ln\left(\frac{I_{E2}}{I_{E1}}\right) = 0.70 - 25.8 \times 10^{-3} \ln 100 \simeq 0.60\ V$$

$$V_{B1} = V_E + V_{BE2} + V_{BE1} \simeq 2.30\ V$$

$$R_{b1} = \frac{V_{CC} - V_{B1}}{I_{B1}} = \frac{6.0 - 2.3}{10^{-7}} = 37\ M\Omega$$

(b)

$$Z_{in} = \frac{Z_{in\ 1} R_{b1}}{Z_{in\ 1} + R_{b1}} = \frac{10 \times 37}{10 + 37} = 7.9\ M\Omega$$

(c)

$$2\pi f C_b Z_{in} = 1; \quad f = \frac{1}{2\pi C_b Z_{in}} = \frac{1}{2\pi \times 10^{-10} \times 7.9 \times 10^6} = 200 \text{ Hz}$$

8.3d. Transistor Amplifier with Unbypassed Emitter Resistor

Figure 8.9(a) shows a transistor amplifier with unbypassed emitter resistor. Figure 8.9(b) is its equivalent circuit.

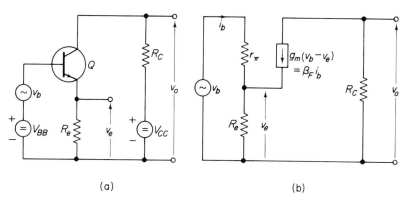

(a)   (b)

Fig. 8.9. (a) Transistor circuit with unbypassed emitter resistor. (b) Equivalent circuit of (a).

Since $g_m(v_b - v_e) = \beta_F i_b$, if $R_e \ll R_0$ and $R_c \ll R_0$, which are usually satisfied, we have

$$v_b - v_e = i_b r_\pi \tag{8.34}$$

$$v_e = (\beta_F + 1)i_b R_e \tag{8.35}$$

$$v_b = i_b[r_\pi + (\beta_F + 1)R_e] \tag{8.36}$$

so that

$$v_e = \frac{(\beta_F + 1)R_e}{r_\pi + (\beta_F + 1)R_e} v_b \tag{8.37}$$

$$v_0 = -g_m(v_b - v_e)R_c = -\beta_F i_b R_c = \frac{-\beta_F R_c}{r_\pi + (\beta_F + 1)R_e} v_b \tag{8.38}$$

If $(\beta_F + 1)R_e \gg r_\pi$, this reduces to

$$\frac{v_0}{v_b} = -\frac{\beta_F}{\beta_F + 1}\frac{R_c}{R_e} \simeq -\frac{R_c}{R_e} \tag{8.38a}$$

Since β_F is large and independent of operating conditions and the condition $(\beta_F + 1)R_e \gg r_\pi$ is usually satisfied, v_0/v_b is practically independent of operating conditions. This means that this amplifier gives practically undistorted amplification. *An unbypassed emitter resistor therefore reduces the gain but strongly reduces the distortion.*

EXAMPLE 3: If $R_c = 2000\ \Omega$, $R_e = 200\ \Omega$, $r_\pi = 2000\ \Omega$, and $\beta_F = 100$, find v_0/v_b.

ANSWER:

$$\frac{v_0}{v_b} = -\frac{\beta_F R_c}{r_\pi + (\beta_F + 1)R_e} = -\frac{2 \times 10^5}{2000 + 100 \times 200}$$

$$= -\frac{2 \times 10^5}{2.2 \times 10^3} = -9$$

Equation (8.38a) gives $v_0/v_b = -9.9$, so that the approximation (8.38a) is reasonably well satisfied.

8.4. DIFFERENTIAL AMPLIFIERS AND OPERATIONAL AMPLIFIERS

A differential amplifier is an amplifier that responds to the *difference* of two input signals. The differential amplifier is the basic circuit used in linear integrated transistor circuits, since its biasing can easily be put in integrated form.

An *operational amplifier* is a feedback amplifier consisting of one or more stages of differential amplification with external feedback.

8.4a. Differential Amplifier

Figure 8.10(a) shows a transistor amplifier that responds to the difference between two input signals v_{b1} and v_{b2}. It consists of two identical transistors connected in a balanced circuit that have an emitter resistance R_e in common.

To understand the operation of the circuit, we first find the common emitter voltage v_e with the help of the equivalent circuit of Fig. 8.10(b). If $R_e \gg r_\pi/(\beta_F + 1)$, we have

$$v_e = v_{b1}\frac{r_\pi}{r_\pi + r_\pi} + v_{b2}\frac{r_\pi}{r_\pi + r_\pi} = \frac{1}{2}(v_{b1} + v_{b2}) \tag{8.39}$$

Consequently

$$v_{b1} - v_e = \tfrac{1}{2}(v_{b1} - v_{b2}); \qquad v_{b2} - v_e = \tfrac{1}{2}(v_{b2} - v_{b1}) \tag{8.40}$$

To evaluate the output voltages v_{01} and v_{02}, we replace the transistors Q_1 and Q_2 by current generators $\tfrac{1}{2}g_m(v_{b1} - v_{b2})$ and $\tfrac{1}{2}g_m(v_{b2} - v_{b1})$, respectively. This gives the equivalent circuit of Fig. 8.10(c). Consequently

$$v_{01} = -\tfrac{1}{2}g_m(v_{b1} - v_{b2})R_c; \qquad v_{02} = \tfrac{1}{2}g_m(v_{b1} - v_{b2})R_c \tag{8.41}$$

so that each output responds to the *difference* of the two input signals. Note that v_{01} and v_{02} have opposite phase. The advantage of the circuit is that it is so well balanced and responds only to $v_{b1} - v_{b2}$ (differential mode response) and not at all to $v_{b1} + v_{b2}$ (common mode response).

The circuit as given here is not useful for integrated circuits in that the bias resistors R_b are far too large to be put in integrated form. We can remedy the situation by replacing R_e by a transistorized constant current generator;

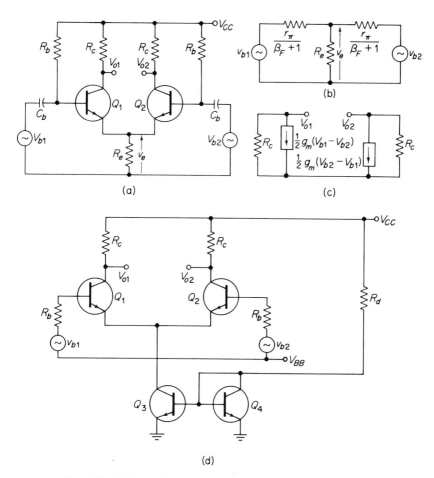

Fig. 8.10. Differential amplifier circuits. (a) Circuit with emitter resistor. (b) Equivalent circuit seen from the emitter side. (c) Equivalent circuit seen from the collector side. (d) Integrated differential amplifier driven by a constant current generator.

the bias resistors R_b of Fig. 8.10(a) can then be removed and the dc potential of the two bases can be taken up by the constant current generator circuit [Fig. 8.10(d)]. The two bases must be returned to a reference potential V_{BB} that is so chosen that the constant current generator circuit operates properly; i.e., $V_{CE3} > 0.50$ V. We thus obtain the circuit shown in Fig. 8.10(d).

We should bear in mind that the two bases carry base current, and therefore a low impedance path (R_b) should be provided to the reference potential V_{BB}. This reference potential lies between V_{CC} and ground and can be obtained by simple diode potentiometer techniques.

By omitting v_{b2} altogether, one can use the circuit as an ac amplifier for the input signal v_{b1}. The balanced circuit then merely serves as a means of stabilizing the operation and of keeping the zero level constant.

The circuit cannot only be used as an ac amplifier but also as a dc amplifier. Matched pairs of transistors are available and excellent stable operation down to relatively low dc voltage levels becomes possible, because the balanced circuit prevents drift to a very great extent.

When used as an amplifier for v_{b1}, it is common practice to eliminate not only the EMF v_{b2} but also the base resistance R_b of Q_2. It is then imperative that the base of Q_1 be returned to the reference voltage level V_{BB} through a low resistance path. For example, if each transistor has a collector current of 1 mA and $\beta_F = 100$, then each transistor has a base current of 10 μA, so that an external base resistance R_b of 100 Ω already gives rise to a dc unbalance of 1 mV.

When used as an ac amplifier, the amplifier suffers from distortion if the input signal level exceeds about 10 mV. We saw that an unbypassed emitter resistor could be used to reduce distortion. To that end equal resistors r_e of about 50–100 Ω are diffused into each emitter lead. However, this reduces the gain that can be obtained, and hence this method must be used with some caution.

When the amplification that can be provided by a single differential amplifier stage is not large enough, it is possible to feed the output of the first differential amplifier pair into the input of the next pair. Usually one does not bother to use v_{01} and v_{02} as input voltages of the new pair, but one uses only one of them, say v_{02}. This has the advantage that v_{02} is *in phase* with v_{b1}. Differential amplifiers can thus always be used in such a way that the output of each stage is *in phase* with the input of each stage.

In some applications one wants an output that is 180 degrees *out of phase* with the input. This can be achieved by simply taking the output from the *other half* of the differential output pair. The same effect can also be achieved by using *both* bases of the first differential pair as inputs. We shall see this in the discussions of the operational amplifier.

EXAMPLE 1: Design a differential amplifier according to Fig. 8.10(d) with $V_{CC} = 6.00$ V, $I_{C1} = I_{C2} = 1.0$ mA, and $V_{CE1} = V_{CE2} = 1.00$ V, V_{BB} being supplied by a diode power supply carrying a 2.0-mA diode current.

ANSWER: Since V_{BB} must be an integral number times 0.70 V and $V_{CE3} > 0.50$ V to avoid saturation, V_{BB} must be 1.40 V. Therefore $V_{CE3} = 0.70$ V and v_{01} and v_{02} have a dc component of 1.70 V above ground. This means that

$$R_c = \frac{V_{CC} - 1.70}{10^{-3}} = 4.3 \times 10^3 \ \Omega$$

Since $g_m = 38.6 \times 10^{-3}$ mho at $I_C = 1.0$ mA, the voltage gain is

$$\tfrac{1}{2} g_m R_c = \tfrac{1}{2} \times 38.6 \times 10^{-3} \times 4.3 \times 10^{-6} = 83x$$

Because the current flowing through R_d is 2.0 mA and $V_{BE4} = 0.70$ V,

$$R_d = \frac{V_{CC} - 0.70}{2.0 \times 10^{-3}} = \frac{5.30}{2.0 \times 10^{-3}} = 2650 \ \Omega$$

The diode power supply for V_{BB} must have two diodes in series. Let them carry a current of 2.0 mA. The currents flowing to the bases of Q_1 and Q_2 are negligible in comparison. Hence the resistance R in Fig. 8.2 must be

$$R = \frac{V_{CC} - 1.40}{2.0 \times 10^{-3}} = \frac{4.60}{2.0 \times 10^{-3}} = 2300 \ \Omega$$

8.4b. Operational Amplifier with Resistive Feedback

We assume first that the output voltage and the input voltage are 180 degrees out of phase and that a feedback resistance R_f is connected between output and input. We further assume that the input impedance of the amplifier is sufficiently large, and we shall investigate later what this assumption means. We thus have the equivalent circuit of Fig. 8.11(a), where the output voltage is $-K_v v_i$, K_v being a positive number.

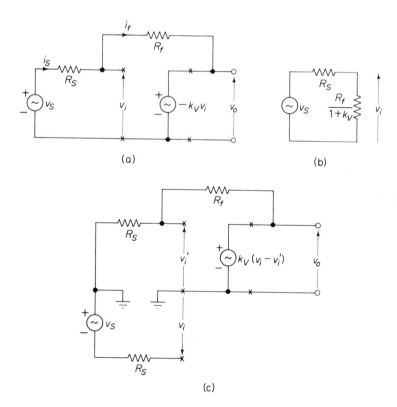

Fig. 8.11. Operational amplifier circuit. (a) Operational amplifier with resistive feedback. (b) Equivalent circuit, as seen from the input. (c) Differential amplifier version of operational amplifier.

We now have

$$v_0 = -K_v v_i \tag{8.42}$$

$$i_s = \frac{v_s - v_i}{R_s} = i_f = \frac{v_i - v_0}{R_f} \tag{8.43}$$

since the current flowing into the input of the amplifier is negligible if the input impedance of the amplifier is sufficiently large. Substituting for v_0 and solving for v_i yields

$$\frac{v_s - v_i}{R_s} = v_i \frac{(1 + K_v)}{R_f} \tag{8.44}$$

so that

$$v_i = v_s \frac{R_f}{R_s(1 + K_v) + R_f} = v_s \frac{R_f/(1 + K_v)}{R_s + R_f/(1 + K_v)} \tag{8.45}$$

Substituting the first half of Eq. (8.45) into the expression for v_0 yields

$$v_0 = -K_v v_i = -\frac{K_v R_f}{R_s(1 + K_v) + R_f} v_s \tag{8.46}$$

If now $K_v \gg 1$ and $R_s(1 + K_v) \gg R_f$, then $v_i \ll v_s$, and v_0/v_s may be written as

$$\frac{v_0}{v_s} = -\frac{R_f}{R_s} \tag{8.46a}$$

which is practically independent of the operating conditions of the amplifier, i.e., independent of K_v. It may thus be assumed that the amplifier is quite linear and has negligible distortion. This is borne out by a more detailed analysis. This type of feedback is called *negative* feedback, since the output voltage is 180 degrees out of phase with respect to the input voltage.

Because of the shunting of the interstage circuits by the device capacitances, K_v becomes complex at high frequencies and $|K_v|$ decreases with increasing frequency at high frequencies. However, $|v_0/v_s|$ is practically independent of K_v as long as $|R_s(1 + K_v)| > R_f$ and $|K_v| \gg 1$. Therefore the feedback also improves the high-frequency response of the amplifier.

We can now also discuss what the assumption of a "sufficiently large input impedance" of the amplifier means. To that end we take the second half of Eq. (8.45) and represent it by the equivalent circuit of Fig. 8.11(b). We then see that the source v_s in series with its internal resistance "sees" an input impedance $R_f/(1 + K_v)$ of the feedback amplifier. If the input impedance R_i of the amplifier itself is large in comparison with $R_f/(1 + K_v)$, this input impedance has no effect on the operation of the amplifier.

We shall now illustrate this with the help of examples.

EXAMPLE 2: If $R_f = 10^5\ \Omega$, $R_s = 1000\ \Omega$, and $K_v = 10^4$, find the voltage gain.

ANSWER:

$$\frac{v_0}{v_s} = -\frac{R_f}{R_s} = -100$$

EXAMPLE 3: Investigate whether the condition $R_i \gg R_f/(1 + K_v)$ is satisfied. You may assume that $R_i = 2000 \, \Omega$.

ANSWER:

$$\frac{R_f}{1 + K_v} = \frac{10^5}{10^4} = 10 \, \Omega$$

which is small in comparison with R_i.

One final word of caution. K_v is real and positive at low frequencies, but because of the shunting effects of the device capacitances, K_v becomes complex at higher frequencies. There may then be a frequency f_1 where K_v is real and negative. At that frequency the output voltage v_0 is in phase with the input voltage v_i; if $|K_v|$ at that frequency is sufficiently high, the circuit may become unstable. Since a single interstage network provides a maximum phase shift of 90 degrees, a phase shift in K_v by more than 180 degrees cannot be provided by an operational amplifier with one or two interstage networks. Such amplifiers are therefore perfectly stable. However, for an amplifier with three interstage networks the maximum phase shift is 270 degrees, and here one must be very careful.

We shall now consider the case that $K_v = K_{v1}$ at $f = f_1$ is real and negative in greater detail. We must then rewrite the first half of Eq. (8.45) as follows:

$$v_i = v_s \frac{R_f}{R_s + R_f - |K_{v1}| R_s} \tag{8.47}$$

so that v_i becomes infinitely large if

$$R_s + R_f - |K_{v1}| R_s = 0 \tag{8.47a}$$

For $R_s + R_f - |K_{v1}| R_s < 0$, the circuit oscillates. This must be avoided.

This type of feedback is called *positive* feedback, since v_0 and v_i are in phase. We see that too large an amount of positive feedback leads to instability and should therefore be avoided unless one wants to build oscillators. Moreover, this type of feedback enhances distortion and deteriorates the high-frequency response of the feedback amplifier.

We thus see that the feedback circuit is unconditionally stable if $|K_{v1}| < 1$. For $|K_{v1}| > 1$ the circuit is stable if

$$R_f > (|K_{v1}| - 1)R_s \tag{8.47b}$$

If the last condition is not satisfied, the feedback circuit will oscillate at the frequency f_1.

To remedy the situation one must remedy the frequency response of the operational amplifier to such an extent that the feedback circuit remains stable. There are several techniques for doing this, but it is beyond the scope of this discussion to go into details.

We shall now investigate the case where v_0 and v_i are in phase and negative feedback is obtained by feeding back from the output to the other input terminal of the first differential amplifier stage. For the sake of simplicity we shall assume once again that the amplifier has a high input impedance. The circuit is shown in Fig. 8.11(c). We have inserted a resistance R_s in series with *each* input to prevent any unbalance of the amplifier. We then see by inspection, if K_v is the voltage gain of the differential amplifier, that

$$v_0 = K_v(v_i - v_i') \tag{8.48}$$

$$v_i' = v_0 \frac{R_s}{R_s + R_f} = \frac{K_v R_s}{R_s + R_f}(v_i - v_i') \tag{8.49}$$

Solving for v_i' yields

$$v_i' = \frac{K_v R_s}{R_f + (1 + K_v)R_s} v_i \tag{8.50}$$

or, since $v_i = v_s$,

$$v_i - v_i' = \frac{R_f + R_s}{R_f + (1 + K_v)R_s} v_s \tag{8.51}$$

Consequently

$$v_0 = K_v(v_i - v_i') = \frac{K_v(R_f + R_s)}{R_f + (1 + K_v)R_s} v_s \tag{8.52}$$

If $K_v \gg 1$ and $(1 + K_v)R_s \gg R_f$, Eq. (8.52) reduces to

$$\frac{v_0}{v_s} = 1 + \frac{R_f}{R_s} \tag{8.52a}$$

This is again independent of operating conditions and hence the amplifier is linear and distortion-free.

We must now investigate what the condition "sufficiently high input impedance R_i" means. For the input half to which the feedback is applied, the requirement is $R_i \gg R_f/(1 + K_v)$ as in the previous case. For the other input half we must require $R_i \gg R_s$, which is not necessarily satisfied. The separate discussion that is required here is beyond the scope of this book.

Finally, we shall investigate how the circuit reduces distortion. We shall assume that the distortion is mainly generated in the last stage and that the distortion, when referred back to the input of that stage, can be represented

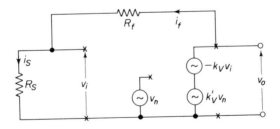

Fig. 8.12. Equivalent circuit of operational amplifier used to calculate distortion.

by an EMF v_n, K_v' being the amplification of that stage. We assume that v_n depends on v_0 only.

We shall calculate this with the help of Fig. 8.12. If the input impedance of the amplifier itself is sufficiently large, we have, if v_n produces an input voltage v_i,

$$v_0 = K_v'v_n - K_v v_i \tag{8.53}$$

$$i_f = \frac{(v_0 - v_i)}{R_f} = i_s = \frac{v_i}{R_s} \tag{8.54}$$

Substitution of Eq. (8.53) into Eq. (8.54) gives

$$\frac{K_v'v_n - v_i(1 + K_v)}{R_f} = \frac{v_i}{R_s} \tag{8.55}$$

or, when solving for v_i,

$$v_i = \frac{K_v'R_s}{R_f + R_s(1 + K_v)}v_n \tag{8.56}$$

Consequently

$$v_0 = K_v'v_n - K_v v_i = K_v'v_n\left[1 - \frac{R_sK_v}{R_f + R_s(1 + K_v)}\right]$$

$$= K_v'v_n\frac{R_f + R_s}{R_f + R_s(1 + K_v)} \simeq \frac{1 + R_f/R_s}{1 + K_v}K_v'v_n \tag{8.57}$$

if $R_f \ll R_s(1 + K_v)$.

We now assume that v_0 is kept constant when the feedback is applied; this can be achieved by increasing the input amplitude of the *wanted* signal appropriately. When this has been done, v_n is unaltered by the feedback and hence the distortion is reduced by the factor

$$\frac{1 + R_f/R_s}{1 + K_v} \tag{8.58}$$

which is appreciable. If the input amplitude is not changed when feedback is applied, v_0 and hence v_n decrease with increasing feedback and the discussion becomes more complicated.

EXAMPLE 4: If the amplifier without feedback has $K_v = 10^4$, the output voltage is 1.0 V, and the gain K_v' of the last stage is 10^2, find the distortion without feedback.

ANSWER: Since the output voltage is 1.0 V, the input voltage of the last stage is 10 mV. However, according to Section 7.2, this gives a distortion of about 10% if the load resistance of the previous stage is not too large*.

EXAMPLE 5: If the amplifier without feedback has $K_v = 10^4$, $R_f/R_s = 10^2$, $K_v' = 10^2$, and v_0 *with* feedback is 1.0 V, find the distortion *with* feedback.

---

*This must be modified somewhat if the last stage is a differential amplifier (see Problem 8.10).

ANSWER:

$$\frac{1 + R_f/R_s}{1 + K_v} \simeq \frac{R_f/R_s}{K_v} = \frac{10^2}{10^4} = 10^{-2}$$

Therefore the distortion with feedback is $10^{-2} \times 10\% = 0.1\%$.

EXAMPLE 6: What is the input voltage v_s needed to produce an output voltage of 1.0 V in the feedback amplifier of Example 5?

ANSWER:

$$v_s = \frac{v_0}{R_f/R_s} = \frac{1.0}{100} = 10 \text{ mV}$$

8.4c. Operational Amplifier with Capacitive Feedback

Just as the case of resistive feedback produced an input impedance $R_f/(1 + K_v)$, so feedback via a capacitance C_f produces an input impedance [Fig. 8.13(a) and (b)]

$$Z_{\text{in}} = \frac{1/(j\omega C_f)}{1 + K_v} = \frac{1}{j\omega C} \tag{8.59}$$

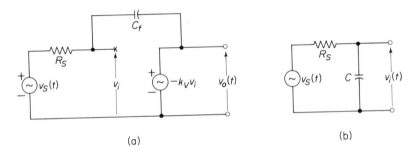

(a)   (b)

Fig. 8.13. (a) Operational amplifier with capacitive feedback. (b) Equivalent circuit as seen from the input.

corresponding to an apparent input capacitance

$$C = C_f(1 + K_v) \tag{8.60}$$

As is seen from Fig. 8.13(b), the input voltage $v_i(t)$ is thus determined by the differential equation

$$C\frac{dv_i}{dt} = \frac{v_s(t) - v_i(t)}{R_s} \simeq \frac{v_s(t)}{R_s} \tag{8.61}$$

as long as $v_i(t) \ll v_s(t)$.

If $v_i(t) = 0$ at $t = 0$, the solution of this equation is

$$v_i(t) = \frac{1}{CR_s} \int_0^t v_s(u) \, du \tag{8.62}$$

so that

$$v_0(t) = -K_v v_i(t) = -\frac{K_v}{1 + K_v} \frac{1}{C_f R_s} \int_0^t v_s(u)\, du \qquad (8.62a)$$

The solution is valid for $t \ll CR_s$; often this condition can be narrowed down to $t < \frac{1}{10} CR_s$, since that gives an inaccuracy of only a few percent. By proper choice of CR_s one can thus perform an integration of $v_s(t)$ over quite a long time interval. For that reason the circuit can be used as an integrator in analog computers and in measuring equipment.

8.5. TRANSISTOR POWER AMPLIFIERS

If one takes a transistor operating at relatively low power and plots the collector current I_C as a function of the base current I_B [Fig. 8.14(a)], a reasonably straight line is obtained as long as β_F is not very current-dependent. One can then operate the transistor at the quiescent operating point (I_{C1}, I_{B1}), feed the base from a sinusoidal current source, and use a current amplitude such that the device stays within its linear range.

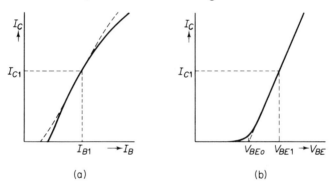

(a)   (b)

Fig. 8.14. (a) (I_C, I_B) characteristic of a power transistor. (b) (I_C, V_{BE}) characteristic of a power transistor.

However, at very high currents β_F decreases appreciably with increasing current and then this approach does not work. Since one wants high ac power production and this requires large ac currents, a transistor power amplifier with the base fed from an ac current generator cannot produce a large, distortion-free ac power.

Fortunately, in those cases there is a better solution. If one plots the collector current I_C of an *n-p-n* power transistor as a function of the base-emitter voltage V_{BE}, one obtains a reasonably straight line for $V_{BE} > V_{BE0}$, where V_{BE0} is of the order of 0.70 V in silicon power transistors [Fig. 8.14(b)]. The reason for this behavior is that the base resistance r_x linearizes the external characteristic. Here one uses an operating point (I_{C1}, V_{BE1}) and can then utilize a base voltage amplitude almost equal to $V_{BE1} - V_{BE0}$.

8.5a. Power Transistor with Resistive Output

Figure 8.15 shows a power transistor circuit with a dc base bias V_{BE1} and a base EMF $V_1 \cos \omega t$, where $V_1 \simeq V_{BE1} - V_{BE0}$. The collector current then swings from $2I_{c1}$ to zero, since the current amplitude $\simeq I_{c1}$ [compare Fig. 8.14(b)]. The transistor must be able to provide this current.

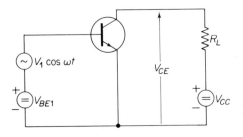

Fig. 8.15. Power transistor circuit.

We must now investigate what happens in the collector circuit and determine how the load resistor R_L must be chosen. If $I_C = 0$, $V_{CE} = V_{CC}$; the transistor must therefore be able to withstand the collector voltage V_{CC} without suffering breakdown. At $I_C = 2I_{c1}$ the voltage V_{CE} has its minimum value

$$V_{CE \, min} = V_{CC} - 2I_{c1}R_L \tag{8.63}$$

R_L must be so chosen that the transistor is never saturated; that is, $V_{CE \, min}$ must be larger than 0.5–1.0 V. Hence, since $V_{CE \, min} \ll V_{CC}$,

$$R_L = \frac{V_{CC} - V_{CE \, min}}{2I_{c1}} \simeq \frac{V_{CC}}{2I_{c1}} \tag{8.64}$$

The value of V_{CE} then swings from V_{CC} (for $I_C = 0$) to about zero (for $I_C = 2I_{c1}$). Consequently the quiescent collector operating point is $(I_{c1}, \frac{1}{2}V_{CC})$ and the ac collector voltage amplitude is $\frac{1}{2}V_{CC}$.

The power operated under quiescent conditions is therefore

$$(P_{diss})_0 = \frac{1}{2}V_{CC}I_{c1} < P_{max} \tag{8.65}$$

where P_{max} is the maximum collector power rating of the transistor.

EXAMPLE 1: If $P_{max} = 10$ W and $V_{CC} = 15$ V, find I_{c1} and R_L.

ANSWER:

$$(I_{c1})_{max} = \frac{2P_{max}}{V_{CC}} = \frac{20}{15} A = 1.33 \text{ A}$$

$$R_L \simeq \frac{V_{CC}}{2I_{c1}} = \frac{15}{2.67} = 5.6 \, \Omega$$

The power supplied to the circuit by the supply voltage V_{CC} is

$$P_{supp} = V_{CC}I_{c1} \tag{8.66}$$

since the circuit is linear. The dc power dissipated in R_L is

$$P_{dc} = I_{c1}^2 R_L = \frac{1}{2}V_{CC}I_{c1} = \frac{1}{2}P_{supp} \tag{8.67}$$

Half the supplied power thus ends up in the resistor R_L; under quiescent conditions the remainder is dissipated in the transistor as $(P_{\text{diss}})_0$.

What happens if an ac voltage is applied? Since the ac current amplitude is I_{c1}, the ac power dissipated in R_L is

$$P_{\text{ac}} = \tfrac{1}{2}I_{c1}^2 R_L = \tfrac{1}{4}V_{cc}I_{c1} = \tfrac{1}{4}P_{\text{supp}} \tag{8.68}$$

so that the power dissipated under normal operating conditions is

$$P_{\text{diss}} = P_{\text{supp}} - P_{\text{dc}} - P_{\text{ac}} = \tfrac{1}{4}V_{cc}I_{c1} = \tfrac{1}{4}P_{\text{supp}} \tag{8.69}$$

This is only half of what is dissipated under quiescent conditions. If the circuit is so designed that the transistor does not become overheated under quiescent conditions, it will certainly not be overheated when an ac signal is applied. One can also put it this way: The circuit transforms dc power normally dissipated in the transistor into ac power dissipated in the load resistor R_L.

The efficiency η of the power amplifier is defined as

$$\eta = \frac{P_{\text{ac}}}{P_{\text{supp}}} = \frac{1}{4} \tag{8.70}$$

so that the circuit has an efficiency of 25%.

EXAMPLE 2: If $P_{\text{max}} = 10$ W, find the undistorted ac power supplied to the load resistance R_L in Example 1.

ANSWER: $P_{\text{ac}} = \tfrac{1}{4}P_{\text{supp}} = 5$ W, since $P_{\text{supp}} = V_{cc}I_{c1} = 15 \times 1.33 = 20$ W.

8.5b. Power Transistor Circuit with Transformer-Coupled Load

A more efficient power amplifier can be obtained by coupling the load resistance R_L to the circuit by means of a transformer of turns ratio n, for the simple reason that no dc power is dissipated in the load resistor. We shall see that this raises the efficiency to 50%. Figure 8.16(a) shows the circuit.

The apparent ac resistance R'_L seen at the primary terminals of the transformer is

$$R'_L = n^2 R_L \tag{8.71}$$

This must be properly chosen.

If I_{c1} is the quiescent collector current, as before, then the collector current swings from $2I_{c1}$ to zero and the collector voltage swings from $(V_{CE})_{\text{min}}$ to $2V_{cc} - V_{CE\,\text{min}}$. The value of V_{cc} must be so chosen that the transistor can stand the voltage $2V_{cc}$ without suffering breakdown. Since the ac collector voltage amplitude is approximately V_{cc} and the ac collector current amplitude is I_{c1}, the apparent load resistance R'_L must be so chosen that

$$R'_L \simeq \frac{V_{cc}}{I_{c1}} \tag{8.72}$$

The dc power supplied by V_{cc} is

$$P_{\text{supp}} = I_{c1}V_{cc} \tag{8.73}$$

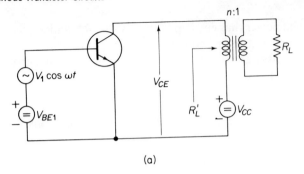

(a)

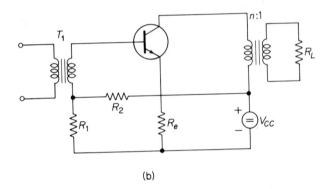

(b)

Fig. 8.16. (a) Power transistor circuit with transformer output coupling. (b) Power transistor circuit with transformer output coupling and dc base supply.

The dc power dissipated under quiescent conditions is

$$(P_{\text{diss}})_0 = I_{C1}V_{CC} < P_{\text{max}} \tag{8.74}$$

where P_{max} is the maximum power that can be dissipated by the transistor. The ac power dissipated by R_L is

$$P_{\text{ac}} = \tfrac{1}{2}I_{C1}^2R_L' = \tfrac{1}{2}I_{C1}V_{CC} = \tfrac{1}{2}P_{\text{supp}} \tag{8.75}$$

so that the efficiency η of the amplifier is

$$\eta = \frac{P_{\text{ac}}}{P_{\text{supp}}} = \frac{1}{2} \quad \text{or} \quad 50\% \tag{8.76}$$

as had to be proved.

The drawback of the circuit of Fig. 8.16(a) is that a separate bias battery V_{BE1} must be applied. In Fig. 8.16(b) an improved circuit is shown in which the base bias voltage is supplied by a potentiometer circuit. Furthermore, a small unbypassed resistor R_e is provided to improve the linearity and to stabilize the dc operation of the circuit. Finally, the input is also fed from a transformer to provide a low-impedance dc path for the base current.

EXAMPLE 3: A transistor power amplifier with transformer coupling has $V_{CC} = 10$ V and $P_{max} = 10$ W. Find the value of I_{C1}, the apparent load resistance R_L, and the ac power delivered to the load resistance R_L.

ANSWER:

$$P_{max} = V_{CC}I_{C1}; \quad I_{C1} = \frac{P_{max}}{V_{CC}} = \frac{10}{10} = 1 \text{ A}$$

$$R'_L = \frac{V_{CC}}{I_{C1}} = \frac{10}{1.0} = 10 \text{ }\Omega$$

$$P_{ac} = \tfrac{1}{2}I_{C1}V_{CC} = \tfrac{1}{2} \times 1.0 \times 10 = 5 \text{ W}$$

EXAMPLE 4: If $R_L = 2.5$ Ω, find the required turns ratio n in Example 3.

$$n = \left(\frac{R'_L}{R_L}\right)^{1/2} = \left(\frac{10}{2.5}\right)^{1/2} = 2$$

EXAMPLE 5: What is the maximum value of V_{CE} that the transistor must be able to withstand?

ANSWER: $V_{CE \text{ max}} \simeq 2V_{CC} = 20$ V.

EXAMPLE 6: Design a transistor power amplifier with transformer output, potentiometer base voltage supply, and $R_e = 0$ that delivers 10 W into a 5.0-Ω load. The transistor has $P_{max} = 20$ W, $V_{CE \text{ max}} = 60$ V, and $I_{C1 \text{ max}} = 6.0$ A. The circuit must be designed for minimum base current. The transistor has $\beta_F = 20$ and you may assume that the base operates at $V_{BE1} = 1.0$ V.

ANSWER: To minimize the base current, we must make I_{C1} as small as possible and V_{CC} as large as possible. We therefore choose $V_{CC} = \tfrac{1}{2}V_{CE \text{ max}} = 30$ V and hence

$$I_{C1}V_{CC} = P_{max} = 20 \text{ W} \quad \text{or} \quad I_{C1} = \tfrac{20}{30} = 0.67 \text{ A} < 6.0 \text{ A}$$

$$R'_L = \frac{V_{CC}}{I_{C1}} = \frac{30}{0.67} = 45 \text{ }\Omega; \quad n = \left(\frac{R'_L}{R_L}\right)^{1/2} = \left(\frac{45}{5}\right)^{1/2} = 3.2$$

$$P_{supp} = I_{C1}V_{CC} = 20 \text{ W}; \quad \eta = 50\%; \quad P_{ac} = \eta P_{supp} = 10 \text{ W}$$

Since $\beta_F = 20$, the quiescent base current is $(0.67/20)$ A $= 33$ mA and the maximum base current is twice as large. We therefore assume that a current $I_2 = 150$ mA passes through R_2. Hence

$$R_2 = \frac{V_{CC} - V_{BE1}}{I_2} \simeq \frac{30}{0.150} = 200 \text{ }\Omega$$

Since under quiescent conditions 33 mA goes to the base, 117 mA passes through R_1. Hence

$$R_1 = \frac{1.0}{0.117} = 8.5 \text{ }\Omega$$

The power amplifiers discussed so far draw current during a full cycle.

Such power amplifiers are called *class A* power amplifiers. The current swings from $2I_{c1}$ to zero and the voltage swings from zero to $2V_{CC}$; the quiescent operating point is (I_{c1}, V_{CC}), the supplied power is $I_{c1}V_{CC}$, and the maximum ac power is $\frac{1}{2}I_{c1}V_{CC}$, so that the maximum efficiency cannot be higher than 50%. A 10-W transistor can thus give a maximum ac power of only 5 W.

The class A amplifiers discussed here have several drawbacks that should be remedied:

1. The dissipated power is a maximum under quiescent conditions, whereas one would like to have zero dissipated power under quiescent conditions.
2. One would like to have an efficiency higher than 50%.
3. One would like to eliminate the power transformers.

The first two problems are solved by going to the class B push-pull circuit (Section 8.5c) and the third problem is solved by going to a complementary symmetry transistor combination (Section 8.5d).

8.5c. Class B Push-Pull Transistor Power Amplifier

In a class B push-pull transistor circuit, each transistor passes current only half the time; one transistor provides half of the sine wave and the other transistor provides the other half. Figure 8.17(a) shows the circuit. In this circuit the transistors are so biased that they are barely on under the quiescent condition, that is, $V_{BE} \simeq 0.70$ V, so that the quiescent power dissipated is very small. In the circuit the base bias is provided by a potentiometer circuit. R_1 can be replaced by a diode; if that is done, R_2 must be so chosen that it can provide a current larger than the maximum base current.

Figure 8.17(b) shows the currents $I_1(t)$ and $I_2(t)$ flowing through each half of the circuit and the current $I_1(t) + I_2(t) = I(t)$ flowing through V_{CC}. Each transistor carries a current amplitude I_{max} and for maximum power the voltage amplitude is approximately equal to V_{CC}; hence the load resistance "seen" by each transistor half must be

$$R'_L = n^2 R_L = \frac{V_{CC}}{I_{max}} \tag{8.77}$$

For the average current we now have

$$I_{av} = \frac{I_{max}}{2\pi} \int_0^{2\pi} |\cos \omega t| \, d(\omega t) = \frac{2}{\pi} I_{max} \tag{8.78}$$

The power supplied is therefore

$$P_{supp} = V_{CC}I_{av} = \frac{2}{\pi} I_{max}V_{CC} \tag{8.79}$$

The ac power is

$$P_{ac} = \frac{1}{2}I_{max}V_{CC} \tag{8.80}$$

since the current amplitude is I_{max} and the voltage amplitude is V_{CC}. Therefore

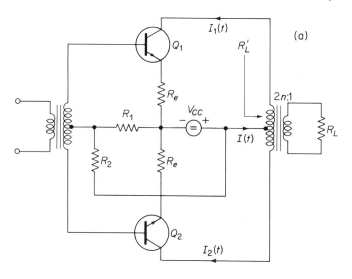

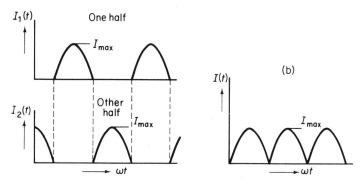

Fig. 8.17. (a) Class B power transistor circuit. (b) Wave form of one half and other half, and wave form seen by the voltage supply.

the efficiency is

$$\eta = \frac{P_{ac}}{P_{supp}} = \frac{\pi}{4} = 78.5\% \tag{8.81}$$

which is much better than a class A amplifier.

Under quiescent conditions practically no current is flowing so that $(P_{diss})_0 \simeq 0$. For full power operation the dissipated power is

$$P_{diss} = P_{supp} - P_{ac} = \left(\frac{1}{\eta} - 1\right) P_{ac} \leq 2P_{max} \tag{8.82}$$

where P_{max} is the maximum power that can be dissipated by each transistor. Consequently the maximum ac power that can be generated is

$$(P_{ac})_{max} = \frac{2\eta}{1-\eta} P_{max} \tag{8.83}$$

This is much larger than the value $2P_{max}$ found for the class A amplifier.

EXAMPLE 7: If $\eta = 78.5\%$ and $P_{max} = 10$ W, find $(P_{ac})_{max}$.

ANSWER:

$$(P_{ac})_{max} = \frac{2 \times 0.785}{1 - 0.785} \times 10 = 73 \text{ W}$$

EXAMPLE 8: If $V_{CC} = 20$ V, find I_{max} for that case.

ANSWER:

$$(P_{ac})_{max} = \tfrac{1}{2}V_{CC}I_{max} = 10I_{max} = 73 \text{ W}$$

$$I_{max} = 7.3 \text{ A}$$

It must be investigated whether the transistors can carry this current.

EXAMPLE 9: The transistors of Example 6 are used to build a class B power amplifier with 78.5% efficiency. R_1 is replaced by a diode and the transistors have $\beta_F = 20$ at the maximum current. The circuit must be designed for the smallest base current.

ANSWER: Since the base current must be as small as possible, we take I_{max} as small as possible. This requires V_{CC} to be as large as possible, or $V_{CC} = \tfrac{1}{2}V_{CE\,max} = 30$ V. In analogy with Example 7,

$$(P_{ac})_{max} = 146 \text{ W}$$

or

$$\tfrac{1}{2}V_{CC}I_{max} = 15I_{max} = 146 \text{ W} \quad \text{or} \quad I_{max} = 9.7 \text{ A}$$

Since this is larger than $I_{C1\,max}$, we choose $I_{max} = 6$ A, which means that the maximum power cannot be achieved:

$$P_{ac} = \frac{1}{2}I_{max}V_{CC} = 90 \text{ W}; \qquad R'_L = \frac{V_{CC}}{I_{max}} = \frac{30}{6} = 5\,\Omega$$

If I_2 is the current flowing through R_2 and $I_{max} = 6$ A, the maximum base current is 300 mA. We therefore choose $I_2 = 500$ mA and require that the diode be able to withstand that current. Hence

$$R_2 = \frac{V_{CC} - 0.70}{I_2} \simeq \frac{30}{0.5} = 60\,\Omega$$

8.5d. Complementary Symmetry Class B Power Amplifier

Finally, we shall discuss a circuit that does away with the transformer. It consists of a driver stage with transistor Q_1 feeding two power transistors Q_2 and Q_3 of complementary symmetry. The circuit arrangement is shown in Fig. 8.18(a); Q_2 is an *n-p-n* transistor, and Q_3 is a *p-n-p* transistor. The diodes D_1 and D_2 provide the dc bias of the bases (each diode has a voltage drop of 0.70 V, corresponding to the voltage drop in each transistor). Again, one transistor provides one half of the sine wave and the other transistor provides the other half. When no ac voltage is applied, half the voltage V_{CC} is developed across Q_2 and the other half is developed across Q_3; hence the maximum ac

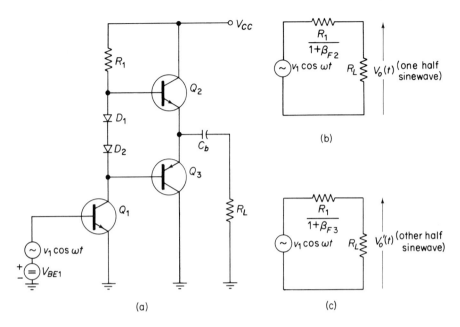

Fig. 8.18. (a) Transformerless class B power transistor circuit. (b) Equivalent circuit seen by one half. (c) Equivalent circuit seen by other half.

output amplitude is $\frac{1}{2}V_{CC}$. The capacitor C_b stores the charge coming from Q_2 and uses it to provide the current for Q_3.

If I_{max} is the current amplitude, then

$$R_L = \frac{V_{CC}}{2I_{max}} \tag{8.84}$$

The ac power developed is

$$P_{ac} = \frac{1}{2}I_{max}\frac{1}{2}V_{CC} = \frac{1}{4}I_{max}V_{CC} \tag{8.85}$$

The current $I(t)$ flows during the half-cycle that Q_2 is on. This means that the supplied power is half the supplied power of the previous case, or

$$P_{supp} = \frac{1}{\pi}I_{max}V_{CC} \tag{8.86}$$

so that the efficiency η is again $\pi/4$ or 78.5%.

Since both input and output are single-ended, it is possible to provide a large amount of negative feedback. The capacitor C_b must be so chosen that it passes the ac signal and stores enough dc charge to provide for the current of Q_3 during the time that Q_3 is on.

The circuit has one small drawback: Q_2 and Q_3 are usually not perfectly matched. If Q_2 has a current amplification factor β_{F2} and Q_3 a current amplification factor β_{F3} and if a voltage $v_1 \cos \omega t$ is developed across R_1, we see that the equivalent circuit of Q_2 for the one half-cycle and of Q_3 for the other half-cycle is as shown in Fig. 8.18(b) and (c), respectively. These equivalent circuits express the simple fact that the circuit consists essentially of two emitter followers passing current alternately. We thus see that there is a small difference in voltage amplitude during the two half-cycles if β_{F2} and β_{F3} are not equal. The difference is negligible if β_{F2} and β_{F3} are sufficiently large.

If more current gain is needed in the output stage than a single transistor can provide, Q_2 and Q_3 should be replaced by Darlington pairs. This raises β_{F2} and β_{F3} sufficiently to make any unbalance between the two half sine waves negligible.

Finally, we shall return to Fig. 8.18(a) and show that the output power is somewhat smaller than previously indicated, because the collector of Q_1 must always be forward-biased. Since the two diodes have a voltage drop of about 0.70 V each, the peak-to-peak ac voltage in the collector is $V_{CC} - 1.40$, so that the ac amplitude at the output has a maximum amplitude of $\frac{1}{2}V_{CC} - 0.70$ instead of $\frac{1}{2}V_{CC}$, so that Eq. (8.85) reduces to

$$P_{ac} = \tfrac{1}{4}I_{max}(V_{CC} - 1.40) \tag{8.87}$$

whereas P_{supp} has not changed. Therefore

$$\eta = \frac{\pi}{4}\frac{V_{CC} - 1.40}{V_{CC}} \tag{8.87a}$$

This difficulty can be overcome by feeding Q_1 from a voltage of $V_{CC} + 1.40$ V, rather than from the voltage V_{CC}.

If the two transistors Q_2 and Q_3 are each replaced by a Darlington pair, one needs four diodes in series to compensate for the voltage drops in the four transistors. One must now replace 1.40 V by 2.80 V in the above formulas.

8.6. *TRANSISTOR CHOPPER CIRCUITS*

In a chopper circuit a transistor is used as a switch that alternately connects and disconnects a signal source to or from a load. To understand its behavior, first we shall consider a mechanical chopper.

8.6a. Mechanical Choppers

Suppose that one has a small dc voltage V_d developed over a resistance R_d and that one wants to measure this signal accurately. One could then amplify the signal with the help of a dc amplifier. However, this has the disadvantage that the small signal drowns in the zero drift of the circuit.

A much better solution is to transform the dc voltage into an ac voltage by means of a mechanical chopper, i.e., a mechanical switch S that alternately opens and closes (Fig. 8.19) and thereby periodically short-circuits the signal source. The output voltage $V_0(t)$ of the chopper is alternately V_d (switch open)

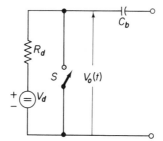

Fig. 8.19. Mechanical chopper at
the input of an amplifier.

and zero (switch closed), so that the chopper transforms the dc voltage V_d
into a square wave of peak-to-peak amplitude V_d with a period $1/f_c$, where f_c
is the frequency with which the switch opens and closes. Putting

$$V_0(t) = A_0 + A_1 \cos \omega_c t + A_2 \cos 2\omega_c t + \cdots \qquad (8.88)$$

where $\omega_c = 2\pi f_c$, we have

$$A_1 = \frac{1}{\pi} \int_{-\pi}^{\pi} V_0(t) \cos \omega_c t \, d(\omega_c t) = \frac{2V_d}{\pi} \qquad (8.88a)$$

since $V_0(t) = V_d$ for $-(\pi/2) < \omega_c t < \pi/2$ and zero for

$$-\pi < \omega_c t < -\frac{\pi}{2} \quad \text{and} \quad \frac{\pi}{2} < \omega_c t < \pi$$

This signal of frequency f_c can now be amplified by standard ac techniques and be detected without being bothered by amplifier drift. If the signal
$V_0(t)$ is fed directly into the amplifier, one can detect signals of the order of
10^{-9} V. If one uses a transformer of turns ratio n, one can measure signals
down to $10^{-9}/n$ V.

8.6b. Transistor Chopper

In a transistor chopper one uses the transistor as a switch that alternately
connects or disconnects a signal source V_d to or from a load. First we snall
discuss the single transistor chopper of Fig. 8.20(a). The resistance R_b is of
the order of 10,000 Ω to limit the flow of base current I_B.

If the base is made *positive* with respect to the collector, the transistor
behaves as a closed switch and $V_0(t) = V_d$. If the base is made *negative* with
respect to the collector, the transistor behaves as an open switch and $V_0(t) =$
0.

However, the transistor is not a perfect switch. If the switch is closed,
it has an offset voltage

$$V_{EC} = \frac{kT}{e}(1 - \alpha_F) + I_B r_{sc} \qquad (8.89)$$

where I_B is the base current, r_{sc} the series resistance of the collector, and α_F
the current amplification factor of the transistor. To make this offset voltage
small, one should make $1 - \alpha_F$ and r_{sc} as small as possible.

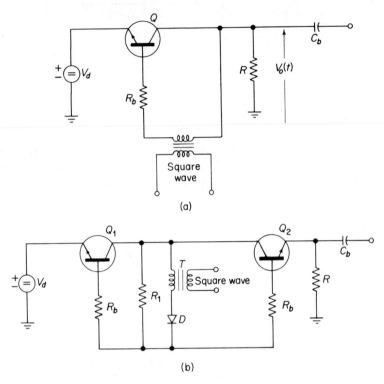

Fig. 8.20. (a) Electronic transistor chopper. (b) Balanced electronic transistor chopper.

If the switch is open, there is a small leakage current flowing. By proper choice of the transistor and by making the output resistance R not too large, one can eliminate this source of error to a considerable extent.

To prove the expression for the offset voltage, we observe that the dc voltage drop in the resistance r_{sc} is equal to $I_{B}r_{sc}$, which corresponds to the second half of Eq. (8.89). To prove the first part, we observe that if V_d is very small in comparison with the voltage driving the switch, and $eV_{BE}/kT \gg 1$ and $eV_{BC}/kT \gg 1$, nearly all the current flows from the base to the collector, and $I_E \simeq 0$. Therefore, from the Ebers-Moll equations,

$$-I_{ES}\exp\left(\frac{eV_{BE}}{kT}\right) + \alpha_R I_{CS}\exp\left(\frac{eV_{BC}}{kT}\right) = 0$$

so that, since $V_{BE} - V_{BC} = -V_{EC}$,

$$\exp\left(-\frac{eV_{EC}}{kT}\right) = \exp\left[\frac{eV_{BE} - V_{BC})}{kT}\right] = \frac{\alpha_R I_{CS}}{I_{ES}} = \alpha_F \qquad (8.90)$$

since $\alpha_R I_{CS} = \alpha_F I_{ES}$. Therefore

$$V_{EC} = -\frac{kT}{e} \ln \alpha_F = \frac{kT}{e}(1 - \alpha_F) \qquad (8.90a)$$

since $\ln(1 - x) = -x$ for small x. This proves the first half of Eq. (8.89).

EXAMPLE: If $I_B = 0.5$ mA, $r_{cs} = 0.5\ \Omega$, and $1 - \alpha_F = 0.010$, find V_{EC}.

ANSWER:

$$\frac{kT}{e}(1 - \alpha_F) = 25.8 \times 10^{-3} \times 10^{-2} = 258\ \mu V$$

$$I_B r_{cs} = 0.5 \times 10^{-3} \times 0.5 = 250\ \mu V$$

Hence $V_{EC} \simeq 500\ \mu V$.

By using the balanced circuit shown in Fig. 8.20(b), the offset voltages of the two transistors cancel each other and so do the effects of the leakage currents. Therefore offset voltages of a few microvolts become attainable.

8.7. HIGH-FREQUENCY CIRCUITS

8.7a. High-Frequency Response of Transistors

The equivalent π-circuit of a transistor is easily extended to higher frequencies by connecting a capacitance C_π in parallel with r_π and a capacitance C_μ in parallel with r_μ. If we then neglect r_μ because its effect is small, we obtain the circuit of Fig. 8.21(a).

The capacitance C_μ is simply the space-charge capacitance of the collector-base junction. If the space-charge region has a width d and the junction has an area A, this capacitance is

$$C_\mu = \frac{\epsilon\epsilon_0 A}{d} \qquad (8.91)$$

It is therefore important to keep the junction area A small and the width d of the space-charge region large. The latter is achieved by making the part of the collector adjacent to the base weakly p-type. The space-charge region of the junction then extends mainly into the collector and d is relatively large.

The capacitance C_π is mainly caused by the minority carriers stored in the base region and is usually considerably larger than C_μ. As long as the effect of the stored charge predominates, which is the case as long as the emitter current is not too small, the product $\tau_\pi = C_\pi r_\pi$ is independent of current. For very small currents, however, the effect of the stored charge is negligible and C_π reduces to the space-charge capacitance C_{je} of the emitter-base junction which does not depend very strongly on current. Since r_π is inversely proportional to the emitter current I_E, $\tau_\pi = C_\pi r_\pi$ is then inversely proportional to I_E. This is discussed in greater detail in Appendix A.

If now an ac current i_b is forced into the base junction, and the output

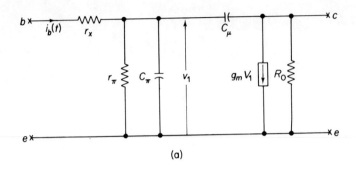

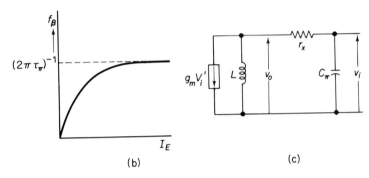

Fig. 8.21. (a) Hybrid π circuit of transistor. (b) Beta cutoff frequency as a function of I_E. (c) High-frequency equivalent circuit of tuned transistor interstage network.

is short-circuited, and if the effect of C_μ is neglected because $C_\mu \ll C_\pi$, the voltage v_1 across r_π is

$$v_1 = \frac{i_b}{1/r_\pi + j\omega C_\pi} = \frac{i_b r_\pi}{1 + j\omega C_\pi r_\pi} \qquad (8.92)$$

so that the ac output current becomes

$$g_m v_1 = \frac{g_m r_\pi}{1 + j\omega C_\pi r_\pi} i_b = \frac{\beta_F}{1 + j\omega C_\pi r_\pi} i_b \qquad (8.93)$$

It thus makes considerable sense to call the factor in front of i_b the *high-frequency current amplification factor β for the common emitter connection,*

$$\beta = \frac{\beta_F}{1 + j\omega C_\pi \pi_\pi} = \frac{\beta_F}{1 + jf/f_\beta} \qquad (8.94)$$

where

$$\frac{1}{f_\beta} = 2\pi C_\pi r_\pi = 2\pi \tau_\pi \qquad (8.95)$$

and τ_π is the time constant of the input circuit. The frequency f_β is called the *beta cutoff frequency* of the transistor; for most transistors it lies in the megahertz range. Because of what was said about the time constant $\tau_\pi = C_\pi r_\pi$, the beta cutoff frequency is practically independent of current for reasonably large currents but varies inversely with the emitter current I_E for very small currents. This must be borne in mind when transistors are used at very low currents [Fig. 8.21(b)].

Besides the frequency f_β, another frequency, the *cutoff frequency* f_T, is introduced as the frequency for which $|\beta| = 1$. Since $\beta_F \gg 1$, this is the case if in Eq. (8.94)

$$\beta_F \simeq \omega_T C_\pi r_\pi = g_m r_\pi \quad \text{or} \quad f_T = \frac{g_m}{2\pi C_\pi} = \beta_F f_\beta \tag{8.96}$$

At relatively large currents C_π and g_m are both proportional to the current, and f_T is then independent of current. At low emitter currents, however, $C_\pi = C_{je}$ is independent of current, and hence f_T is proportional to the current I_E.

We thus see that the current amplification factor becomes complex at high frequencies and that its absolute value decreases with increasing frequency. There should therefore be a frequency at which the transistor ceases to operate as an amplifier. In most cases this frequency is of the order of, or at most somewhat larger than, the cutoff frequency f_T, but it is usually determined by r_x (see Examples 2 and 3).

It is not difficult to make transistors with cutoff frequencies f_T in the range 500–1000 MHz. With somewhat more care one can design transistors that can operate in the microwave range.

EXAMPLE 1: A transistor operating at a current of 1 mA has $f_\beta = 10$ MHz and $\beta_F = 100$. Find f_T.

ANSWER: $f_T = \beta_F f_\beta = 100 \times 10$ MHz $= 1000$ MHz.

There is another effect that must be taken into account. At very high frequencies the capacitance C_π shunts the resistance r_π. The input circuit then consists of the resistance r_x and the capacitance C_π in series. This heavily loads a preceding amplifier stage and thus reduces the gain of that stage. The input impedance is thus $r_x + 1/(j\omega C_\pi)$ and hence the input admittance Y_i is

$$Y_i = g_i + jb_i = \frac{1}{r_x + 1/(j\omega C_\pi)} = \frac{j\omega C_\pi}{1 + j\omega C_\pi r_x}$$
$$= \frac{j\omega C_\pi (1 - j\omega C_\pi r_x)}{1 + \omega^2 C_\pi^2 r_x^2} \tag{8.97}$$

The susceptive part can be eliminated by tuning, but the input conductance g_i of the transistor due to r_x cannot. From Eq. (8.97)

$$g_i = \frac{\omega^2 C_\pi^2 r_x^2}{1 + \omega^2 C_\pi^2 r_x^2} \frac{1}{r_x} \tag{8.97a}$$

This can be much larger than the low-frequency input conductance $g_{be} = 1/r_\pi$.

EXAMPLE 2: If $r_x = 20\ \Omega$ and $C_\pi = 5$ pF, find the input conductance due to r_x at 1000 MHz.

ANSWER:

$$\omega C_\pi r_x = 2\pi \times 10^9 \times 5 \times 10^{-12} \times 20 = 0.63$$

$$g_i = \frac{1}{r_x} \frac{(\omega C_\pi r_x)^2}{1 + (\omega C_\pi r_x)^2} = \frac{1}{20} \frac{(0.63)^2}{1 + (0.63)^2} = 0.0143 \text{ mho}$$

corresponding to an input impedance of 70 Ω. We thus see that $g_i \gg g_{be}$, as previously stated.

EXAMPLE 3: The circuit of Fig. 8.21(c) is part of a cascaded set of stages with tuned interstage networks. If the previous stage is represented by a current generator $g_m v_i'$ connected in parallel with the input of the stage in question, find the values $|v_0/v_i'|$ and $|v_i/v_i'|$ at 1000 MHz if $g_m = 40 \times 10^{-3}$ mho and r_x, C_π have the same values as in Example 2; L tunes the network.

ANSWER:

$$\left| \frac{v_0}{v_i'} \right| = \frac{g_m}{g_i} = 40 \times 10^{-3} \times 70 = 2.8$$

$$\frac{v_i}{v_0} = \frac{1/(j\omega C_\pi)}{r_x + 1/(j\omega C_\pi)} = \frac{1}{1 + j\omega C_\pi r_x}$$

$$\left| \frac{v_i}{v_0} \right| = \frac{1}{[1 + (\omega C_\pi r_x)^2]^{1/2}} = \frac{1}{(1.40)^{1/2}} = 0.85$$

so that $|v_i/v_i'| = 2.4$. Note that $|v_i/v_i'|$ is the cascaded gain per stage. The resistance r_x therefore seriously deteriorates the cascaded gain at high frequencies.

8.7b. Effects of the Capacitance C_μ

The capacitance C_μ provides feedback between output and input. For low-frequency amplifier stages in cascade this reduces the high-frequency gain and for tuned amplifiers it can give rise to oscillations.

For low frequencies C_μ gives rise to a Miller capacitance (compare Chapter 4)

$$C' = (1 + g_m R_c)C_\mu \simeq g_m R_c C_\mu \tag{8.98}$$

in parallel with the capacitance C_π, where R_c is the resistance in the collector lead. The total input capacitance C_T of the device is therefore

$$C_T = C_\pi + g_m R_c C_\mu \tag{8.99}$$

and this can be considerably larger than C_π, leading to a reduction in the upper cutoff frequency of an amplifier stage if amplifier stages are connected in cascade.

The time constant of the input of the transistor is thus changed from the value $\tau_\pi = C_\pi r_\pi$ for short-circuited output ($R_c = 0$) to the value

$$\tau_T = C_T r_\pi = C_\pi r_\pi + g_m r_\pi R_c C_\mu = \tau_\pi + \beta_F R_c C_\mu \tag{8.100}$$

This is important for transistors used in pulse circuits.

EXAMPLE 4: If $C_\pi = 15$ pF, at $g_m = 50 \times 10^{-3}$ mho, $C_\mu = 3$ pF, $\beta_F = 100$, and $R_c = 2000\ \Omega$, find C_T, τ_π, and τ_T.

ANSWER:

$$C_T = C_\pi + g_m R_c C_\pi = 15 + 50 \times 10^{-3} \times 2 \times 10^3 \times 3 = 315 \text{ pF}$$

$$r_\pi = \frac{1}{g_{be}} = \frac{\beta_F}{g_m} = 2000\ \Omega$$

$$\tau_\pi = C_\pi r_\pi = 15 \times 10^{-12} \times 2 \times 10^3 = 3 \times 10^{-8} \text{ s}$$

$$\tau_T = C_T r_\pi = 315 \times 10^{-12} \times 2 \times 10^3 = 6.3 \times 10^{-7} \text{ s}$$

In tuned amplifiers the capacitance C_μ may cause instability of the circuit unless the interstage networks are heavily loaded. Fortunately, transistors have a large transconductance, about 40 millimhos for $I_C = 1$ mA and proportionally larger values for higher currents. Therefore considerable amplification per stage is possible, even if the interstage networks are heavily loaded. We refer to the end of Chapter 4 for details.

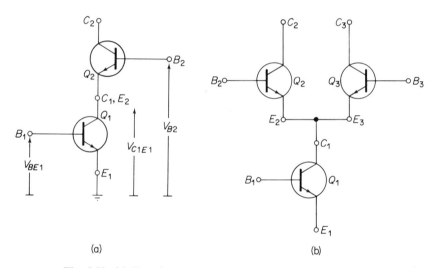

(a)   (b)

Fig. 8.22. (a) Transistor cascode circuit. (b) Transistor cascode circuit with volume control operation.

The instability can be improved by going to a *cascode circuit* [Fig. 8.22(a)]. Here a common emitter connection is followed by a common base connection, both operating at approximately the same current. The first half gives a voltage gain of about unity and the second stage can give a large voltage gain. The improvement in stability comes about because the interstage between Q_1 and Q_2 is untuned and heavily damped, whereas the direct capacitance between C_2 and B_1 can be kept very small.

Obviously the combination must be so biased that a considerable current flows. Therefore $V_{BE1} \simeq 0.70$ V, and since V_{CE1} must be larger than 0.50 V to avoid saturation and V_{B2} is about 0.70 V above V_{CE1}, V_{B2} should be larger than 1.20 V. This can be arranged in practice.

In Fig. 8.22(b) the second transistor is replaced by a differential pair. By changing the bias of B_2, one can shift more or less ac current coming from Q_1 through Q_3. The circuit can thus be used for *volume control purposes*, just as the corresponding FET circuit discussed in Chapter 6. Because of the similarity of the circuits it is not necessary to go into details. The circuit is quite popular and very useful.

Another possibility is to apply a small signal of frequency f_1 to B_1 and a large signal of frequency $f_2 > f_1$ between B_2 and B_3. A signal of frequency $f_3 = f_2 - f_1$ can then be taken from the collectors of Q_2 or Q_3. Such a circuit is called a *mixer circuit*. It is beyond the scope of this book to go into details.

8.8. TRANSISTOR CASCADED AMPLIFIERS

Here we shall apply the theory of cascaded amplifiers, as described in Chapter 4, to transistor amplifier stages.

8.8a. Untuned Interstage Network

To apply the theory of Chapter 4, we shall return to the interstage network of Fig. 4.10 and identify the network elements. To simplify matters, we shall first ignore the effect of the series base resistance r_x.

The three parameters g_m, R_i, and R_0 were defined in Chapter 7. For n-p-n transistors

$$g_m = \frac{\partial I_C}{\partial V_{BE}}; \qquad \frac{1}{R_i} = \frac{1}{r_\pi} = \frac{\partial I_B}{\partial V_{BE}}; \qquad \frac{1}{R_0} = \frac{\partial I_C}{\partial V_{CE}} \qquad (8.101)$$

The resistance R_b corresponds to the parallel connection of the two bias resistances R_1 and R_2, that is,

$$R_b = \frac{R_1 R_2}{R_1 + R_2} \qquad (8.102)$$

where R_1 and R_2 were identified in Fig. 7.9. The capacitance C_0 is negligible, $C_i = C_\pi$, and the Miller effect capacitance $C_f = C_\mu(1 + |g_{v0}|)$, where g_{v0} is the midband gain of the cascaded stages. Hence, if C_w is the wiring capacitance, the total interstage capacitance is

$$C_{par} = C_w + C_\pi + C_\mu(1 + |g_{v0}|) \qquad (8.103)$$

Often the effect of C_w is quite small. The total parallel resistance of the interstage network is given by

$$\frac{1}{R_{par}} = \frac{1}{R_0} + \frac{1}{R_L} + \frac{1}{R_1} + \frac{1}{R_2} + \frac{1}{r_\pi} \qquad (8.103a)$$

The midband gain is defined as $g_{vo} = -g_m R_{par}$, and

$$|g_{vo}|f_1 = g_m R_{par} f_1 = \frac{g_m}{2\pi C_{par}} \quad \text{or} \quad f_1 = \frac{1}{2\pi C_{par} R_{par}} \tag{8.104}$$

An alternative way of writing this is as

$$|g_{vo}|f_1 = \frac{g_m}{2\pi C_\pi} \frac{C_\pi}{C_{par}} = f_T \frac{C_\pi}{C_{par}} \tag{8.104a}$$

where f_T is the cutoff frequency defined in Eq. (8.96); it is of the order of 300–1500 MHz in usual transistors and up to 4000–8000 MHz in microwave transistors. The factor C_π/C_{par} is called the Miller effect deterioration factor; it has a value of 5–10 for high values of the midband gain, whereas its value may drop below 2 for low-gain stages.

The midband gain $|g_{vo}| = g_m R_{par} < g_m r_\pi = \beta_F$. Values of $|g_{vo}|$ between 0.4 and $0.7\beta_F$ are possible in high-gain stages. The gain is limited because of the dc voltage drop in the collector resistance R_c; one must require that V_{CE} does not drop below 0.5 V under worst-case conditions.

EXAMPLE 1: Transistors having $\beta_F = 100$, $f_T = 1000$ MHz, $g_m = 40 \times 10^{-3}$ mho, $R_0 = \infty$, and $C_\mu = 1.0$ pF are used in a circuit having $R_L = r_\pi$, $R_1 = 10,000\ \Omega$, and $R_2 = 5000\ \Omega$. If the wiring capacitance is 2.0 pF, find the midband gain and the upper cutoff frequency of cascaded amplifier stages.

ANSWER:

$$r_\pi = \frac{\beta_F}{g_m} = \frac{100}{40 \times 10^{-3}} = 2500\ \Omega$$

$$C_\pi = \frac{g_m}{2\pi f_T} = \frac{40 \times 10^{-3}}{2\pi \times 10^9} = 6.4\ \text{pF}$$

$$\frac{1}{R_{par}} = \frac{1}{2500} + \frac{1}{10,000} + \frac{1}{5000} + \frac{1}{2500} = 1.1 \times 10^{-3}\ \text{mho}$$

$$|g_{vo}| = g_m R_{par} = \frac{40 \times 10^{-3}}{1.1 \times 10^{-3}} = 36$$

$$C_{par} = [2.0 + 6.4 + 1.0(1 + 36)]\ \text{pF} = 45.4\ \text{pF}$$

$$f_1 = \frac{1}{2\pi C_{par} R_{par}} = \frac{1.1 \times 10^{-3}}{2\pi \times 45.4 \times 10^{-12}} = 3.9\ \text{MHz}$$

For low-gain stages the Miller effect capacitance is less and hence the bandwidth is much larger. If we again neglect r_x for the time being, we have $R_{par} \simeq R_L$; if we also neglect C_w, we have $C_{par} = C_\pi + C_\mu(1 + |g_{vo}|)$.

EXAMPLE 2: Calculate the load resistance and the upper cutoff frequency of an interstage network in a cascaded amplifier chain if the midband gain per stage is 4 and the transistor is as in Example 1.

ANSWER:

$$C_\pi = 6.4 \text{ pF}; \qquad C_{\text{par}} = 6.4 + 1.0 \times 5 = 11.4 \text{ pF}$$

$$R_L = \frac{4}{g_m} = 100 \, \Omega$$

$$f_1 = \frac{1}{2\pi \times 100 \times 11.4 \times 10^{-12}} = 140 \text{ MHz}$$

We shall now take the effect of r_x into account. To simplify matters, we assume that $r_x + R_L \ll r_\pi$, so that the effect of r_π can be ignored. We shall also neglect the wiring capacitance C_w and so obtain the circuit of Fig. 8.23.

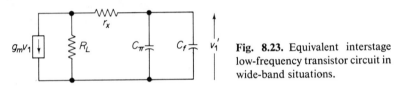

Fig. **8.23.** Equivalent interstage low-frequency transistor circuit in wide-band situations.

As seen by inspection,

$$v_1' = -\frac{g_m R_L v_1 / [j\omega(C_\pi + C_f)]}{R_L + r_x + 1/[j\omega(C_\pi + C_f)]} = \frac{g_m R_L v_1}{1 + j\omega(R_L + r_x)(C_\pi + C_f)}$$

$$= -\frac{g_{v0}}{1 + jf/f_1} v_1 \qquad (8.105)$$

so that

$$2\pi f_1 (R_L + r_x)(C_\pi + C_f) = 1 \qquad (8.106)$$

EXAMPLE 3: For the interstage network of Example 2, evaluate the effect of r_x on the midband gain and on the upper cutoff frequency f_1.

ANSWER: The effect of r_x on $|g_{v0}|$ is negligible, whereas

$$f_1 = \frac{1}{2\pi(R_L + r_x)(C_\pi + C_f)} = \frac{1}{2\pi \times 120 \times 11.4 \times 10^{-12}} = 116 \text{ MHz}$$

8.8b. Tuned Interstage Network

The case of the tuned amplifier, depicted in Fig. 4.14(a), can be treated in the same way. If one uses a cascode transistor circuit, the voltage gain of the first half of the cascode is -1 and hence the input capacitance $C_i = C_\pi + 2C_\mu$.

If one does not use a cascode circuit, one has to limit the gain per stage to prevent oscillations. According to Eq. (4.51a), since $C_{oi} = C_\mu$, the requirement for unconditional stability is

$$|g_{v0}| = g_m R_{\text{par}} = \left(\frac{g_m}{\pi f C_\mu}\right)^{1/2} \qquad (8.107)$$

We shall illustrate this with an example.

EXAMPLE 4: If $g_m = 4 \times 10^{-2}$ mho and $C_\mu = 1.0$ pF, find the maximum gain per stage for which the circuit is stable at 100 MHz.

ANSWER:

$$|g_{vo}| = \left(\frac{g_m}{\pi f C_\mu}\right)^{1/2} = \left(\frac{40 \times 10^{-3}}{\pi \times 10^8 \times 10^{-12}}\right)^{1/2} = 11.3$$

If one wants high selectivity, one must "tap" both the output of the previous stages and the input of the next stage to the tuned circuit in such a way that R_0 and r_π do not affect the total tuned circuit impedance too much. This is illustrated in Fig. 8.24.

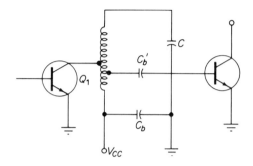

Fig. 8.24. Tuned interstage transistor circuit with tapped collector and base to provide better selectivity of the circuit.

For frequencies close to f_T the above considerations are not sufficient. A more careful analysis of the equivalent circuit of the transistor is needed before the stability condition of the circuit can be evaluated.

8.8c. Emitter Follower at High Frequencies

Section 8.3b gave the low-frequency theory of the emitter follower circuit. We shall now investigate what happens at high frequencies.

At high frequencies the resistance r_π in Fig. 8.7(b) must be replaced by $r_\pi/(1 + j\omega C_\pi r_\pi)$ and the low-frequency current amplification factor β_F must be replaced by the high-frequency current amplification factor $\beta = \beta_F/(1 + j\omega C_\pi r_\pi)$. Instead of Eq. (8.28), we thus have

$$v_e = v_b \frac{R_e(1 + \beta)}{R_b + r_\pi/(1 + j\omega C_\pi r_\pi) + R_e(1 + \beta)} \tag{8.108}$$

This is not very strongly frequency-dependent as long as $R_e|1 + \beta| > R_b$. The emitter follower circuit is therefore a wide-band circuit.

PROBLEMS*

8.1. If $V_{CC} = 7.5$ V, design a constant current supply that feeds a current of 1.0 mA into the differential pair of Fig. 8.1(b). $V_{BE0} = 0.70$ V.

*Problems with an asterisk are more difficult.

8.2. In the circuit of Fig. 8.2, Q has $\beta_F = 100$ and $V_{sup} = 7.5$ V. All diodes have $V_{D0} = 0.70$ V at $I_D = 1.0$ mA and Q has $V_{BE0} = 0.70$ V at $I_E = 1.0$ mA. (a) Choose R so that $I_D = 2.0$ mA at $I_E = 0$, and determine V_{BB} accurately. (b) If a current $I_E = 100$ mA is drawn, find the voltage V_E between emitter and ground. Do not ignore the change in voltage drop in the diode chain.

8.3. If in the circuit of Fig. 8.3 $R_c = 2000\ \Omega$ and $R_e = 2000\ \Omega$ and the transistors have $\beta_F = 100$, find the input resistance of Q_2 and determine whether this significantly affects the gain of Q_1.

8.4. Figure 8.25 shows a differential amplifier in which the bases are returned to a reference level of 2.8 V. (a) Calculate the collector currents of Q_1 and Q_2. (b) Choose R_c so that $V_0 = 2.8$ V. (c) Calculate the voltage gain of the amplifier stage. (d) Calculate total current drawn from the supply voltage $V_{CC} = 4.9$ V. $V_{BE0} = 0.70$ V. Neglect the small effect of the value of R_e on the gain.

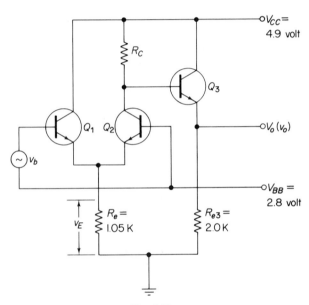

Fig. 8.25.

8.5. Figure 8.26 shows a two-stage differential amplifier. $V_{BB} = 2.1$ V is the dc reference voltage to which the bases of Q_1, Q_2, and Q_5 are returned. $V_{CC} = 4.2$ V. The resistors are as indicated. (a) Calculate the collector currents of Q_2 and Q_5 and show that the base of Q_4 is at 2.1 V. (b) Calculate the transconductance of Q_2 and Q_5. (c) Calculate the overall gain, neglecting the small decrease in gain because the emitter resistances are not very large. (d) Calculate the total current of the circuit.

8.6. The circuit of Fig. 8.4(a) is used as a logic circuit. Find the breaks in the V_{CC}, V_{BB} characteristic if $V_{BE0} = 0.70$ V, $V_{CC} = 10$ V, $V_{CE\ sat} \simeq 0$ V, $R_b = 10^4\ \Omega$, $R_c = 2 \times 10^3\ \Omega$, and $\beta_F = 100$.

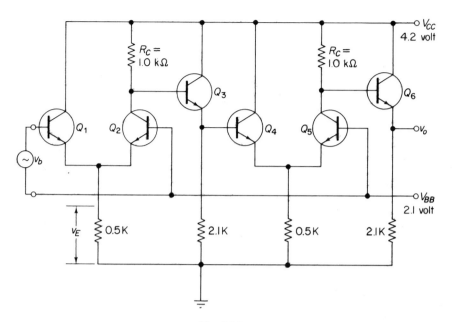

Fig. 8.26.

8.7. The circuit of Fig. 8.27 is used to measure a small current I of the order of 10^{-10} A. (a) If each transistor has $\beta_F = 100$, find the output voltage v_{out}. (b) Find the input resistance presented to the current generator I.

8.8. The circuit of Fig. 8.27 is modified by returning the base of Q_1 to V_{CC} by means of resistance R_b that is so chosen that the current through R_e is 10 mA. R_e is changed to 100 Ω. (a) Find the value of R_b needed. (b) The amplifier

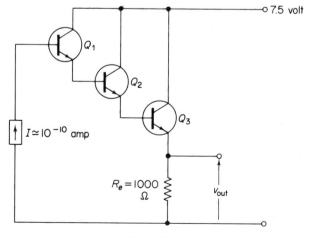

Fig. 8.27.

is connected to a capacitor microphone having a signal EMF $v_b = 1.0$ mV in series with a capacitor of 100 pF. Find the lower cutoff frequency of the circuit. $\beta_F = 100$; each transistor has $V_{BE0} = 0.70$ V at $I_E = 1.0$ mA. See also Fig. 8.8(b) for notation.

*8.9. (a) Find the second harmonic distortion in a transistor amplifier with unbypassed emitter resistor R_e. (b) Now adjust the base signal $v_{b0} \cos \omega t$ so that you get the *same* output voltage across R_c as without emitter feedback. Show that the noise reduction factor due to feedback is then $1 + g_m R_e$. *Hint for* (a): If I_E is the dc emitter current, $v_b \doteq v_{b0} \cos \omega t$ is the ac base voltage, and $v_e = v_{e1} \cos \omega t + v_{e2} \cos 2\omega t + \cdots$ is the ac emitter voltage, then the ac emitter current is

$$\Delta I_E = I_E \left\{ \exp \left[\frac{e(v_b - v_e)}{kT} \right] - 1 \right\}; \qquad v_e = \Delta I_E R_e$$

Make a Taylor expansion of the exponent, neglecting higher than second-order terms and substitute for v_b and v_e.

*8.10. In a differential amplifier the input signal is applied to one base and the output signal is taken from the other collector. Show that in first approximation the circuit cancels the second harmonic distortion. You may assume that $g_m R_e \gg 1$. *Hint:* Put

$$\Delta I_{E1} = I_{E0} \left\{ \exp \left[\frac{e(v_b - v_e)}{kT} \right] - 1 \right\}$$

$$\Delta I_{E2} = I_{E0} \left[\exp \left(- \frac{e v_e}{kT} \right) - 1 \right]$$

where I_{E0} is the dc current, and $v_b = v_{b0} \cos \omega t$, and $v_e = (\Delta I_{E1} + \Delta I_{E2}) R_e = v_{e1} \cos \omega t + v_{e2} \cos 2\omega t + \cdots$. Make a Taylor expansion and neglect all terms higher than the second order.

8.11. In the differential amplifier of Fig. 8.10(a), show that the input resistance seen by v_{b1} or v_{b2} is $2r_\pi$, where $r_\pi = kT/eI_B$, if R_e is sufficiently large. *Hint:* Transform Fig. 8.10(b) back to the inputs after assuming that $R_e \gg r_\pi/(\beta_F + 1)$.

8.12. Design a power transistor amplifier with transformer output coupling and $R_e = 0$ that delivers 5 W into a 5.0-Ω load. The transistor has $P_{\max} = 10$ W, $V_{CE \max} = 50$ V, and $I_{C1 \max} = 3.0$ A. Design the circuit such that the ac base current is as small as possible. Find the ac base current if $\beta_F \simeq 20$.

8.13. The transistor of Problem 8.12 is operated at $V_{BE1} = 1.0$ V, and this base voltage is supplied by a potentiometer circuit as shown in Fig. 8.16(b). The resistance R_1 is bypassed by a large capacitor to keep V_{BE1} practically independent of the ac base current. (a) Choose C_1 such that V_{BE1} varies less than 10% for a frequency of 50 cycles. (b) Find R_1 and R_2 if the current through R_2 is 40 mA.

8.14. In an ideal class *B* transistor amplifier the efficiency η is $\pi/4$, since the collector voltage swings from $2V_{CC}$ to zero, where V_{CC} is the supply voltage. (a) Show that if the collector voltage swings down to $V_{CE \min}$, then

$$\eta = \frac{\pi}{4} \left(1 - \frac{V_{CE \min}}{V_{CC}} \right)$$

(b) In a particular class B amplifier operated that way, the efficiency is 70%. Find $V_{CE\ min}$ if $V_{CC} = 20$ V. (c) If each transistor can withstand 10-W dissipation, what is the maximum undistorted power?

8.15. In the circuit of Fig. 8.17(a) the resistance R_1 is replaced by a diode. If the maximum base current passing alternately through Q_1 and Q_2 is 300 mA and the resistance R_2 supplies a dc current of 600 mA, find the voltage fluctuation across the diode.

8.16. Find the efficiency of the circuit of Fig. 8.18(a). Evaluate ac power dissipated by R_L if $I_{max} = 5$ A and $V_{CC} = 15$ V. Find also the power dissipated by the two transistors.

8.17. A transistor has $\beta_F = 100$. At low currents $C_\pi = C_{je} = 2.0$ pF. Express f_β in terms of I_C at those currents.

8.18. (a) Find the high-frequency cascaded gain of transistor amplifier stages for frequencies for which $\omega C_\pi r_\pi \gg 1$. (b) Show that for large r_x $(\omega^2 C_\pi^2 r_x^2 \gg 1)$

$$\left| \frac{v_i}{v_i'} \right| = \frac{g_m}{\omega C_\pi} \simeq \frac{f_T}{f}$$

so that the voltage gain per stage is unity for $f \simeq f_T$.

8.19. Show that if V_{B3} in Fig. 8.22(b) is fixed and $V_{B2} = V_{B3} + \Delta V_{B2}$, then the current through Q_3 changes by a factor of 99 if ΔV_{B2} changes from -0.12 to $+0.12$ V.

9

Vacuum Tubes

While solid-state devices have replaced most vacuum tubes, there are still some applications where the latter are important. This is especially the case for vacuum tube circuits involving high temperature, high voltage, and/or high power. Therefore we shall discuss these devices and such applications.

9.1. VACUUM DIODES, TRIODES, TETRODES, AND PENTODES

9.1a. Vacuum Diode

A vacuum diode consists of a hot filament or equipotential cathode emitting electrons into a vacuum. In many cases the hot filament or equipotential cathode is coated with a (Ba, Sr)O layer, since it emits at much lower temperatures than metal cathodes. In tubes that must withstand high voltages, the cathode usually consists of a tungsten or thoriated tungsten filament (metal cathode).

An opposing electrode, the *anode*, collects the emitted electrons when its voltage V_A is positive and repels them when it is negative [Fig. 9.1(a)]. A considerable current I_A flows if V_A is positive, whereas $I_A = 0$ for $V_A < 0$ [Fig. 9.1(b)].

The current flow in the diode for $V_A > 0$ is limited by the space charge

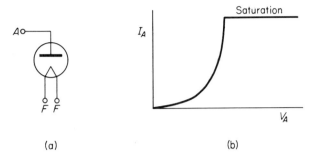

Fig. 9.1. (a) Schematic diagram of a vacuum diode. (b) Characteristic showing space-charge limited and saturated regime.

of the electrons moving between the cathode and anode; a calculation shows
that I_A varies as $V_A^{3/2}$

$$I_A = PV_A^{3/2} \qquad (9.1)$$

(Child's law); the coefficient P is expressed in amperes per volt$^{3/2}$ and is
called the *perveance* of the diode.

For very high anode voltages the anode collects all the electrons emitted by the cathode and the diode is said to be *saturated*. For oxide-coated
cathodes this condition must be avoided, since it destroys the emitting layer,
but it can be tolerated for metal cathodes.

Because of the high vacuum, the anode can withstand large negative
voltages; for that reason properly designed vacuum diodes can be used as
high-voltage rectifiers. Stacks of solid-state diodes can also be used for these
applications, however.

9.1b. Vacuum Triode

The vacuum triode is a three-electrode device, consisting of a hot filament or
equipotential cathode and a positively biased anode, with a negatively biased
grid in between [Fig. 9.2(a)]. Since the grid is negative, no electrons can
arrive at the grid, so that the grid current I_G is zero. Because of the positive
anode voltage V_A, the electric field in the vicinity of the cathode is so directed
that the emitted electrons are drawn away from the cathode so that they
arrive at the anode. The anode current I_A is therefore a function of both the
grid voltage V_G and the anode voltage V_A.

To calculate I_A, we replace the grid by a solid electrode through its
center, kept at the effective potential V_E, and choose V_E such that the equivalent diode and the actual triode have the same current. A calculation then
shows that

$$V_E = C\left(V_G + \frac{V_A}{\mu}\right) \qquad (9.2)$$

where C is a constant slightly smaller than unity and μ is a factor, usually

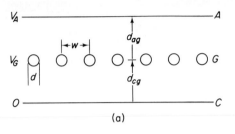

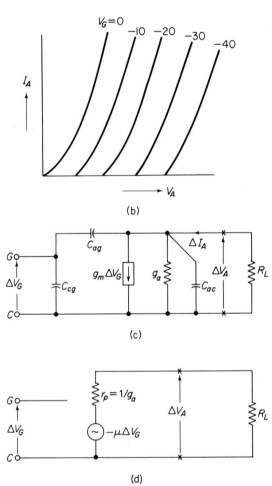

Fig. 9.2. (a) Cross section of a triode. (b) (I_A, V_A) characteristics of triode. (c) Small-signal equivalent circuit. (d) Alternate low-frequency, small-signal equivalent circuit.

between 5 and 100, that depends on the geometry of the triode but very little on the operating conditions. In analogy with Eq. (9.1) we thus have

$$I_A = PV_E^{3/2} = A\left(V_G + \frac{V_A}{\mu}\right)^{3/2} \tag{9.3}$$

where $A = PC$. This is a good approximation if

1. The ratio w/d between the distance w of the grid wires and the diameter d of these wires is relatively small, say 4–5.
2. w is relatively small in comparison with the distance d_{cg} between cathode and grid, say w/d_{cg} less than unity. In that case the field distribution near the cathode is more or less uniform.

However, if w/d is relatively large and w is relatively large in comparison with d_{cg}, the field distribution near the cathode is very inhomogeneous and the device current comes mostly from the cathode regions near the middle of the holes between the grid wires. It is found that the anode current can then be represented as

$$I_A = A'\left(V_G + \frac{V_A}{\mu}\right)^2 \tag{9.4}$$

Figure 9.2(b) shows the (I_A, V_A) characteristics of the triode with the grid voltage V_G as a parameter. The characteristics are somewhat curved, as expected from Eq. (9.3); we shall later approximate them by linear equidistant characteristics [Fig. 9.4(b)].

If the voltages V_G and V_A are varied by small amounts ΔV_G and ΔV_A, respectively, the variation ΔI_A in anode current is

$$\Delta I_A = \frac{\partial I_A}{\partial V_G}\,\Delta V_G + \frac{\partial I_A}{\partial V_A}\,\Delta V_A = g_m\,\Delta V_G + g_a\,\Delta V_A \tag{9.5}$$

where

$$g_m = \frac{\partial I_A}{\partial V_G}; \qquad g_a = \frac{\partial I_A}{\partial V_A} \tag{9.5a}$$

These parameters must be evaluated at the operating point. The small-signal equivalent circuit is thus identical to that of the FET; it is shown in Fig. 9.2(c). It is easily seen from Eqs. (9.2) and (9.5a) that

$$g_a = \frac{g_m}{\mu} \tag{9.6}$$

In Fig. 9.2(c) we have added the interelectrode capacitances C_{cg}, C_{ag}, and C_{ac}. By proper design of the triode the capacitance, C_{ac} can be quite small, but the other two capacitances are significant.

We now change the low-frequency equivalent circuit of Fig. 9.2(c) into the equivalent circuit of Fig. 9.2(d) by replacing the current generator $g_m\,\Delta V_G$

in parallel with g_a by an EMF

$$-\frac{g_m}{g_a}\Delta V_G = -\mu\,\Delta V_G \tag{9.7}$$

in series with the resistance $r_p = 1/g_a$. It is seen by inspection that the triode has a voltage gain of

$$\frac{\Delta V_A}{\Delta V_G} = -\mu \tag{9.7a}$$

if the load resistance R_L is large in comparison with r_p. For this reason the factor μ is called the *amplification factor* of the triode; as mentioned before, it depends on the tube geometry but very little on the operating conditions. The resistance r_p is often called the *plate resistance* of the triode.

　　The vacuum triode is hardly ever used as a small-signal amplifier except at high temperatures, but it is still the only way to develop very large powers.

9.1c. Tetrode and Pentode

It would be convenient if a vacuum tube could be so designed that the anode current I_A depends much less on V_A than in the triode. This can be achieved by inserting a positive electrode, called the *screen grid* G_2, between the grid and the anode [Fig. 9.3(a)]. The configuration thus obtained has four electrodes and is therefore called a *tetrode*.

　　In analogy with Eqs. (9.3) and (9.4) one would thus expect that

$$I_A = A\left(V_G + \frac{V_2}{\mu_2}\right)^{3/2} \tag{9.8}$$

for w/d relatively small and where w/d_{cg} is not too large, whereas

$$I_A = A'\left(V_G + \frac{V_2}{\mu_2}\right)^2 \tag{9.8a}$$

if w/d and w/d_{cg} are both large. Here V_2 is the screen grid voltage and μ_2 is the amplification factor, measured when the screen grid and anode are connected together to form a triode. The dependence of I_A on the anode voltage V_A is quite small and can often be neglected.

　　Because the screen grid voltage is positive, it draws some grid current I_2. The ratio I_2/I_A is of the order of the ratio of the screen grid wire diameter over the distance between the screen grid wires, as expected from geometrical considerations.

　　If the anode voltage V_A becomes smaller than the screen grid voltage V_2, slow secondary electrons, which are emitted by the anode because of the electron bombardment, can arrive at the screen grid. This gives rise to a considerable drop in the anode current and a considerable increase in the

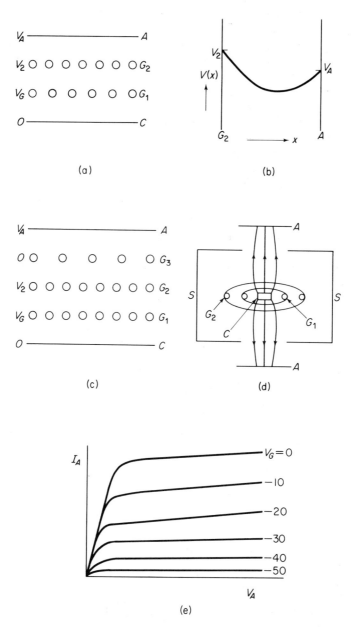

Fig. 9.3. (a) Cross section of a tetrode. (b) Potential distribution between screen grid and anode. (c) Cross section of a pentode. (d) Pentode with beam plates replacing suppressor grid. (e) (I_A, V_A) characteristics of pentode.

screen grid current. This is undesirable, and hence steps must be taken to reduce the effect. Three such measures are discussed here. They all have the following in common: a potential minimum is introduced between screen grid and anode that prevents the screen grid from collecting the slow secondary electrons emitted by the anode [Fig. 9.3(b)–(d)].

1. If the distance between the anode and the screen grid is made relatively large, the space charge due to electrons moving from screen grid to anode gives rise to a potential minimum between the two electrodes [Fig. 9.3(b)].
2. A coarsely wound grid, kept at cathode potential, is inserted between the screen grid and anode. Again, this gives rise to a potential minimum between the screen grid and anode. This third grid is called *suppressor grid* G_3, since it suppresses the effects of secondary electron emission. Because the tube has five electrodes, it is called a *pentode* [Fig. 9.3(c)].
3. The tube is so constructed that the electrons flow in a relatively narrow beam. A metal screen S with a hole sufficiently large to pass the electron beam is inserted between the screen grid and anode. By keeping this metal screen at cathode potential, the required potential minimum is introduced. A tube constructed in this manner is called a *beam power tetrode* [Fig. 9.3(d)].

Since the screen grid, the suppressor grid, and the metal screen all reduce the capacitive coupling between the anode and grid, the capacitance C_{ag} between anode and grid can be much smaller for the tetrode, beam power tetrode, and pentode than for the triode. This has the additional advantage of better stability when used as an HF amplifier (see Section 4.5b).

Figure 9.3(e) shows the (I_A, V_A) characteristic of the tetrode. Because the anode still has a small effect on the equivalent potential V_E in the grid plane, I_A increases slightly with increasing V_A. At low values of V_A not all electrons can arrive at the anode because of their deflection by the grid wires; therefore I_A decreases rapidly with decreasing V_A for small V_A.

9.2. VACUUM TUBE POWER AMPLIFIERS

9.2a. Triode Power Amplifier

Figure 9.4(a) shows a triode power amplifier with transformer output. The triode is indicated by the symbol shown, V_{BB} is the anode supply voltage, and V_{G0} (< 0) is the dc grid voltage, which is usually obtained by a bypassed resistor in the cathode lead (comparable to the FET circuit). An ac voltage $v_g \cos \omega t$ is also applied; it must be chosen so that the device draws current at all times and so that the grid does not become positive at any time, since it

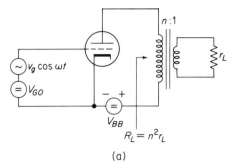

(a)

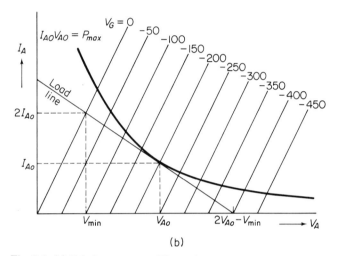

(b)

Fig. 9.4. (a) Triode power amplifier. (b) Linearized characteristics with load line.

will then draw grid current. The transformer has a turns ratio n, so that for a load resistance r_L the anode "sees" an apparent load resistance

$$R_L = n^2 r_L \tag{9.9}$$

The value of R_L must be so chosen that the ac output power is optimized.

To discuss the operation in detail, the (I_A, V_A) characteristics are linearized, as shown in Fig. 9.4(b), so that for each given value of the grid voltage V_G the (I_A, V_A) characteristic is a straight line with slope $1/r_p$. The different characteristics, for given steps ΔV_G in V_G, are then parallel, equidistant lines. Also shown is the quiescent operating point (I_{A0}, V_{A0}) at V_{G0}, where $V_{A0} = V_{BB}$, and the power limitation is

$$P_{\max} = I_{A0} V_{A0} = (P_{\text{diss}})_0 \tag{9.10}$$

representing the power dissipated by the triode under quiescent conditions, whereas $P_{\max}$ is the maximum power that can be dissipated by the tube. The line $I_{A0} = P_{\max}/V_{A0}$ is called the *power limitation line*.

The operating point (I_{A0}, V_{A0}) is so chosen that it lies on this line. This is done by choice of I_{A0},

$$I_{A0} = \frac{P_{\max}}{V_{A0}} \tag{9.10a}$$

We read from the characteristic that this occurs at $V_G = V_{G0}(< 0)$.

We now draw a load line so that $I_A(t)$ can swing from $2I_{A0}$ to zero when V_G swings from zero to $2V_{G0}$; according to Fig. 9.4(b), this determines the load line completely. Since $I_A = 2I_{A0}$ at $V_G = 0$, where $V_A = V_{\min}$, the apparent load resistance R_L follows from

$$R_L = \frac{V_{A0} - V_{\min}}{I_{A0}} \tag{9.11}$$

and the load line satisfies

$$I_A = I_{A0} + \frac{V_{A0} - V_A}{R_L} \tag{9.12}$$

By choosing V_{A0} one thus defines I_{A0} and R_L. If V_{A0} is made larger, I_{A0} decreases and $V_{\min}$ decreases so that the anode amplitude $(V_{A0} - V_{\min})$ increases. It is shown below that the optimum ac power increases if V_{BB} is made larger. Unfortunately, this is not a procedure that should be carried too far, since one then enters the curved parts of the characteristics where the linear approximation is not adequate. One often makes the compromise that the load resistance R_L is chosen so that $R_L = 2r_p$. In this case $V_{\min} = \frac{1}{2}V_{A0}$, as can be seen from Fig. 9.4(b), for since $r_p = \frac{1}{2}V_{\min}/I_{A0}$ and R_L is given by Eq. (9.11),

$$R_L = \frac{V_{A0} - V_{\min}}{I_{A0}} = 2r_p = 2\frac{V_{\min}}{2I_{A0}} \quad \text{or} \quad V_{\min} = \frac{1}{2}V_{A0} \tag{9.13}$$

Since the maximum ac amplitude is $V_{A0} - V_{\min}$, the maximum anode voltage is $V_{A0} + V_{A0} - V_{\min} = 2V_{A0} - V_{\min}$. The triode must be able to withstand this anode voltage.

The ac current thus has an amplitude I_{A0} and the ac output amplitude for maximum power is $V_{A0} - V_{\min}$. Therefore the ac power dissipated in r_L is

$$P_{ac} = \frac{1}{2}I_{A0}(V_{A0} - V_{\min}) \tag{9.14}$$

Because the circuit is linear, the supplied power is

$$P_{supp} = I_{A0}V_{A0} \tag{9.15}$$

so that the efficiency of the power amplifier is

$$\eta = \frac{P_{ac}}{P_{supp}} = \frac{1}{2}\left(1 - \frac{V_{\min}}{V_{A0}}\right) \tag{9.16}$$

Hence $\eta = 25\%$ if $V_{\min} = \frac{1}{2}V_{A0}$, whereas η increases with decreasing $V_{\min}/V_{A0}$, as mentioned before.

Because of the curvature of the characteristic, the ac signal at the anode contains harmonics, mostly second harmonic. For full power output the distortion is about 5%.

EXAMPLE 1: A power triode has a maximum power dissipation of 5 kW and a plate resistance of 1250 Ω. Design a power amplifier with such a triode that has 25% efficiency.

ANSWER: Since $\eta = 25\%$, $V_{\min} = \frac{1}{2}V_{A0}$. Furthermore,

$$P_{\max} = I_{A0}V_{A0} = 5000 \text{ W}; \qquad r_p = \frac{V_{\min}}{2I_{A0}} = \frac{V_{A0}}{4I_{A0}} = 1250 \ \Omega$$

Solving for I_{A0} and V_{A0} yields $I_{A0} = 1$ A, $V_{A0} = 5000$ V. The anode voltage thus swings from 2500 to 7500 V, the anode current swings from 2 A to zero, the load resistance $R_L = 2r_p = 2500 \ \Omega$, and $P_{ac} = 1250$ W.

The above mode of operation corresponds to class A power amplification, since the triode draws current for the full cycle. Higher efficiency can be obtained for class B push-pull operation; here each triode passes half a sine wave for one half-cycle and the other supplies the other half of the sine wave, as in the transistor case.

For high-power triode amplifiers the triode must be effectively cooled. For medium-high powers, forced air cooling is sufficient; for very high power, water cooling is needed.

9.2b. High-Frequency Power Amplification; Class C Operation

For high-frequency power amplification one uses tuned circuits at both input and output. Since the tuned circuits filter out the unwanted harmonics, one can now drive the grid positive and so increase the peak value of the anode current. Because of their better stability, tetrode or pentode power amplifiers are often used.

Figure 9.5(a) shows a HF tetrode power amplifier. The symbol for the tetrode is as shown. The screen grid G_2 is fed from the supply voltage V_{BB} by means of a properly chosen bypassed series resistor R_2. The grid bias V_{G0} (< 0) is usually provided by a properly chosen bypassed resistor in the cathode lead. The input signal $v_i \cos \omega t$ drives the grid positive during part of the cycle. One chooses the load resistance R_L so that a maximum power output is obtained.

Figure 9.5(b) shows the linearized characteristics of the tetrode; they consist of equidistant parallel straight lines, terminating in a limiting characteristic at low anode voltages that arises because not all electrons reach the anode at those voltages. A quiescent operating point V_{A0} is chosen, and it is determined from the characteristic that if V_G swings to $V_{G\max}$, then the full current is collected by the anode as long as $V_A > V_{\min}$. Therefore $V_{A0} - V_{\min}$ is the maximum ac amplitude.

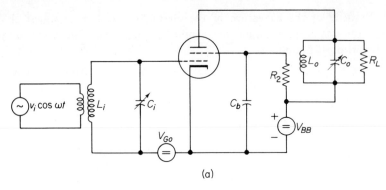

(a)

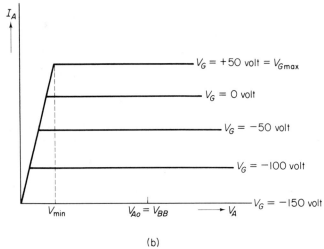

(b)

Fig. 9.5. (a) Tuned tetrode power amplifier. (b) (I_A, V_A) character-istic, showing range of linear operation.

If we make a Fourier analysis of the anode current $I_A(t)$,

$$I_A(t) = I_{A1} + i_{a1} \cos \omega t + i_{a2} \cos 2\omega t + \cdots \qquad (9.17)$$

then the ac amplitude is i_{a1}, the supplied power is

$$P_{\text{supp}} = V_{A0} I_{A1} \qquad (9.18)$$

the ac power is

$$P_{\text{ac}} = \tfrac{1}{2}(V_{A0} - V_{\min}) i_{a1} \qquad (9.19)$$

and the load resistance R_L is

$$R_L = \frac{V_{A0} - V_{\min}}{i_{a1}} \qquad (9.20)$$

so that the efficiency η may be expressed as

$$\eta = \frac{P_{ac}}{P_{supp}} = \frac{(V_{A0} - V_{min})}{V_{A0}} \frac{i_{a1}}{2I_{A1}} \qquad (9.21)$$

If the current flows in larger and larger peaks of shorter and shorter duration, i_{a1} can become larger than I_{A1} and reaches a limiting value $2I_{A1}$ for very short, sharp current peaks. The circuit is then said to be operating in *class C*, and the efficiency becomes

$$\eta_c = \frac{V_{A0} - V_{min}}{V_{A0}} \qquad (9.22)$$

This can be as large as 90%, as shown below.

EXAMPLE 2: Find η_c if $V_{A0} = 5000$ V, $V_{min} = 500$ V, and $i_{a1} \simeq 2I_{A1}$.

ANSWER:

$$\eta_c = \frac{5000 - 500}{5000} = 90\%$$

EXAMPLE 3: If the tetrode in Example 2 has $P_{max} = 5000$ W and $\eta_c = 90\%$, find the maximum value of P_{ac} under the most favorable conditions.

ANSWER:

$$\eta_c = \frac{P_{ac}}{P_{supp}} = 0.90; \qquad P_{diss} = P_{supp} - P_{ac} = P_{ac}\left(-1 + \frac{1}{\eta_c}\right) = 5000 \text{ W}$$

$$P_{ac} = \frac{\eta_c}{1 - \eta_c} \times 5000 = \frac{0.90}{0.10} \times 5000 = 45 \text{ kW}$$

It must be investigated, of course, whether the tetrode can supply large enough peak currents to generate such a large amount of power. In most cases it would be advisable to raise V_{A0} as much as possible, so that smaller peak currents can be used.

In HF power triodes the capacitance C_{ag} between anode and grid gives rise to feedback that can cause instability. To overcome this, a push-pull circuit is used and each grid is connected to the other anode by means of a capacitor C_{ag}. The internal feedback in each tube is then just compensated by the external feedback [Fig. 9.6(a)]. The load resistor R_L must be so chosen that maximum ac power is generated.

9.2c. Second Harmonic Power Generators

In the push-pull circuit the current components of frequency ω passing through the two tubes are 180 degrees out of phase, whereas the second harmonic currents are in phase. Therefore, if the anodes of a push-pull triode circuit are connected together and the anode circuit is tuned to a frequency 2ω, the harmonic currents reinforce each other so that a large output power

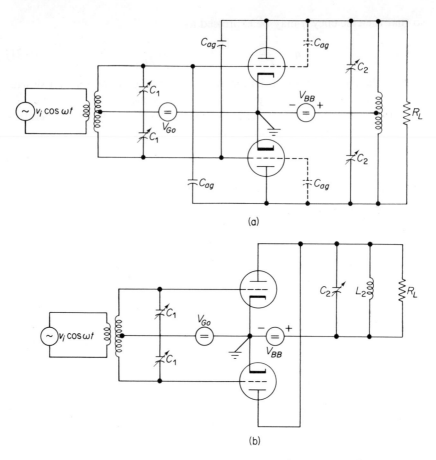

Fig. 9.6. (a) Tuned push-pull triode power amplifier with neutralization. (b) Push-pull triode frequency doubler.

of frequency 2ω can be obtained. The circuit is called a *second harmonic generator* and is shown in Fig. 9.6(b).

The load resistance R_L and the operating conditions must be so chosen that maximum second harmonic power is generated. The output power of frequency 2ω can be very much larger than the power of frequency ω needed to drive the input. Since input and output are tuned to different frequencies, feedback effects are unimportant.

PROBLEMS

9.1. You are to design an electronic amplifier with 120-dB voltage gain that can operate at 400°C. Since silicon transistors and FETs cease to function above 250°C because of excessive leakage currents, you are forced to use vacuum tubes. You have the choice between triodes and pentodes, but because the latter are capable of higher gain, you choose the latter. You have at your disposal a voltage supply V_{BB} of 150 V dc. Your pentodes have $V_G = -2.0$ V, $V_2 = 100$ V, $I_2 = 1.0$ mA, and $I_A = 4.0$ mA for $V_A \geq 30$ V. $g_{ma} = 5.0 \times 10^{-3}$ mho; $g_{m2} = 1.25 \times 10^{-3}$ mho. Here g_{ma} and g_{m2} are the transconductances in the anode and screen lead, respectively. (a) Design a single-stage pentode so that the voltage gain of the stage is 40 dB, and investigate whether you stay within the limits stated for V_A. (b) How many stages are needed? (c) What anode voltage amplitude is allowed in the last stage before the condition $V_A \geq 30$ V is violated. (d) What is the total current and the total power supplied by V_{BB}? *Hint:* Use a cathode resistor R_c and bypass it with a capacitance C_c so that $\omega C_c \geq g_{ma} + g_{m2}$ for all frequencies of practical interest. Use a screen grid resistor R_2 and bypass it with a capacitor C_2, so that $5\omega C_2 \geq g_{m2}$ for all frequencies of practical interest. Choose $f_2 = 10$ Hz as the lowest frequency. The suppressor grid is connected to the cathode.

9.2. Repeat Problem 9.1 for a triode amplifier operating at $V_G = -2.0$ V, $V_A = 100$ V, $I_A = 4.0$ mA, $g_m = 3.0 \times 10^{-3}$ mho, and $r_p = 10,000\ \Omega$. (a) Design a single triode stage and evaluate the voltage gain. (b) How many stages are needed? (c) What is the total current and the total power supplied by V_{BB}? (d) Compare the results with Problem 9.1. *Hint:* Use a cathode resistor R_c and bypass it with a capacitance C_c so that $\omega C_c \geq g_m$ for $f \geq f_2 = 10$ cycles.

9.3. Can the absolute value of the voltage gain g_v in a triode amplifier be larger than the amplification factor μ?

9.4. Show that in a triode with a quadratic characteristic

$$ I_A = C\left(v_G + \frac{v_A}{\mu}\right)^2 $$

the negative feedback from the anode multiplies the distortion by a factor of $[1 - (|g_v|/\mu)]^2 \ll 1$. Can this factor become zero? *Hint:* Substitute $V_G = V_{G0} + v_{g0}\cos\omega t$ and $V_A = (V_{A0} + v_{a0}\cos\omega t)$, collect the terms with $\cos^2\omega t$, and bear in mind that $g_v = v_{a0}/v_{g0}$ and that v_{a0} and v_{g0} have opposite phase. Compare the case $|g_v| > 0$ with the case of the short-circuited output ($v_{a0} = 0$).

9.5. Repeat the problem of Example 1, Section 9.2a, for the case $I_{A0} = 0.5$ A and $V_{A0} = 10,000$ V to show the improvement in ac power that can be obtained. Find η, P_{ac}, and R_L.

9.6. Repeat the problem of Example 1, Section 9.2a, for a tetrode and compare the results. $P_{max} = 5000$ W, $V_{A0} = 5000$ V, $I_{A0} = 1.0$ A, and $V_{min} = 200$ V. Find η, P_{ac}, and R_L.

9.7. In the second harmonic generator of Fig. 9.6(b), $V_{BB} = 5000$ V and $V_{min} = 2500$ V; each device gives a second harmonic current of 0.5 A amplitude. Determine R_L for maximum output and the second harmonic power generated.

10

Pulse Response
in FETs and Transistors

An exact calculation of the pulse response of FETs is rather difficult, since one must not only take into account the response of the input and output circuits but also the feedback effect caused by the drain-gate capacitance C_{dg} and the response of the FET itself.

In Section 10.1a we shall discuss the pulse response of the FET and show that the turn-on response is much faster than the turn-off response. In Section 10.1b we shall discuss a circuit that has a much faster turn-off response.

10.1a. Pulse Response of MOSFETs

To illustrate the problem, we shall turn to Fig. 10.1(a), which shows an enhancement mode n-channel MOSFET ($V_T > 0$) driven by a pulse signal $V(t)$, with a resistance R_g in the gate lead, a resistance R_d in the drain lead, and a load capacitance C_d, which is mostly due to the input capacitances of subsequent FET stages. The equivalent circuit is shown in Fig. 10.1(b), which also shows the device capacitances C_{gs} and C_{dg}; the capacitance C_{ds} has been absorbed into the load capacitance C_d.

For sake of simplicity we shall now assume that R_d is taken quite small so that the feedback via the capacitance C_{dg} can be neglected. For the response of the input circuit, if $V(t) = 0$ for $t < 0$ and $V(t) = V_1$ for $t > 0$,

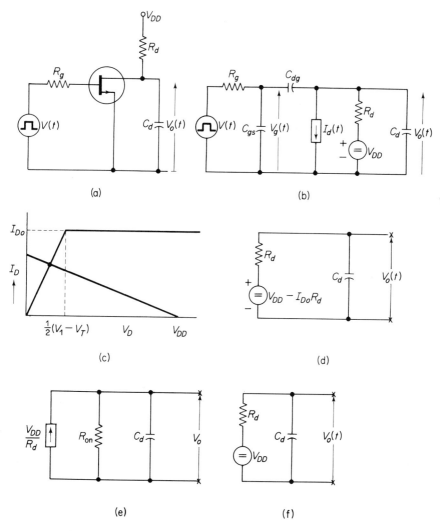

Fig. 10.1. (a) MOSFET pulse circuit. (b) Equivalent circuit of (a). (c) Piecewise linear characteristic of MOSFET. (d) Equivalent output circuit in pinch-off regime. (e) Equivalent output circuit in ohmic regime. (f) Equivalent output circuit for turned-off MOSFET.

we then have

$$V_g(t) = V_1\left[1 - \exp\left(-\frac{t}{\tau_g}\right)\right] \tag{10.1}$$

where

$$\tau_g = R_g(C_{gs} + C_{dg}) \tag{10.1a}$$

This is the turn-on response of the MOSFET.

Next we must take into account the *intrinsic* response of the FET, i.e., the time it takes to form a conducting channel between source and drain after the gate voltage has been turned on. This time is usually so short ($< 10^{-9}$ s) that it can be neglected. We may thus assume that $I_d(t)$ follows $V_g(t)$ instantaneously, so that for $V_g(t) > V_T$

$$I_d(t) = K[V_g(t) - V_T]^2 \tag{10.2}$$

This current generator now operates on the output circuit. If the time constant $\tau_d = C_d R_d$ is much smaller than the time constant τ_g, we have in good approximation

$$V_0(t) = V_{DD} - I_D(t)R_d = V_{DD} - KR_d[V_g(t) - V_T]^2 \tag{10.2a}$$

so that the input pulse causes a decrease in the output voltage, as expected.

We shall now add a small refinement. We see from the circuit of Fig. 10.1(b) that the capacitance C_{dg} gives rise to a feed-through effect. As long as the device has not been turned on, this feed-through effect gives rise to a positive spike in the output caused by $V_g(t)$. After the device has been turned on, the current generator $I_d(t)$ takes over and drives $V_0(t)$ lower, ending up as before.

Next we shall discuss another simple case. Here R_g is quite small and R_d much larger, so that

$$\tau_g = R_g(C_{gs} + C_{dg}) \ll C_d R_d = \tau_d$$

We shall first consider the turn-on transient of the circuit by assuming that $V(t) = 0$ for $t < 0$ and that $V(t) = V_1$ for $t > 0$, where V_1 is so large that the device is driven deeply into the ohmic regime. Since τ_g is so small, we may assume that $V_g(t) = V(t)$.

We shall now represent the (I_D, V_{DS}) characteristic of the device by a linear model, consisting of a pinch-off regime for $V_0 \geq \frac{1}{2}(V_1 - V_T) = V_{p0}$, where $I_D = I_{D0} = $ constant, and an ohmic regime for which $V_0 = I_D R_{on}$, where $1/R_{on} = 2K(V_1 - V_T)$ [Fig. 10.1(c)] if $V_0 \leq \frac{1}{2}(V_1 - V_T)$.

First then, the device is pinched off and the equivalent circuit is as shown in Fig. 10.1(d). For $t > 0$ the differential equation of the circuit is

$$C_d \frac{dV_0}{dt} = \frac{V_{DD} - I_{D0}R_d - V_0}{R_d} \tag{10.3}$$

or, by dividing by C_d and putting $\tau_d = C_d R_d$,

$$\frac{dV_0}{dt} + \frac{V_0}{\tau_d} = \frac{V_{DD} - I_{D0}R_d}{\tau_d} \tag{10.3a}$$

These equations hold as long as the device is pinched off, or $V_0 \geq V_{p0} = \frac{1}{2}(V_1 - V_T)$. The initial condition is $V_0(t) = V_{DD}$ at $t = 0$. Here $I_{D0}R_d$ is usually large in comparison with V_{DD}.

The solution of this equation is (Compare also Appendix B)

$$V_0(t) = V_{DD} - I_{D0}R_d + A \exp\left(-\frac{t}{\tau_d}\right) \tag{10.3b}$$

since $V_0(t) = V(\infty) = V_{DD} - I_D R_d$ is a solution of Eq. (10.3a). Substitution of the initial condition at $t = 0$ yields $A = I_{D0}R_d$, or

$$V_0(t) = V_{DD} - I_{D0}R_d\left[1 - \exp\left(-\frac{t}{\tau_d}\right)\right] \tag{10.3c}$$

This part of the solution terminates at $V_0 = V_{p0}$. Let this be at $t = t_1$. Substitution of $V_0(t) = V_{p0}$ at $t = t_1$ yields

$$V_{p0} = V_{DD} - I_{D0}R_d + I_{D0}R_d \exp\left(-\frac{t_1}{\tau_d}\right) \tag{10.4}$$

from which it follows that

$$t = t_1 = \tau_d \ln \frac{I_{D0}R_d}{I_{D0}R_d - V_{DD} + V_{p0}} \tag{10.4a}$$

We now bear in mind that $I_{D0}R_d \gg V_{DD} - V_{p0}$. We may then rewrite Eq. (10.4a) as

$$t_1 = -\tau_d \ln\left[1 - \frac{V_{DD} - V_{p0}}{I_{D0}R_d}\right] \simeq \tau_d \frac{V_{DD} - V_{p0}}{I_{D0}R_d} = \frac{C_d(V_{DD} - V_{p0})}{I_{D0}} \tag{10.4b}$$

since $\ln(1 - x) \simeq -x$ for small x. This has a very simple explanation: $C_d(V_{DD} - V_{p0})$ is the charge lost by C_d during the time t_1. Since the capacitor is discharged by a constant current I_{D0}, t_1 is simply the ratio of the lost charge over the discharging current.

We shall now investigate the response in the ohmic regime. If we replace the EMF V_{DD} by the current generator V_{DD}/R_d and assume that $R_d \gg R_{on}$, we obtain the equivalent circuit of Fig. 10.1(e). The differential equation of the circuit is

$$C_d \frac{dV_0}{dt} + \frac{V_0}{R_{on}} = \frac{V_{DD}}{R_d} \tag{10.5}$$

or, if $\tau_{on} = C_d R_{on}$,

$$\frac{dV_0}{dt} + \frac{V_0}{\tau_{on}} = \frac{V_{DD}}{\tau_d} \tag{10.5a}$$

with the initial condition $V_0 = V_{p0}$ at $t = t_1$. Since

$$V_0 = V_0(\infty) = V_{on} = V_{DD}\frac{\tau_{on}}{\tau_d} = V_{DD}\frac{R_{on}}{R_d} \tag{10.5b}$$

is a solution of the equation and $A \exp\left[-(t - t_1)/\tau_{on}\right]$ is a solution of the homogeneous equation, the solution of (10.5a) satisfying the initial condition is (Compare also Appendix B)

$$V_0(t) = V_{on}\left[1 - \exp\left(-\frac{t - t_1}{\tau_{on}}\right)\right] + V_{p0} \exp\left(-\frac{t - t_1}{\tau_{on}}\right) \tag{10.5c}$$

We shall now define the turn-on time as the total time t_2 that it takes to let $V_0(t)$ drop over 90% of its range $V_{DD} - V_{on}$. Let this occur at $t = t_2$; then

$V_0(t_2) = V_{DD} - 0.90(V_{DD} - V_{on})$. Substitution into Eq. (10.5c) yields

$$0.10(V_{DD} - V_{on}) = (V_{p0} - V_{on}) \exp\left(-\frac{t_2 - t_1}{\tau_{on}}\right) \qquad (10.5d)$$

or

$$t_2 - t_1 = \tau_{on} \ln\left[\frac{V_{p0} - V_{on}}{0.10(V_{DD} - V_{on})}\right] \qquad (10.5e)$$

We shall now illustrate this with the help of an example.

EXAMPLE 1: Let $V_{DD} = 20$ V, $V_1 = 12$ V, $V_T = 2$ V, $I_{D0} = 20$ mA, $R_d = 5000\ \Omega$, and $C_d = 20$ pF. Find the response time t_2 for the turn-on transient.

ANSWER:

$$V_{p0} = \tfrac{1}{2}(V_1 - V_T) = 5 \text{ V}; \qquad I_{D0}R_d = 100 \text{ V}$$

$$\tau_d = C_d R_d = 20 \times 10^{-12} \times 5 \times 10^3 = 100 \times 10^{-9} \text{ s}$$

Then from Eq. (10.4a)

$$t_1 = 10^{-7} \ln \tfrac{100}{85} = 16.4 \times 10^{-9} \text{ s}$$

If we had used Eq. (10.4b), we would have obtained

$$t_1 = \frac{20 \times 10^{-12} \times 15}{20 \times 10^{-3}} = 15 \times 10^{-9} \text{ s}$$

which is quite close. Next we evaluate $t_2 - t_1$ as follows:

$$I_{D0} = K(V_1 - V_T)^2; \qquad 20 \times 10^{-3} = 100K, \ K = 0.2 \text{ mA/V}^2$$

$$R_{on} = \frac{1}{2K(V_1 - V_T)} = \frac{1}{4 \times 10^{-3}} = 250\ \Omega$$

$$V_{on} = V_{DD}\frac{R_{on}}{R_d} = 20 \times \frac{250}{5000} = 1.0 \text{ V}$$

$$\tau_{on} = C_d R_{on} = 20 \times 10^{-12} \times 250 = 5 \times 10^{-9} \text{ s}$$

$$t_2 - t_1 = 5 \times 10^{-9} \ln\left(\frac{4}{1.9}\right) = 3.7 \times 10^{-9} \text{ s}$$

$$t_2 = 16.4 \times 10^{-9} + 3.7 \times 10^{-9} = 20.1 \times 10^{-9} \text{ s}$$

The actual response time is slightly larger, for the piecewise linear characteristic of Fig. 10.1(c) is higher than the actual characteristic for the full voltage range. This difference is not very large, however; the value of t_2 obtained here is probably accurate within 10%.

We shall now discuss the turn-off response. Since now $V(t) = V_1$ for $t < 0$ and $V(t) = 0$ for $t > 0$, we have $V_0(0+) = V_{on}$ and $I(t) = 0$ for $t > 0^+$. Hence the output equivalent circuit is as shown in Fig. 10.1(f) and the output response is

$$V_0(t) = V_{on} \exp\left(-\frac{t}{\tau_d}\right) + V_{DD}\left[1 - \exp\left(-\frac{t}{\tau_d}\right)\right]$$

$$= V_{DD} - (V_{DD} - V_{on})\exp\left(-\frac{t}{\tau_d}\right) \tag{10.6}$$

since $V_0(t) = V_0(\infty) = V_{DD}$ is a solution of Eq. (10.3a) if $I_{D0} = 0$.

We shall now investigate how long it takes $V_0(t)$ to change from V_{on} to $V_{on} + 0.90(V_{DD} - V_{on})$. Let this occur at $t = t_3$; then

$$V_{on} + 0.90(V_{DD} - V_{on}) = V_{DD} - (V_{DD} - V_{on})\exp\left(-\frac{t_3}{\tau_d}\right) \tag{10.6a}$$

or

$$\exp\left(-\frac{t_3}{\tau_d}\right) = 0.10; \qquad t_3 = \tau_d \ln 10 = 2.3\tau_d \tag{10.6b}$$

This is much larger than the turn-on time t_2, as is seen from the following example.

EXAMPLE 2: Find the turn-off time t_3 for the circuit discussed in Example 1.

ANSWER: Since $\tau_d = 10^{-7}$ s, $t_3 = 2.3\tau_d = 230 \times 10^{-9}$ s, which is about 10 times larger than t_2.

If both R_g and R_d are large, one must take the full feedback effect into account. It may be assumed, however, that $I_D(t)$ follows $V_g(t)$ instantaneously; that is

$$I_D(t) = K[V_{g1}(t) - V_T]^2 \tag{10.7}$$

as before. It is beyond the scope of this discussion to solve the problem in detail.

10.1b. Complementary Symmetry MOSFET (COSMOS) Circuit

The complementary symmetry MOSFET circuit is shown in Fig. 10.2(a). It consists of a p-channel and an n-channel enhancement mode MOSFET pair connected in series for dc, with the gates connected in parallel. C_0 is the load capacitance of the circuit, due to subsequent MOS stages. We shall see that the circuit has practically zero current, except during switching.

We see by inspection that the n-channel device has gate-source and drain-source voltages

$$V_{GSn} = v_i; \qquad V_{DSn} = v_0 \tag{10.8}$$

respectively, whereas the p-channel device has voltages

$$V_{GSp} = v_i - V_{DD}; \qquad V_{DSn} = v_0 - V_{DD} \tag{10.9}$$

If v_i is at a logical 1, e.g., $v_i \simeq V_{DD}$, then Q_p, since it is an enhancement mode device, is turned off; i.e., it is operating at a leakage current of about

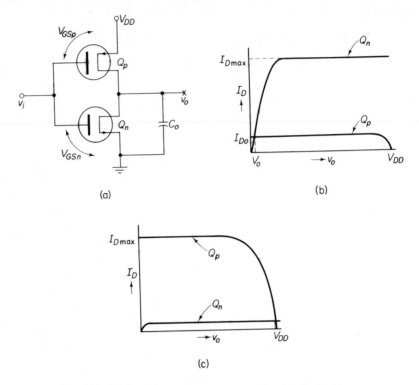

Fig. 10.2. (a) Complementary symmetry pulse circuit. (b) Characteristics and operating point if *n*-channel device is on and *p*-channel device is off. (c) Same, but *n*-channel device is off, and *p*-channel device is on.

10^{-12} A, whereas Q_n is turned on. However, since the current is determined by Q_p, Q_n operates deeply in the ohmic regime. We now have the (I_D, v_0) characteristics as shown in Fig. 10.2(b). The Q_n characteristic passes through the origin and the Q_p characteristic passes through the point $(I_D = 0, v_0 = V_{DD})$. The operating point is the point of intersection of the two characteristics; i.e., v_0 is very close to 0 V, so that the output is a logical 0.

If v_i is a logical 0, e.g., $v_i \simeq 0$, then Q_n is turned off, i.e., operating at a leakage current of the order of 10^{-12} A, and Q_p is turned on. However, since the current is determined by Q_n, Q_p operates deeply in the ohmic regime. We now have the (I_D, v_0) characteristics shown in Fig. 10.2(c). The Q_n characteristic again passes through the origin and the Q_p characteristic passes through the point $(I_D = 0, v_0 = V_{DD})$. The operating point of intersection of the two characteristics, i.e., v_0, is very close to V_{DD}, so that the output is a logical 1. If the two devices are completely symmetric, they have the same current I_D when off and the same pinch-off current $I_{D \max}$ when on.

Since a logical 1 input produces a logical 0 output and a logical 0 input produces a logical 1 output, the circuit is an *inverter*.

Suppose that v_i switches from a logical 1 to a logical 0; then Q_n is switched off and Q_p is switched on. Since v_0 is first at 0 V, Q_p has a strongly negative drain voltage $V_{DS} \simeq -V_{DD}$, so that it is pinched off; i.e., $I_D \simeq I_{D\ max}$. Hence a large current $I_D(t)$ flows that charges C_0 until it reaches the voltage V_{DD}.

We shall now define the charging time as the time t_1 needed to charge C_d to $0.90V_{DD}$. The total charge displaced during t_1 is thus $0.90C_dV_{DD}$, and hence t_1 follows from the relation

$$\int_0^{t_1} I_D(t)\, dt = 0.9C_0V_{DD} \qquad (10.10)$$

If we make the slightly optimistic assumption that $I_D(t) = I_{D\ max}$ for $0 < t < t_1$, then

$$I_{D\ max}t_1 = 0.90C_0V_{DD} \quad \text{or} \quad t_1 = \frac{0.90C_0V_{DD}}{I_{D\ max}} \qquad (10.10a)$$

Since $I_D(t) < I_{D\ max}$ part of the time, Eq. (10.10a) somewhat underestimates t_1 but not by an important factor. Suppose that v_1 switches from a logical 0 to a logical 1; then Q_p is switched off and Q_n is switched on. Since at first $v_0 \simeq V_{DD}$, a large current flows through Q_n, discharging C_0 until it reaches zero voltage. If the devices are fully symmetric, the currents are equal at identical values of $|V_{DS}|$ and hence the discharge time is exactly equal to the charging time; this overcomes one of the drawbacks of the regular MOSFET circuit.

10.2. PULSE RESPONSE OF TRANSISTORS

We shall now show that the equivalent circuit of Fig. 8.21(a) allows us to evaluate the large-signal pulse response of transistor circuits. We do so for an *n-p-n* transistor.

10.2a. Charge Control Equations for Nonsaturated Transistors

We start from the small-signal equivalent circuit of Fig. 10.3(a) and inject a current $i_b(t)$ into the base. The differential equation of the base circuit, neglecting the effects of the capacitance C_μ and of r_x, is then

$$C_\pi \frac{dv_1}{dt} + \frac{v_1}{r_\pi} = i_b(t) \qquad (10.11)$$

We shall now define

$$q_b(t) = C_\pi v_1(t) \qquad (10.12)$$

as the charge stored in the base. Introducing the time constant $\tau_\pi = C_\pi r_\pi$,

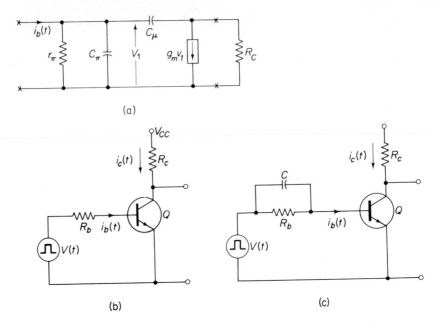

Fig. 10.3. (a) Hybrid π-circuit of transistor. (b) Transistor pulse circuit. (c) Transistor pulse circuit with speedup capacitor.

Eq. (10.11) may be written as

$$\frac{dq_b}{dt} + \frac{q_b}{\tau_\pi} = i_b(t) \tag{10.13}$$

This is derived as a small-signal equation, but since τ_π is practically independent of the operating conditions, Eq. (10.13) should be equally valid for large signals. This follows simply from steady-state considerations, for let Q_B be the dc charge stored in the base; then the recombination current is Q_B/τ_π and hence the dc base current $I_B = Q_B/\tau_\pi$, in accordance with Eq. (10.13).

If applied to the large-signal pulse response, $q_b(t)$ is a positive charge, but for *n-p-n* transistors the minority carriers stored in the base are electrons, which give a negative charge. However, this stored charge is compensated by an equal and opposite hole charge, and it is this *positive* hole charge that makes up $q_b(t)$.

Since q_b/τ_π is the recombination current, the collector current $i_c(t)$ is β_F times as large, or

$$i_c(t) = \frac{\beta_F}{\tau_\pi} q_b(t) \tag{10.14}$$

since the collected hole current is β_F times the recombination current.

This holds as long as the collector current $i_c(t)$ is less than the saturated current

$$I_{Cs} = \frac{V_{CC} - V_{CE \text{ sat}}}{R_c} \qquad (10.14a)$$

[compare Fig. 10.3(b)], that is,

$$i_c(t) < I_{Cs} \qquad (10.14b)$$

We assume that this condition is satisfied. If it is not satisfied, $i_c(t)$ will ultimately reach the value I_{Cs} and stay there.

Now let a voltage $V(t)$ be applied to the base by means of a resistor R_b [compare Fig. 10.3(b)] and let $V(t) = 0.70$ V, the turn-on voltage of the transistor for $t = 0$, and $V(t) = V_1 > 0.70$ V for $t > 0$. The base current $i_b(t)$ for $t > 0$ may thus be written as

$$i_b(t) = I_{B1} = \frac{V_1 - 0.70}{R_b} \qquad (10.15)$$

so that Eq. (10.13) becomes

$$\frac{dq_b}{dt} + \frac{q_b}{\tau_\pi} = I_{B1} \qquad (10.16)$$

with the initial condition $q_b(0^+) = 0$. To make sure that the transistor is never saturated, we must require that

$$\beta_F I_{B1} < I_{Cs} \quad \text{or} \quad I_{B1} < \frac{I_{Cs}}{\beta_F} = I_{Bs} \qquad (10.15a)$$

We assume that this is the case.

To solve Eq. (10.16), we observe that $q_b(\infty) = I_{B1}\tau_\pi$, so that the full solution is (compare the end of Appendix B)

$$q_b(t) = q_b(\infty)\left[1 - \exp\left(-\frac{t}{\tau_\pi}\right)\right] = I_{B1}\tau_\pi\left[1 - \exp\left(-\frac{t}{\tau_\pi}\right)\right] \qquad (10.16a)$$

or

$$i_c(t) = \beta_F I_{B1}\left[1 - \exp\left(-\frac{t}{\tau_\pi}\right)\right] \qquad (10.16b)$$

After a time $t_1 = 2.3\tau_\pi$, $i_c(t)$ has reached 90% of its final value.

Having evaluated the turn-on transient, we shall now investigate the turn-off transient. To that end we assume that $V(t)$ switches from $+V_1$ to $-V_2$ at $t = 0$. Since at first there is still charge stored in the base, the base voltage remains at about 0.70 V, and hence a current

$$i_b(t) = -I_{B2} = \frac{-V_2 - 0.70}{R_b} \qquad (10.17)$$

flows into the base. Equation (10.13) must now be written as

$$\frac{dq_b}{dt} + \frac{q_b}{\tau_\pi} = -I_{B2} \qquad (10.18)$$

with the initial condition $q_b(0^+) = I_{B1}\tau_\pi$.

To solve Eq. (10.18), we observe that $q_b(\infty) = -I_{B2}\tau_\pi$; hence the full solution of Eq. (10.18) is

$$q_b(t) = q_b(0^+)\exp\left(-\frac{t}{\tau_\pi}\right) + q_b(\infty)\left[1 - \exp\left(-\frac{t}{\tau_\pi}\right)\right]$$

$$= I_{B1}\tau_\pi \exp\left(-\frac{t}{\tau_\pi}\right) - I_{B2}\tau_\pi\left[1 - \exp\left(-\frac{t}{\tau_\pi}\right)\right] \qquad (10.19)$$

If $V_2 = -0.70$ V, $I_{B2} = 0$ and the last term in Eq. (10.19) disappears.

However, $q_b(t)$ is a positive charge, and hence the turn-off transient stops if $q_b(t) = 0$. Let this occur at $t = t_2$; then

$$(I_{B1} + I_{B2})\tau_\pi \exp\left(-\frac{t_2}{\tau_\pi}\right) - I_{B2}\tau_\pi = 0$$

or

$$\exp\left(\frac{t_2}{\tau_\pi}\right) = \frac{I_{B1} + I_{B2}}{I_{B2}} \quad \text{or} \quad t_2 = \tau_\pi \ln\left(\frac{I_{B1} + I_{B2}}{I_{B2}}\right) \qquad (10.20)$$

If $I_{B1} = I_{B2}$, we have

$$t_2 = \tau_\pi \ln 2 = 0.69\tau_\pi \qquad (10.20a)$$

so that the turn-off transient is faster than the turn-on transient. If $I_{B2} = 0$, the transient would have been 90% completed at $t = 2.3\tau_\pi$. Switching from $+V_1$ to $-V_2$, rather than from $+V_1$ to 0.70 V, thus speeds up the response.

To be more exact, $V(t)$ should also switch from $-V_2$ to $+V_1$ for the turn-on transient. One then has to wait a certain time before $V_{be}(t)$ reaches the value $V_{BE0} = 0.70$ V needed to turn the transistor on. This is called the *turn-on time delay*; it is discussed in Section 10.2c.

EXAMPLE 1: If $C_\pi = 10$ pF, $r_\pi = 2000\ \Omega$, $R_b = 5000\ \Omega$, $V_1 = 5.0$ V, $V_2 = -5.0$ V, and $v_{be0} = 0.70$ V, find (a) the time when the turn-on transient is 90% completed and (b) the full time for the turn-off response.

ANSWER:

(a)

$$\tau_\pi = C_\pi r_\pi = 2.0 \times 10^{-8}\ \text{s}$$

$$t_1 = 2.3\tau_\pi = 4.6 \times 10^{-8}\ \text{s}$$

(b)

$$I_{B1} = \frac{V_1 - 0.70}{R_b} = \frac{5.0 - 0.70}{5000} = 0.86\ \text{mA}$$

$$I_{B2} = \frac{5.0 + 0.70}{5000} = 1.14\ \text{mA}$$

$$t_2 = \tau_\pi \ln\left(\frac{I_{B1} + I_{B2}}{I_{B2}}\right) = 2.0 \times 10^{-8} \ln\left(\frac{2.00}{1.14}\right) = 1.12 \times 10^{-8}\ \text{s}$$

Next we shall take the effect of C_μ into account. Since C_μ gives rise to a Miller capacitance,

$$C_f = (1 + g_m R_c)C_\mu \simeq g_m R_c C_\mu \qquad (10.21)$$

in parallel with C_π, the total capacitance C_T in parallel with r_π is now

$$C_T = C_\pi + g_m R_c C_\mu \qquad (10.21a)$$

We must now replace C_π in Eq. (10.11) by C_T, so that

$$C_T \frac{dv_1}{dt} + \frac{v_1}{r_\pi} = i_b(t) \qquad (10.22)$$

However, the stored charge is still as given by Eq. (10.12), that is,

$$q_b(t) = C_\pi v_1$$

Therefore, switching to $q_b(t)$ as the new variable,

$$\frac{C_T}{C_\pi} \frac{dq_b}{dt} + \frac{q_b}{C_\pi r_\pi} = i_b(t) \qquad (10.22a)$$

Introducing the time constants $\tau_\pi = C_\pi r_\pi$ and $\tau_T = C_T r_\pi = \tau_\pi + \beta_F R_c C_\mu$, Eq. (10.22a) may be written as

$$\frac{dq_b}{dt} + \frac{q_b}{\tau_T} = \frac{\tau_\pi}{\tau_T} i_b(t) \qquad (10.23)$$

with $i_b(t) = I_{B1}$ for the turn-on transient and $i_b(t) = -I_{B2}$ for the turn-off transient.

Since for the turn-on transient

$$q_b(0^+) = 0; \qquad q_b(\infty) = \tau_\pi I_{B1} \qquad (10.23a)$$

and for the turn-off transient

$$q_b(0^+) = \tau_\pi I_{B1}; \qquad q_b(\infty) = -\tau_\pi I_{B2} \qquad (10.23b)$$

as in the previous cases, we see that Eqs. (10.16a) and (10.19) remain correct except that the expression $\exp(-t/\tau_\pi)$ must be replaced by $\exp(-t/\tau_T)$ everywhere. The feedback effect via the capacitance C_μ thus slows down the response.

One word of caution should be added here. The capacitance C_μ depends on the collector voltage and is thus a function of time. The above solution remains reasonably correct, however, if a suitable intermediate value is chosen for C_μ.

EXAMPLE 2: Find by what factor the turn-on and turn-off transients are changed by the Miller effect if $\tau_\pi = 10^{-8}$ s, $\beta_F = 50$, $R_c = 2000\ \Omega$, and $C_\mu = 1.5$ pF.

ANSWER:

$$\tau_T = \tau_\pi + \beta_F R_c C_\mu = 10^{-8} + 50 \times 2 \times 10^3 \times 1.5 \times 10^{-12} = 16 \times 10^{-8}\ \text{s}$$

so that the switching times are increased by a factor of $\tau_T/\tau_\pi = 16$.

It would be useful if the transient response of the circuit could be speeded up. It is easily seen from Fig. 10.3(c) how this can be achieved if the effect of C_μ is ignored: One puts a capacitance C in parallel with the resistance R_b such that

$$CR_b = C_\pi r_\pi = \tau_\pi \tag{10.24}$$

because the circuit then acts as a simple voltage divider, transmitting the input step voltage directly to the base.

EXAMPLE 3: If $R_b = 5000\ \Omega$ and $\tau_\pi = 2 \times 10^{-8}$ s, find the value of C needed to eliminate the turn-on response.

ANSWER:

$$C = \frac{\tau_\pi}{R_b} = \frac{2 \times 10^{-8}}{5000} = 4\ \text{pF}$$

It is beyond the scope of this discussion to investigate how condition (10.24) must be altered if the effect of the capacitance C_μ is taken into account. Roughly speaking, compensation should be achieved if $CR_b = C_T r_\pi = \tau_T$.

10.2b. Charge Control Equations for Saturated Transistors

If, for the turn-on transient, $I_{B1} > I_{Bs}$, then the collector current becomes saturated after some time. For the turn-off transient it takes the so-called *storage delay time* before the collector current becomes unsaturated; after that the collector current goes rather rapidly to zero. We shall see that the unsaturated transient response is considerably speeded up, but this is bought at the price of introducing the storage delay time.

The unsaturated part of the transient is easily evaluated. If the voltage $V(t)$ switches from 0.70 V to V_1, the differential equation for the turn-on transient is

$$\frac{dq_b}{dt} + \frac{q_b}{\tau_T} = I_{B1}\frac{\tau_\pi}{\tau_T} \tag{10.25}$$

with $q_b(0^+) = 0$. Since $q_b(\infty) = I_{B1}\tau_\pi$, the solution is

$$q_b(t) = I_{B1}\tau_\pi\left[1 - \exp\left(-\frac{t}{\tau_T}\right)\right] \tag{10.26}$$

as in the previous case. However, when $q_b(t)$ now reaches the value $I_{Bs}\tau_\pi$, $i_c(t)$ reaches the value I_{Cs} and the collector saturates. Let this occur at $t = t_1$; then

$$I_{Bs}\tau_\pi = I_{B1}\tau_\pi - I_{B1}\tau_\pi \exp\left(-\frac{t_1}{\tau_T}\right); \qquad \exp\left(-\frac{t_1}{\tau_T}\right) = \frac{I_{B1} - I_{Bs}}{I_{B1}}$$

or

$$t_1 = \tau_T \ln\left(\frac{I_{B1}}{I_{B1} - I_{Bs}}\right) \simeq \tau_T\frac{I_{Bs}}{I_{B1}} \tag{10.27}$$

if $I_{Bs} \ll I_{B1}$, so that a considerable speedup of the response occurs. For example, if $I_{Bs} = 0.1 I_{B1}$, we have $t_1 \simeq 0.1\tau_T$, whereas for the unsaturated transistor the transient is only 90% completed at $t_1 = 2.3\tau_T$.

We shall now investigate the turn-off transient. If $V(t)$ switches again from $+V_1$ to $-V_2$, it takes some time before the transistor becomes unsaturated. Let this occur at $t = 0$; then for $t > 0$

$$\frac{dq_b}{dt} + \frac{q_b}{\tau_T} = -I_{B2}\frac{\tau_\pi}{\tau_T} \tag{10.28}$$

with the initial condition $q_b(0^+) = I_{Bs}\tau_\pi$. Since $q_b(\infty) = -I_{B2}\tau_\pi$, the full solution of Eq. (10.28) is

$$q_b(t) = I_{Bs}\tau_\pi \exp\left(-\frac{t}{\tau_T}\right) - I_{B2}\tau_\pi\left[1 - \exp\left(-\frac{t}{\tau_T}\right)\right] \tag{10.29}$$

The transient stops when $q_b(t) = 0$. Let this be at $t = t_2$; then

$$0 = (I_{B2} + I_{Bs})\tau_\pi \exp\left(-\frac{t_2}{\tau_T}\right) - I_{B2}\tau_\pi$$

or

$$\exp\left(\frac{t_2}{\tau_T}\right) = \frac{I_{B2} + I_{Bs}}{I_{B2}} \quad \text{or} \quad t_2 = \tau_T \ln\left(\frac{I_{B2} + I_{Bs}}{I_{B2}}\right) \simeq \tau_T\frac{I_{Bs}}{I_{B2}} \tag{10.30}$$

if $I_{Bs} \ll I_{B2}$, resulting again in a speedup of the response.

EXAMPLE 4: If $V_{CC} = 10$ V, $R_c = 2000\ \Omega$, $\beta_F = 50$, $V_{CE\ sat} \simeq 0$ V, $R_b = 5000\ \Omega$, $V_1 = 5.0$ V, $V_2 = 5.0$ V, and $\tau_T = 30 \times 10^{-8}$ s, find t_1 and t_2.

ANSWER: As in Example 1, $I_{B1} = 0.86$ mA and $I_{B2} = 1.14$ mA. According to Eq. (10.14a),

$$I_{CS} = \frac{10}{2000} = 5\ \text{mA}; \qquad I_{BS} = \frac{I_{CS}}{50} = 0.10\ \text{mA}$$

$$t_1 = \tau_T \ln\left(\frac{I_{B1}}{I_{B1} - I_{Bs}}\right) = 30 \times 10^{-8}\ln\left(\frac{0.86}{0.76}\right) = 3.7 \times 10^{-8}\ \text{s}$$

$$t_2 = \tau_T \ln\left(\frac{I_{B2} + I_{Bs}}{I_{B2}}\right) = 30 \times 10^{-8}\ln\left(\frac{1.24}{1.14}\right) = 2.5 \times 10^{-8}\ \text{s}$$

Finally, we shall investigate the response during saturation. Equations (10.25) and (10.28) are then not valid, but we can write a similar equation for the excess stored charge $q_s(t) = q_b(t) - I_{Bs}\tau_\pi$ if the following points are taken into account:

1. It is reasonable to assume that q_s will rise exponentially in the turn-on case with a time constant τ_s and decay exponentially in the turn-off case with the same time constant.
2. Since the collector voltage remains constant at saturation, there is no Miller effect capacitance. That is, τ_T must be replaced by τ_s everywhere.

For the turn-on transient this yields

$$\frac{dq_s}{dt} + \frac{q_s}{\tau_s} = I_{B1} - I_{Bs} \tag{10.31}$$

For a symmetric transistor, that is, a transistor for which the characteristic does not change if the emitter and collector are interchanged, τ_s should be equal to τ_π; for an unsymmetric transistor, τ_s may be different from τ_π.

Let the saturated part of the turn-on transient start at $t = 0$; then $q_s(0^+) = 0$. Since $q_s(\infty) = (I_{B1} - I_{Bs})\tau_s$, the solution of Eq. (10.31) is

$$q_s(t) = (I_{B1} - I_{Bs})\tau_s\left[1 - \exp\left(-\frac{t}{\tau_s}\right)\right] \tag{10.32}$$

so that it takes a time of $2.3\tau_s$ before the stored charge reaches 90% of its full value. The transistor remains saturated all the time, so that $i_c(t)$ does not change, and the transient does not show up in the response.

For the turn-off transient, under saturated conditions, if the turn-off starts at $t = 0$, we have

$$\frac{dq_s}{dt} + \frac{q_s}{\tau_s} = -I_{B2} - I_{Bs} \tag{10.33}$$

with $q_s(0^+) = (I_{B1} - I_{Bs})\tau_s$. Since $q_s(\infty) = -(I_{B2} + I_{Bs})\tau_s$, the solution of Eq. (10.33) is

$$q_s(t) = (I_{B1} - I_{Bs})\tau_s \exp\left(-\frac{t}{\tau_s}\right) - (I_{B2} + I_{Bs})\tau_s\left[1 - \exp\left(-\frac{t}{\tau_s}\right)\right] \tag{10.34}$$

This part of the transient stops if the excess stored charge $q_s(t)$ becomes zero. Let this be at $t = t_3$; then

$$(I_{B1} + I_{B2})\tau_s \exp\left(-\frac{t_3}{\tau_s}\right) - (I_{B2} + I_{Bs})\tau_s = 0$$

or

$$\exp\left(\frac{t_3}{\tau_s}\right) = \frac{I_{B2} + I_{B1}}{I_{B2} + I_{Bs}}; \qquad t_3 = \tau_s \ln\left(\frac{I_{B2} + I_{B1}}{I_{B2} + I_{Bs}}\right) \tag{10.35}$$

If $I_{B2} \gg I_{Bs}$ and $I_{B1} = I_{B2}$, this results in

$$t_3 \simeq \tau_s \ln 2 = 0.69\tau_s \tag{10.35a}$$

The storage delay time of the transistor is thus equal to t_3.

EXAMPLE 5: If $\tau_s = 2 \times 10^{-8}$ s, find t_3.

ANSWER: $t_3 = 0.69\tau_s = 1.4 \times 10^{-8}$ s.

10.2c. Turn-on Delay in Transistors

If the base voltage $V(t)$ of a transistor is switched from $-V_2$ to $+V_1$ at $t = 0$, it takes some time before V_{BE} reaches the value $V_{BE0} = 0.70$ V needed to turn the transistor on. This time is called the *turn-on delay time*.

The equivalent circuit is shown in Fig. 10.4(a), with C_{in} being the input capacitance of the transistor

$$C_{in} = C_{je} + C_\mu \tag{10.36}$$

where C_{je} is the differential capacitance of the emitter-base space-charge

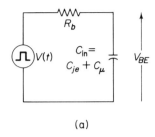

(a)

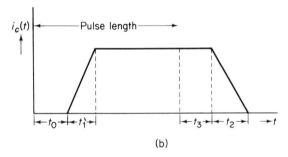

(b)

Fig. 10.4. (a) Equivalent circuit of pulsed transistor before turn-on. (b) Output wave form of transistor pulse circuit showing turn-on delay t_0, turn-on time t_1, turn-off delay t_3 and turn-off time t_2.

region* and C_μ the differential capacitance of the collector-base space-charge region. Both capacitances depend on the applied voltage; to simplify calculations, an appropriately chosen effective value for C_{in} must be used. We assume that this has been done.

We then have for $t > 0$

$$C_{\text{in}}\frac{dV_{BE}}{dt} = \frac{V_1 - V_{BE}}{R_b}; \qquad \frac{dV_{BE}}{dt} + \frac{V_{BE}}{\tau_b} = \frac{V_1}{\tau_b} \tag{10.37}$$

with $\tau_b = C_{\text{in}}R_b$. The initial condition is $V_{BE}(0^+) = -V_2$, whereas $V_{BE}(\infty) = V_1$; according to Eq. (10.37), we thus have

$$V_{BE} = -V_2\exp\left(-\frac{t}{\tau_b}\right) + V_1\left[1 - \exp\left(-\frac{t}{\tau_b}\right)\right] \tag{10.38}$$

We now substitute $V_{BE}(t_1) = V_{BE0} = 0.70$ V and calculate the *turn-on delay time* t_0. This yields

$$V_{BE0} = V_1 - (V_1 + V_2)\exp\left(-\frac{t_0}{\tau_b}\right) \quad \text{or} \quad \exp\left(\frac{t_0}{\tau_b}\right) = \frac{V_1 + V_2}{V_1 - V_{BE0}}$$

or

$$t_0 = \tau_b\ln\left(\frac{V_1 + V_2}{V_1 - V_{BE0}}\right) \tag{10.38a}$$

Figure 10.4(b) shows the four times t_0, t_1, t_2, and t_3 in a saturated transistor.

*Because the transistor is off, the stored charge in the base is zero, or $C_\pi = C_{je}$.

EXAMPLE 6: Find t_0 if $C_{\text{in}} = 3.0$ pF, $R_b = 5000 \, \Omega$, $V_1 = V_2 = 5.0$ V, and $V_{BE0} = 0.70$ V.

ANSWER:

$$\tau_b = C_{\text{in}}R_b = 3 \times 10^{-12} \times 5 \times 10^3 = 1.5 \times 18^{-8} \text{ s}$$

$$t_0 = 1.5 \times 10^{-8} \ln\left(\frac{10}{4.3}\right) = 1.5 \times 10^{-8} \times 0.84 = 1.3 \times 10^{-8} \text{ s}$$

EXAMPLE 7: Combining the results of Examples 4, 5, and 6, find the total turn-on time and the total turn-off time.

ANSWER:

$$t_{\text{on}} = 1.3 \times 10^{-8} + 3.3 \times 10^{-8} = 4.6 \times 10^{-8} \text{ s}$$

$$t_{\text{off}} = 1.4 \times 10^{-8} + 2.5 \times 10^{-8} = 3.9 \times 10^{-8} \text{ s}$$

In the turn-off transient we saw that the transient in $q_b(t)$, and hence in $i_b(t)$, stopped at the time at which $q_b(t) = 0$. Insofar as $i_c(t)$ is concerned, the transient is then completed, but it takes a certain time, the so-called *recovery time*, before $V_{BE}(t)$ has dropped from the value $V_{BE0} = 0.70$ V to $-V_2$. This recovery time can be calculated in a similar way to the turn-on delay. Equation (10.37) now becomes

$$\frac{dV_{BE}}{dt} + \frac{V_{BE}}{\tau_b} = -\frac{V_2}{\tau_b} \tag{10.39}$$

with $V_{BE} = V_{BE0}$ at $t = 0^+$.

10.3. TRANSIENTS IN BASIC TRANSISTOR SWITCHING CIRCUITS

In this section we shall evaluate the transient response of the transistor inverter circuit and the emitter follower circuit. We shall assume that these circuits feed into other circuits and that the load impedance of these circuits can be represented by a capacitance C_0.

10.3a. Transistor Inverter Circuit

Figure 10.5 shows the inverter circuit with a capacitance C_0 (load capacitance) connected between the collector and ground. If V_c is the voltage developed across C_0, the turn-on transient satisfies the differential equation

$$C_0\frac{dV_c}{dt} + \beta_F I_B = \frac{V_{CC} - V_c}{R_c} \tag{10.40}$$

since $I_C = \beta_F I_B$ as long as the collector current is unsaturated. The equation holds as long as

$$V_c > V_{CE \text{ sat}} \simeq 0 \text{ V}$$

Dividing by the capacitance C_0 and introducing the time constant τ

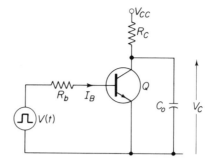

Fig. 10.5. Transistor pulse circuit
with capacitive output load.

$= C_0 R_c$, Eq. (10.40) may be written as

$$\frac{dV_c}{dt} + \frac{V_c}{\tau} = \frac{V_{CC} - \beta_F I_B R_c}{\tau} \tag{10.40a}$$

with the initial condition $V_c(0^+) = V_{CC}$ since the transistor was off for $t < 0$.
However,

$$V_c(\infty) = V_{CC} - \beta_F I_B R_c \tag{10.40b}$$

and hence the solution of Eq. (10.40a) may be written as

$$V_c(t) = V_{CC} \exp\left(-\frac{t}{\tau}\right) + (V_{CC} - \beta_F I_B R_c)\left[1 - \exp\left(-\frac{t}{\tau}\right)\right] \tag{10.41}$$

The turn-on transient ends when $V_c(t) = 0$, or

$$(V_{CC} - \beta_F I_B R_c) + \beta_F I_B R_c \exp\left(-\frac{t}{\tau}\right) = 0$$

or

$$\exp\left(\frac{t}{\tau}\right) = \frac{\beta_F I_B R_c}{\beta_F I_B R_c - V_{CC}} \quad \text{or} \quad t = t_1 = \tau \ln\left(\frac{\beta_F I_B R_c}{\beta_F I_B R_c - V_{CC}}\right) \tag{10.42}$$

Since $\beta_F I_B R_c \gg V_{CC}$, the turn-on transient is quite short.

The turn-off transient lasts much longer. Since the transistor is off for
$t > 0$,

$$C_0 \frac{dV_c}{dt} = \frac{V_{CC} - V_c}{R_c} \quad \text{or} \quad \frac{dV_c}{dt} + \frac{V_c}{\tau} = \frac{V_{CC}}{\tau} \tag{10.43}$$

with $\tau = C_0 R_c$ and $V_c(0^+) = 0$. We read from Eq. (10.43) that $V_c(\infty) = V_{CC}$,
so that the turn-off transient is

$$V_c(t) = V_{CC}\left[1 - \exp\left(-\frac{t}{\tau}\right)\right] \tag{10.44}$$

and 90% of the transient has been completed if

$$t = t_2 = 2.3\tau \tag{10.44a}$$

EXAMPLE 1: If $C_0 = 50$ pF, $R_c = 2000\ \Omega$, $V_{CC} = 10$ V, $I_B = 1$ mA, and β_F
$= 100$, determine t_1 and t_2.

ANSWER:

$$\tau = C_0 R_c = 50 \times 10^{-12} \times 2000 = 10^{-7} \text{ s}$$

$$\beta_F I_B R_c = 100 \times 10^{-3} \times 2000 = 200 \text{ V}$$

$$t_1 = \tau \ln\left(\frac{\beta_F I_B R_c}{\beta_F I_B R_c - V_{CC}}\right) = 10^{-8} \ln\left(\frac{200}{190}\right) = 0.51 \times 10^{-8} \text{ s}$$

so that t_1 is quite short. Furthermore, $t_2 = 2.3 \times 10^{-7}$ s. To this we must add the turn-on delay time of the transistor itself.

10.3b. Emitter Follower Circuit

Figure 10.6(a) shows the circuit. The load is again represented by a capacitance C_0. For the turn-on transient, $V(t)$ switches from $-V_2$ to $+V_1$ at $t = 0$; for the turn-off transient, $V(t)$ switches from $+V_1$ to $-V_2$ at $t = 0$.

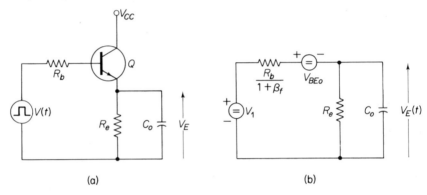

(a)   (b)

Fig. 10.6. (a) Emitter follower pulse circuit with capacitive output load. (b) Equivalent circuit of (a).

Since V_1 usually comes from a previous stage fed from the same supply voltage V_{CC}, we have $V_1 \le V_{CC}$. Since V_E has a maximum value $V_1 - V_{BE0}$, where $V_{BE0} \simeq 0.70$ V, we see that $V_{CE} = V_{CC} - V_E \ge 0.70$ V, so that the collector of the transistor is never saturated.

The equivalent circuit for the turn-on transient is given by Fig. 10.6(b), with $V_E(0^+) = 0$. As seen by inspection, $V_E(\infty) \simeq V_1 - V_{BE0}$ if $R_b/(1 + \beta_F) \ll R_E$. Moreover, the time constant τ_1 of the circuit in this condition is

$$\tau_1 = \frac{C_0 R_b}{1 + \beta_F} \tag{10.45}$$

Hence the turn-on transient is

$$V_E(t) \simeq (V_1 - V_{BE0})\left[1 - \exp\left(-\frac{t}{\tau_1}\right)\right] \tag{10.46}$$

so that the transient is 90% completed if

$$t = t_1 = 2.3\tau_1 \tag{10.47}$$

For the turn-off transient the transistor is off for $t > 0$. Since the voltage on the capacitor at $t = 0^+$ is $V_c(0^+) = V_1 - V_{BE0}$, the voltage decays freely with a time constant $\tau_2 = C_0 R_e$, so that

$$V(t) = (V_1 - V_{BE0}) \exp\left(-\frac{t}{\tau_2}\right) \tag{10.48}$$

The transient is 90% completed if

$$t = t_2 = 2.3\tau_2 \tag{10.49}$$

Since $\tau_1 \ll \tau_2$, the turn-on transient is much faster than the turn-off transient.

EXAMPLE 2: If $R_b = 5000\ \Omega$, $R_e = 1000\ \Omega$, $C_0 = 50$ pF, and $\beta_F = 50$, find t_1 and t_2.

ANSWER:

$$\tau_1 = 50 \times 10^{-12} \times \tfrac{5000}{51} \simeq 5 \times 10^{-9}\ \text{s}; \qquad t_1 = 2.3\tau_1 \simeq 1.1 \times 10^{-8}\ \text{s}$$

$$\tau_2 = 50 \times 10^{-12} \times 1000 = 5 \times 10^{-8}\ \text{s}; \qquad t_2 = 2.3\tau_2 \simeq 11 \times 10^{-8}\ \text{s}$$

10.3c. TTL Gate

A TTL gate is a transistor with a double emitter driving a normal transistor. The circuit is shown in Fig. 10.7. The emitters E_1 and E_2 are driven from previous stages which give voltages $V_1(t)$ and $V_2(t)$, respectively.

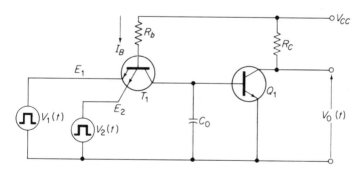

Fig. 10.7. TTL gate driving a transistor.

If $V_1(t) \simeq V_2(t) \simeq V_{CC}$, the emitters are off and the transistor T_1 acts as a (base-collector) diode. Therefore, since the circuit has two voltage drops of $V_{BE0} \simeq 0.70$ V,

$$I_B = \frac{V_{CC} - 1.40}{R_b} \tag{10.50}$$

This current is sufficiently large to drive the transistor Q_1 into saturation, so that $V_0(t) = 0$. Therefore a 1 input at each gate gives 0 output for Q_1.

If $V_1(t)$ or $V_2(t)$ or both are zero, the gate T_1 acts as a normal transistor, discharging the capacitance C_0 until $V_{CE} \simeq 0$; C_0 represents the input capacitance of Q_1 plus possible capacitance between the collector and substrate. That is, Q_1 will now be turned off, so that a zero input either at one or both of the emitters gives a 1 output. We thus have the following logic operations:

0 and 0 gives 1
1 and 0 gives 1
0 and 1 gives 1
1 and 1 gives 0

Since this is just the inverse of the *and* operation, it is called a *nand* (= "not and") operation or circuit.

The turn-off transient is very short, for if $V_1(t) = V_2(t) = V_{CC}$ for $t < 0$ and $V_1(t)$ switches to zero at $t = 0$, then the voltage V_c on the capacitor C_0 has the value $V_c(0^+) = V_{BE0}$ at $t = 0^+$. The current by which the capacitor C_0 is discharged is $\beta_F I_B$. Since the total charge on the capacitor is $C_0 V_{BE0}$, the discharge time is

$$t_d = \frac{C_0 V_{BE0}}{\beta_F I_B} \tag{10.51}$$

The turn-off time t_2 of Q_1 is even considerably shorter than t_d, since Q_1 is already turned off after the base-emitter voltage of Q_1 drops as little as 0.10 V.

We shall now investigate the turn-on time t_1 of Q_1. Let $V_2(t) = V_{CC}$ and $V_1(t) = 0$ for $t < 0$ and let $V_1(t)$ be switched to V_{CC} at $t = 0$. Then $V_c(0^+) = 0$. The transistor Q_1 is turned on as soon as $V_c(t)$ reaches the value V_{BE0}. The current I_B thus gradually drops from the value $(V_{CC} - V_{BE0})/R_b$ at $t = 0$ to the value $(V_{CC} - 2V_{BE0})/R_b$ at $t = t_1$. Neglecting this small drop and bearing in mind that the total charge on C_0 at $t = t_1$ is $C_0 V_{BE0}$, we have

$$t_1 \simeq \frac{C_0 V_{BE0}}{I_B} \tag{10.52}$$

which is much longer than t_d.

EXAMPLE 3: Find t_1 and t_d if $C_0 = 50$ pF, $I_B = 1$ mA, $\beta_F = 50$, and $V_{BE0} = 0.70$ V.

ANSWER:

$$t_1 = \frac{C_0 V_{BE0}}{I_B} = \frac{50 \times 10^{-12} \times 0.70}{10^{-3}} = 3.5 \times 10^{-8} \text{ s}$$

$$t_d = \frac{C_0 V_{BE0}}{\beta_F I_B} = \frac{3.5 \times 10^{-8}}{50} = 0.7 \times 10^{-9} \text{ s}$$

10.3d. Complementary Symmetry Transistor Circuits

Figure 10.8(a) shows a diode *and* gate driving a transistor inverter. In integrated circuits the diodes are generally transistors connected as diodes, but if that is the case, we can replace the diodes D_1 and D_2 by p-n-p transistors,

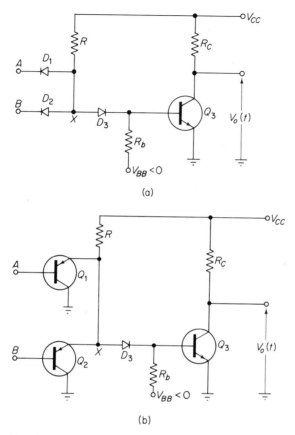

Fig. 10.8. (a) Diode and gate driving a transistor. (b) Two diodes of the *and* gate are replaced by *p-n-p* transistors.

since these transistors have current gain, so that a single previous stage can drive several and stages.

Before doing so, however, we compare Fig. 10.8(a) with Fig. 10.7 and observe that there is a considerable similarity, since each has three diodes: D_1 compares with the E_1 diode, D_2 with the E_2 diode, and D_3 with the collector diode. However, there is also some difference in that in Fig. 10.8(a) the three diodes are independent, whereas in Fig. 10.7 the collector diode forms a transistor with each of the emitter diodes. We saw already how that speeded up the turn-off time of the TTL gate.

We shall now show how D_1 and D_2 can be replaced by *p-n-p* transistors. This is easily achieved in practice if the integrated transistors are triple-diffused transistors. Here one starts with a *p*-type substrate and subsequently diffuses in an *n*-type collector, a *p*-type base, and an *n*-type emitter; this isolates individual transistors. One now uses the base as emitter, the collector as base, and the substrate as collector; one can even omit the last *n*-type diffusion altogether, since one wants a *p-n-p* transistor anyway. However, since the various transistors have the substrate in common, they can be used only in the emitter follower connection. We shall see, however, that this is no handicap.

The *p-n-p* transistors thus introduced are not specially built for that purpose; therefore β_F is relatively small, often only about 3. Since the emitter follower has a current gain $1 + \beta_F$, this corresponds to a gain of about 4.

The circuit thus obtained is shown in Fig. 10.8(b). As mentioned before, since the circuit has a current gain $\beta_F + 1$, only a base current $I_B = I_E/(\beta_F + 1)$ is needed to produce the emitter current I_E. Since $\beta_F + 1 \simeq 4$, the previous stage can drive about 4 times as many gates as when diodes had been used.

The circuits of Fig. 10.8(a) and (b) are combinations of *and* and inverter stages and are thus *nand* circuits.

10.4. TOTEM-POLE CIRCUITS

Totem-pole circuits combine inverter and emitter follower circuits. The resulting combination has the properties of an inverter circuit with very small turn-on and turn-off times and with a very high current capability.

10.4a. Direct-Drive Totem-Pole Circuit

Figure 10.9 shows a direct-drive totem-pole circuit. A logical zero at the input turns Q_1 and Q_2 off, which means that Q_3 has a base current I_{B3} flowing through the resistor R_2, which turns Q_3 on and makes it act as an emitter

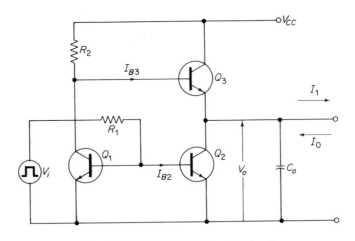

Fig. 10.9. Direct drive totem-pole circuit.

follower. Consequently, the output voltage V_0 is high and the circuit gives a logical 1 output. A logical 1 at the input drives Q_1 and Q_2 into saturation and turns Q_3 off. Therefore V_0 is close to 0 V and the output is a logical 0.

To understand the turn-off time of the output, we switch V_i on at $t = 0$. Q_1 and Q_2 are then on and saturated and the output voltage V_0 drops to zero. If both transistors are identical,

$$I_{B2} = \frac{1}{2} \frac{(V_i - V_{BE0})}{R_1}. \tag{10.53}$$

and hence, as long as the transistor is nonsaturated,

$$I_{C2} = \beta_F I_{B2} \tag{10.54}$$

which is quite large. The current I_{C2} discharges the *load capacitance* C_0, representing subsequent stages, very rapidly to practically 0 V. Therefore the turn-off time is

$$t_2 \simeq \frac{C_0 V_0}{I_{C2}} \tag{10.55}$$

To understand the turn-on time t_1 of the output, we switch V_i off at $t = 0$. Q_1 and Q_2 are then off, and

$$I_{B3} = \frac{V_{CC} - V_0 - V_{BE0}}{R_2} \tag{10.56}$$

can be quite large. Since Q_3 acts as an emitter follower, the time constant τ of the output circuit is

$$\tau = \frac{C_0 R_2}{1 + \beta_{F3}} \tag{10.57}$$

since C_0 "sees" a resistance $R_2/(1 + \beta_{F3})$. The actual turn-on time is now defined as the time needed to complete the transient for 90%. This yields

$$t_1 = 2.3\tau \tag{10.57a}$$

which is also quite short.

Next, something should be said about the output current capability of the circuit. Suppose, when the output is at a logical 0, that the gates into which the output feeds force a current I_0 into the collector of Q_2. Then Q_2 remains saturated as long as

$$I_0 \leq \beta_{F2} I_{B2} \tag{10.58}$$

By proper design of the circuit, this can be a large current.

Suppose, when the output is at a logical 1, that the gates into which the output feeds draw a current I_1 from the output. Obviously, the maximum current that can be drawn is

$$I_1 = (\beta_{F3} + 1)I_{B3} \tag{10.59}$$

but that does not solve our problem, since I_{B3} depends on I_1. We observe that the voltage V_0 will adjust itself so that

$$I_{B3} = \frac{V_{CC} - V_{BE0} - V_0}{R_2} = \frac{I_1}{\beta_{F3} + 1} \qquad (10.60)$$

Hence

$$V_0 = V_{01} = V_{CC} - V_{BE0} - \frac{I_1 R_2}{\beta_{F3} + 1} \qquad (10.60a)$$

Even for relatively large values of I_1 this can be a logical 1 output, since $R_2/(\beta_{F3} + 1)$ can be a small resistance.

10.4b. Phase-Splitter-Drive Totem-Pole Circuit

This circuit, which is shown in Fig. 10.10, is similar to the previous circuit. Q_1 acts as an emitter follower driving Q_2. If V_i is at a logical 0, Q_1 and Q_2 are off, whereas Q_3 is on and acts as an emitter follower; therefore V_0 is at a logical 1. The only difference with the previous circuit is that there is a voltage drop of 0.70 V in diode D.

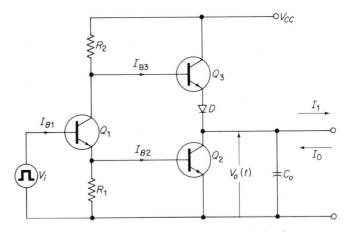

Fig. 10.10. Phase-splitter drive totem-pole circuit.

If V_i is a logical 1, Q_1 is on and saturated, driving Q_2 into saturation. To achieve this, V_i must be larger than two diode voltage drops, or 1.40 V. If I_{B1} is the base current of Q_1, then the emitter current I_{E1}, practically all flowing into the base of Q_2, is

$$I_{E1} = \frac{V_{CC} - V_{BE0}}{R_2} + I_{B1} \qquad (10.61)$$

and hence the output current capability is given by

$$I_0 = \beta_{F2} I_{B2} \qquad (10.62)$$

where

$$I_{B2} = I_{E1} - \frac{V_{BE0}}{R_2} \qquad (10.62a)$$

This is larger than in the previous case.

If V_i is at a logical 0, Q_1 is off, the voltage drop across R_1 is zero, and hence Q_2 is off. Q_3 now acts as an emitter follower, having a current capability similar to the previous case, except that the voltage drop in diode D must be taken into account. Diode D is needed, because otherwise Q_3 can draw current when Q_1 and Q_2 are saturated.

The turn-on and turn-off times of the output circuit are similar to the ones in the previous case. The input circuit of Q_1 must be protected so that I_{B1} does not become too large. This can be done by driving it from a TTL gate or from a diode *and* gate.

10.5. OTHER SWITCHING DEVICES

In this section we shall discuss the *p-n-p-n* switch, the silicon-controlled rectifier, and the double-base diode or unijunction transistor.

10.5a. *p-n-p-n* Switch and Silicon Controlled Rectifier

The *p-n-p-n switch* is a four-layer device with ohmic contacts to the end *p*- and *n*-regions that act as emitters. When the two emitter junctions are forward-biased, the junction between the two bases B_1 and B_2 is back-biased at small currents I and forward-biased for large currents I. The applied voltage to the device is

$$V = V_{E1} + V_{B1B2} + V_{E2} \qquad (10.63)$$

where V_{E1}, V_{B1B2}, and V_{E2} are the voltages applied to the three junctions [Fig. 10.11(a)]. For a back-biased middle junction the current is small and the applied voltage V is large, since V_{B1B2} is positive and large; it can reach the breakdown voltage of the B_1, B_2 junction. For a forward-biased middle junction the current is large and V_{B1B2} is negative and small; V_{E1}, $|V_{B1B2}|$, and V_{E2} are all of the order of 0.70 V. In between is a current range where the differential resistance is negative; this part of the (V, I) characteristic is not easily accessible.

The (V, I) characteristic of the device is thus as presented in Fig. 10.11(b), which shows a large peak at very small currents and is nearly horizontal for large currents. We now draw a load line that intersects the characteristic at three points A, C, and B; A and B are then stable operating points and C is an unstable operating point. Suppose that the operating point is at A and that a positive voltage pulse is applied that lifts the load line above the peak in the (V, I) characteristic; then the device switches from the low-

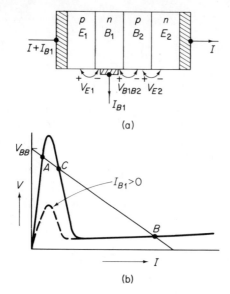

(a)

(b)

Fig. 10.11. (a) *p-n-p-n* switch. (b) (V, I) characteristics of switch with load line.

current to the high-current regime. If the device is operating at point B and a large negative voltage pulse is applied that shifts the load line below the minimum in the (V, I) characteristic, then the device switches from the high-current to the low-current regime.

By providing base B_1 with an external contact and drawing a current I_{B1} out of that contact, one forces the junction $B_1 B_2$ to operate in the forward mode. As a consequence the peak in the (V, I) characteristic becomes much lower and may even disappear altogether [see Fig. 10.11(b)]. Such a device is called a *silicon controlled rectifier*, since the base current I_{B1} controls the operating point; that is, it determines whether the device is operating in point A or point B of the characteristic. The switching from the low-current to the high-current regime is thus achieved by applying a negative voltage pulse to the base B_1, whereas the switching from the high-current to the low-current regime is achieved by applying a negative voltage pulse to the emitter E_1.

10.5b. Double-Base Diode or Unijunction Transistor

Figure 10.12(a) shows the schematic diagram of a double-base diode. It consists of weakly *n*-type material to which two ohmic base contacts B_1 and B_2 are made. A voltage V_{B1B2} forces current to flow from B_1 to B_2. On the side, relatively close to B_1, an emitter junction E is made. If forward-biased, this *p-n* junction injects holes into the *n*-region that flow to the base of B_2. Because no net space charge can be maintained in the conductor between E and B_2, electrons must enter through the ohmic contact B_2 to neutralize the space charge caused by the injected holes. While this does not affect the resistance

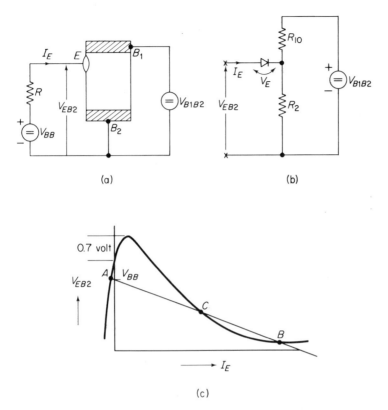

(a)

(b)

(c)

Fig. 10.12. (a) Unijunction transistor (double-base diode). (b) Equivalent circuit of (a). (c)(V_{BE2}, I_E)characteristic of (a) with load line.

R_{10} between the base B_1 and the n-side of the emitter E, it strongly reduces the resistance R_2 between the n-side of the emitter E and the base B_2. Consequently, the voltage V_{EB2} between E and B_2 decreases with increasing emitter current I_E according to the relationship [Fig. 10.12(b)]

$$V_{EB2} = V_{B1B2}\frac{R_2}{R_{10} + R_2} + I_E\frac{R_{10}R_2}{R_{10} + R_2} + V_E \qquad (10.64)$$

where V_E is the turn-on voltage of the emitter diode.

The (V_{EB2}, I_E) characteristic is thus as shown in Fig. 10.12(c). V_{EB2} has the value V_{EB20} at $I_E = 0$, rises rapidly by about 0.70 V when the diode is turned on, and then decreases with increasing I_E because the first term in Eq. (10.64) takes over. If we now draw a load line that intersects the (V_{EB2}, I_E) characteristic at three points A, C, and B, then A and B are stable operating points, whereas C is an unstable operating point.

Suppose that the operating point is at A and that a positive voltage pulse is applied that lifts the load line over the voltage peak; then the operating point must switch from the low-current (point A) to the high-current (point B) regime. If the operating point is at B and a negative pulse is applied that shifts the load line below the falling part of the characteristic, then the operating point must switch from the high-current (point B) to the low-current (point A) regime.

PROBLEMS

10.1. In the circuit of Fig. 10.2(a), $V_{DD} = 8$ V, $K_1 = K_2 = K = 1.0$ mA/V^2, $V_T = -2.0$ V for Q_p, and $V_T = +2.0$ V for Q_n. $C_0 = 20$ pF. (a) Find $I_{D\,max}$ and R_{on}. (b) Replace the characteristic by a linear model and find the breakpoint V_{po} between the parts $I_D = V_D R_{on}$ and $I_D = I_{D\,max}$. (c) Applying the piecewise linear approach of Section 10.1a, properly modified, calculate the turn-off time of the circuit, that is, the time it takes to let v_0 go from V_{DD} to $0.10 V_{DD}$. (d) What is the turn-on time of the circuit?

10.2. The input time constant τ_T of a transistor (including the Miller effect) is 10^{-7} s. If the resistance R_b in the base is 5000 Ω, evaluate the capacitance C needed to eliminate the turn-on transient [see Fig. 10.3(c)].

10.3. In a switching transistor the input voltage switches from -5.0 to $+5.0$ V and back ($V_1 = V_2 = 5.0$ V). (a) Find the turn-on delay of the transistor. (b) Find the turn-on time. (c) Find the turn-off delay. (d) Find the turn-off time. (e) Evaluate the total turn-on and turn-off times. $V_{BE0} = 0.70$ V, $\tau_\pi = 2 \times 10^{-8}$ s, and $\tau_s = 2 \times 10^{-8}$ s = time constant for excess carriers. $C_\mu = 1.0$ pF, $\beta_F = 50$, $R_c = 2000$ Ω, $R_b = 4300$ Ω, $V_{CC} = 8.0$ V, $V_{BE} = 0.70$ V, $V_{CE\,sat} \simeq 0.00$ V, and the input capacitance C_{in} for device off $= 2.0$ pF.

10.4. Repeat Problem 10.3 if the input voltage switches from 0 to 5.0 V ($V_1 = 5.0$ V, $V_2 = 0$) and back and discuss the differences.

10.5. The TTL gate of Fig. 10.7 has $V_{CC} = 4.9$ V, $R_b = 1750$ Ω, and $R_c = 1000$ Ω. $C_0 = 20$ pF, Q_1 has $\beta_F = 100$, $V_{BE0} = 0.70$ V, and $V_{CE\,sat} \simeq 0.00$ V. (a) Find the time it takes to turn Q_1 on by means of T_1 after it was first off. (b) Find the output current capability of Q_1 when on.

10.6. In the totem-pole circuit of Fig. 10.10, $R_2 = 1500$ Ω, $R_1 = 700$ Ω, $I_{B1} = 2.0$ mA, and $C_0 = 50$ pF. All transistors and diodes have a turn-on voltage of 0.70 V, $V_{CE\,sat} \simeq 0.00$ V, and the transistors have $\beta_F = 100$. If $V_{CC} = 4.9$ V, find (a) the current I_{B2} when Q_1 is on and saturated, (b) the current capability of Q_2 in that case, (c) the turn-off time of the output of Q_2 if $V_0(t)$ started at 2.0 V, (d) the current capability of Q_3 when Q_1 is off if $V_0(t)$ is not allowed to drop below 2.0 V, and (e) the turn-on time of the output due to Q_3.

11

Bistable, Monostable,
and Astable Multivibrators

11.1. BISTABLE MULTIVIBRATOR

A *bistable multivibrator* or *flip-flop circuit* is a circuit consisting of two FETs or two transistors directly coupled together and operated in such a way that one device is off and the other one on. This means that there are two stable operating conditions:

1. Device 1 is off; device 2 is on.
2. Device 1 is on; device 2 is off.

Hence the circuit is indeed a *bistable* circuit.

We shall now give several examples of such circuits involving either transistors or FETs.

11.1a. Transistor Bistable Multivibrator

Figure 11.1(a) shows a transistor flip-flop circuit. Here the resistors are so chosen that when Q_1 is off, Q_2 is on and saturated and vice versa. No negative voltage supply is needed since $V_{CE\,sat} < V_{BE0}$, so that when the one transistor is saturated, the other one is off.

Let Q_1 be off; then

$$I_{B2} = \frac{V_{CC} - V_{BE0}}{R_c + R_1}; \qquad I_{C2} = \frac{V_{CC} - V_{CE\,sat}}{R_c} \tag{11.1}$$

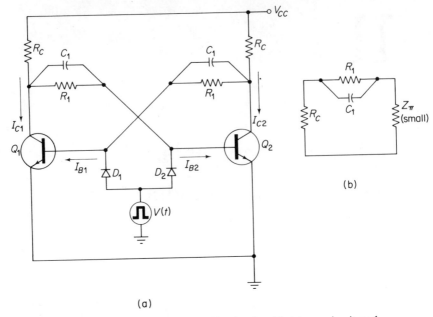

(a)

Fig. 11.1. (a) Transistor bistable circuit with trigger circuit and speedup capacitors. (b) Equivalent circuit demonstrating the effect of the speedup capacitors.

Since Q_2 is saturated, R_c and R_1 must be so chosen that

$$I_{C2} < \beta_F I_{B2}; \qquad \beta_F > \frac{R_c + R_1}{R_c} \qquad (11.2)$$

since $V_{CC} - V_{CE\,sat}$ and $V_{CC} - V_{BE0}$ differ only a little. This condition is basically not difficult to achieve, for R_1 and R_c can easily be made comparable. Often R_c and R_1 are of the order of 1000 Ω; this produces reasonably high currents. It also makes it possible to put the circuit in integrated form.

EXAMPLE 1: In the circuit $R_1 = 1000\ \Omega$, $R_c = 500\ \Omega$, $V_{CC} = 4$ V, $V_{BE0} = 0.70$ V, $V_{CE\,sat} \simeq 0$, and $\beta_F = 50$. Find I_{B2} and I_{C2}.

ANSWER:

$$I_{B2} = \frac{4.0 - 0.7}{1500} = 2.2\ \text{mA}; \qquad I_{C2} = \frac{4.0}{500} = 8\ \text{mA}$$

$$\beta_F > \frac{R_c + R_1}{R_c} = \frac{1500}{500} = 3$$

To switch the bistable circuit from the one stable condition to the other, a trigger pulse must be applied. Figure 11.1(a) shows how this can be done with two diode triggers. One must make sure that the trigger pulse is long enough, so that the transistor that is off is triggered on and short enough so that the circuit is not triggered back to the original state. This means that the trigger pulse must be so long that the switching is well on its way when the pulse stops and so short that the switching is only partly completed.

It should be observed that $V_{BE0} > V_{CE\,sat}$. This means that Q_1 is already off before Q_2 is saturated. There may be a possibility that both transistors are on and nonsaturated during the switching process, but this is not a stable configuration. Let us assume that Q_1 and Q_2 are both on and nonsaturated and that I_{C1} increases by an amount ΔI_{C1}. Then V_{C1} decreases, which causes a decrease in I_{B2} and hence in I_{C2}. This raises V_{C2} and hence increases I_{B1}, which increases I_{C1} still further. If the total gain of the circuit is larger than unity, this will continue until Q_1 is fully on and Q_2 fully off. In the same way a small decrease in I_{C1} will be amplified and ultimately turn Q_1 off and Q_2 on.

To speed up the response, capacitances C_1 are connected in parallel with the resistances R_1, so that

$$C_1 R_1 = \tau_\pi \tag{11.3}$$

If Q_1 is off, the equivalent circuit seen by the base of Q_2 is then as shown in Fig. 11.1(b). When the base current is sufficiently large, the effect of the base impedance Z_π is small and can be neglected. The time constant of the circuit is then

$$\tau' = \frac{R_1 R_c}{R_1 + R_c} C_1 = \frac{R_c}{R_1 + R_c} \tau_\pi \tag{11.4}$$

This gives an idea about the speedup that can be achieved. Actually the situation is slightly more complicated, because of the effect of the Miller capacitance observed when Q_2 is on but not yet saturated.

EXAMPLE 2: If in Example 1 we have $\tau_\pi = 10^{-8}$ s, find τ'.

ANSWER:

$$\tau' = \tfrac{500}{1500} \times 10^{-8} = \tfrac{1}{3} \times 10^{-8} \text{ s.}$$

11.1b. JFET Bistable Multivibrator

As a second example we shall discuss the junction FET flip-flop. Figure 11.2(a) shows the circuit, involving two junction FETs with an n-type channel. Here it is necessary to use two batteries, one with positive polarity (V_{DD}) and one with negative polarity (V_{GG}), because the FETs have a pinch-off voltage $V_P < 0$, so that a negative gate voltage is needed to turn the device off, whereas a positive drain voltage is needed to keep the device on.

We shall now assume that Q_1 is off and Q_2 on and that the circuit has been so designed that $V_{DS2} = V_{on}$ is close to zero, or Q_2 operates in the ohmic regime. For V_{GS2} we then have

$$V_{GS2} = V_{DD}\frac{R_2}{R_d + R_1 + R_2} + V_{GG}\frac{R_d + R_1}{R_d + R_1 + R_2} > V_P \tag{11.5}$$

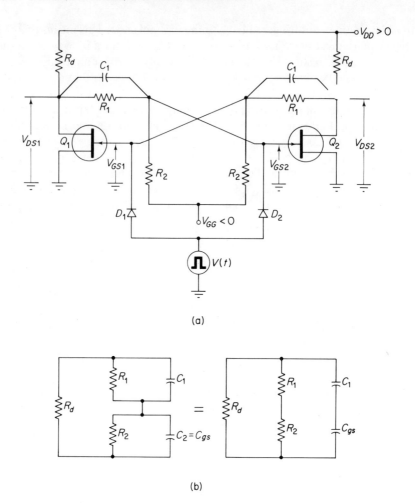

(a)

(b)

Fig. 11.2. (a) JFET bistable circuit with trigger circuit and speedup capacitors. (b) Equivalent circuits demonstrating the effect of the speedup capacitors.

for since Q_1 is off, the supply for G_2 is a simple potentiometer circuit, provided that the gate of Q_2 does not draw too much current. This is certainly the case if $V_{GS2} < 0.60$ V. The required range of V_{GS2} is therefore

$$V_P < V_{GS2} < 0.60 \text{ V} \tag{11.5a}$$

with V_{GS2} preferably close to the highest value. For Q_1, however,

$$V_{GS1} = V_{DS2} \frac{R_2}{R_1 + R_2} + V_{GG} \frac{R_1}{R_1 + R_2} < V_P \tag{11.6}$$

Here $V_{DS2} = V_{on}$ is developed over the resistance R_{on} (compare Chapter 6), which is small in comparison with $R_1 + R_2$. Since V_{DS2} is small, Eq. (11.6) may be approximated as

$$V_{GG} < V_P \frac{R_1 + R_2}{R_1} \qquad (11.6a)$$

Both conditions (11.5) and (11.6) must be satisfied simultaneously. The other stable condition is found by interchanging the subscripts 1 and 2 in V_{GS1}, V_{GS2}, V_{DS1}, and V_{DS2}.

To switch the bistable circuit from one stable condition to the other, a trigger pulse must be applied. Figure 11.2(a) shows how this can be done with two diode triggers. The situation is similar to the transistor case.

We assumed that only one of the devices would be on at the same time. During switching both devices may be on at the same time, but, as in the transistor case, the situation $V_{GS1} > V_P$, $V_{GS2} > V_P$ is unstable if the total gain of the circuit is larger than unity. Let V_{GS1} increase by an amount ΔV_{GS1}; then this is amplified in Q_1 and fed to the gate of Q_2 where it decreases V_{GS2}. This change in V_{GS2} is amplified in Q_2 and fed into the gate of Q_1, where it enhances the original increase ΔV_{GS1}. If the overall gain of the system is larger than unity, then the original variation will grow until Q_1 is fully on and Q_2 is fully off. Conversely, if V_{GS1} decreases by an amount ΔV_{GS1}, then this will increase V_{GS2}, which, in turn, will decrease V_{GS1}, and this will continue until Q_1 is fully off and Q_2 fully on.

EXAMPLE 3: If $V_{DD} = 30$ V, $V_{GG} = -30$ V, $V_P = -3$ V, $R_{on} = 300\ \Omega$, $R_d = 5000\ \Omega$, $R_1 = 10{,}000\ \Omega$, and $R_2 = 15{,}000\ \Omega$, find V_{GS2} and V_{GS1}.

ANSWER:

$$V_{GS2} = 30 \frac{15}{5 + 10 + 15} - 30 \frac{5 + 10}{5 + 10 + 15} = 15 - 15 = 0$$

well satisfying the condition $V_P < V_{GS2} < 0.6$ V.

$$V_{on} = V_{DD} \frac{R_{on}}{R_d + R_{on}} = 30 \frac{300}{5300} = 1.7\text{ V}$$

$$V_{GS1} = 1.7 \frac{15}{10 + 15} - 30.0 \frac{10}{10 + 15} = 1.0 - 12.0 = -11.0\text{ V}$$

well satisfying the condition $V_{GS1} < V_P$.

The input capacitance C_{gs} of the FET shunts R_2 and this slows the switching. The time constant involved in the switching is then

$$\tau = C_{gs} \frac{R_2(R_1 + R_d)}{R_2 + R_1 + R_d} \qquad (11.7)$$

The switching can be speeded up by connecting a capacitance C_1 in parallel with the resistance R_1, so that

$$R_1 C_1 = R_2 C_{gs} \qquad (11.8)$$

The time constant τ' is then given by the equivalent circuit of Fig. 11.2(b) as

$$\tau' = \frac{R_d(R_1 + R_2)}{R_d + R_1 + R_2} \frac{C_{gs}C_1}{C_1 + C_{gs}} \tag{11.9}$$

for since $R_1C_1 = R_2C_{gs}$, the series connection of R_1, C_1 and R_2, C_{gs} corresponds to the parallel connection of $R_1 + R_2$ and C', where C' is the capacitance of the series connection of C_1 and C_{gs}. This speeds up the response. Actually the situation is slightly more complicated, since the capacitance C_{dg} between the drain and gate gives rise to the Miller effect, which increases the input capacitance above C_{gs}.

EXAMPLE 4: If $C_{gs} = 5.0$ pF, find the value of C_1 needed for the speedup of the response and the time constants τ and τ'.

ANSWER:

$$\tau = 5 \times 10^{-12}\frac{15 \times 15}{30}10^3 = 37.5 \times 10^{-9}\text{ s}$$

$$C_1 = \frac{C_{gs}R_2}{R_1} = \frac{5 \times 10^{-12} \times 15}{10} = 7.5\text{ pF}$$

$$\tau' = \frac{5 \times 25}{30}10^3 \times \frac{7.5 \times 5.0}{12.5}10^{-12} = 12.5 \times 10^{-9}\text{ s}$$

11.1c. MOSFET Bistable Multivibrator with Enhancement Mode Devices

MOSFETs operating in the depletion mode, such as n-channel MOSFETs with $V_T < 0$, operate in the same way as JFET circuits and hence no further

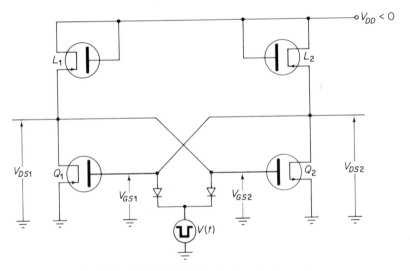

Fig. 11.3. MOSFET bistable circuit with trigger circuit.

discussion is needed. The only difference is that the gate never draws current, so that there is no restriction on V_{GS} when a device is on.

As we saw in Chapter 6, enhancement mode devices opened up the opportunity to build circuits that did not contain any resistors. Two such amplifiers connected directly together can give a simple bistable circuit that is easily put in integrated form. If the gain per stage is large enough, the circuit works with a single battery.

Figure 11.3 shows such a circuit with p-channel MOSFETs. Let it be assumed that all four devices have the same turn-on voltage $V_T < 0$. Let it further be assumed that Q_1 and Q_2 have a K-factor K_1 and that L_1 and L_2 have a K-factor K_2, where $K_1 > K_2$, since each half must give an amplification $(K_1/K_2)^{1/2} > 1$. We shall see in a moment what value of K_1/K_2 is required. Let Q_1 be off, or $V_{GS1} > V_T$, and Q_2 on, or $V_{GS2} < V_T$; then

$$V_{GS2} = V_{DS1} = V_{DD} - V_T; \qquad V_{GS1} = V_{DS2} = V_{on} > V_T \qquad (11.10)$$

Because Q_2 is on and $V_{DS2} = V_{on} \simeq 0$, L_2 has a current

$$I_{DL2} = -K_2(V_{DD} - V_T - V_{DS2})^2 \simeq -K_2(V_{DD} - V_T)^2 \qquad (11.11)$$

and Q_2 has a current

$$I_{D2} = -K_1[2(V_{GS2} - V_T)V_{DS2} - V_{DS2}^2] \simeq -2K_1(V_{GS2} - V_T)V_{DS2} \qquad (11.12)$$

However, since the devices are in series, $I_{DL2} = I_{D2}$, so that

$$K_2(V_{DD} - V_T)^2 = 2K_1(V_{GS2} - V_T)V_{DS2} = 2K_1(V_{DD} - 2V_T)V_{DS2} \qquad (11.13)$$

or

$$V_{DS2} = \frac{K_2(V_{DD} - V_T)^2}{2K_1(V_{DD} - 2V_T)} \qquad (11.14)$$

as $V_{GS2} = V_{DD} - V_T$. However, according to Eq. (11.10), $V_{GS1} = V_{DS2} > V_T$ which leads to the requirement

$$\frac{K_2(V_{DD} - V_T)^2}{2K_1(V_{DD} - 2V_T)} > V_T \quad \text{or} \quad \frac{K_1}{K_2} > \frac{(V_{DD} - V_T)^2}{2(V_{DD} - 2V_T)V_T} \qquad (11.15)$$

(note that $V_{DD} - V_T$ and V_T are negative quantities). The circuit works properly if this condition is satisfied.

Triggering the circuit to effect the switching from the one stable state to the other can be done in the same way as for the JFET. The only difference is that here we used p-type devices, so that the polarity of the diodes and of the triggering pulse must be reversed (compare Fig. 11.3).

EXAMPLE 5: If $V_{DD} = -20$ V and $V_T = -4$ V, find the inequality for K_1/K_2.

ANSWER:

$$\frac{K_1}{K_2} > \frac{(-20 + 4)^2}{2(-20 + 8)(-4)} = \frac{16^2}{2 \times 12 \times 4} = 2\frac{2}{3}$$

which is not difficult to satisfy.

11.2. MONOSTABLE MULTIVIBRATOR

A monostable multivibrator is a circuit that goes through a complete wave form when triggered. We shall now discuss two examples.

11.2a. Transistor Monostable Multivibrator

By replacing the one $R_1 C_1$ combination in Fig. 11.1(a) by a capacitor C and returning the now floating base to V_{CC} via a resistor R, we obtain the circuit of Fig. 11.4. Here the resistors R and R_c are chosen so that Q_2 is normally on and saturated and so that Q_1 is normally off. If Q_1 is triggered by a pulse,

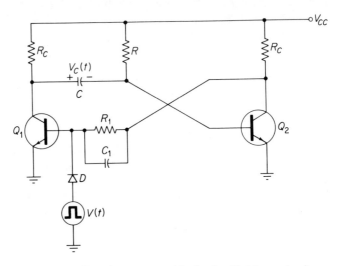

Fig. 11.4. Transistor monostable circuit with trigger circuit.

it is turned on, which turns Q_2 off. The base of Q_2 is then charged through the resistor R until its voltage reaches the value V_{BE}, at which point Q_2 is again turned on.

If the transistor Q_1 is triggered at $t = 0$, then just before triggering

$$V_{C1}(0^-) = V_{CC}; \qquad V_{B2}(0^-) = V_{BE0} \simeq 0.70 \text{ V} \qquad (11.16)$$

Right after triggering and switching, that is, at $t = 0^+$,

$$V_{C1}(0^+) = V_{CEsat} \simeq 0 \quad \text{and hence} \quad V_{B2}(0^+) = V_{BE0} - V_{CC} \qquad (11.17)$$

since the change $-V_{CC}$ in $V_{C1}(t)$ at $t = 0$ cannot be taken up by the capacitor C but must be transmitted to B_2.

Since the capacitor C is charged through the resistance R, the time constant is RC, and $V_{B2}(t)$ would rise to the final value $V_{B2}(\infty) = V_{CC}$ if nothing intervened. Consequently (see Appendix B),

$$V_{B2}(t) = V_{B2}(0^+)\exp\left(-\frac{t}{RC}\right) + V_{B2}(\infty)\left[1 - \exp\left(-\frac{t}{RC}\right)\right]$$

$$= V_{CC} - (2V_{CC} - V_{BE0})\exp\left(-\frac{t}{RC}\right) \tag{11.18}$$

This part of the wave form stops if $V_{B2}(t)$ reaches the value V_{BE0}; at that point Q_2 is turned on and Q_1 is turned off. Let this occur at $t = T_s$, then

$$\exp\left(-\frac{T_s}{RC}\right) = \frac{V_{CC} - V_{BE0}}{2V_{CC} - V_{BE0}}; \qquad T_s = RC\ln\left(\frac{2V_{CC} - V_{BE0}}{V_{CC} - V_{BE0}}\right) \tag{11.19}$$

If $V_{CC} \gg V_{BE0}$, this reduces to

$$T_s \simeq RC\ln 2 = 0.69RC \tag{11.19a}$$

The original situation has not yet been fully restored, however, since V_{C1} has not yet recovered. It takes some time before that is accomplished; this has a bearing on the repetition rate with which the circuit can be triggered.

To investigate this problem, we observe that just before switching, that is, at $t = T_s^-$, Q_1 is still saturated, or $V_{C1}(T_s^-) = V_{CE\,sat} \simeq 0$ and $V_{B2}(T_s^-) = V_{BE0} \simeq 0.70$ V. Right after the switching, that is, at $t = T_s^+$, V_{B2} has hardly changed and hence $V_{C1}((T_s^+)$ remains at about 0 V. Since $V_{B2}(t)$ remains fixed at V_{BE0}, the capacitor C must charge through the resistance R_c, making $V_{C1}(t)$ approach the ultimate value $V_{C1}(\infty) = V_{CC}$. Since the time constant $\tau = CR_c$ and $V_{C1}(T_s^+) \simeq 0$,

$$V_{C1}(t) = V_{CC}\left[1 - \exp\left(-\frac{t - T_s}{CR_c}\right)\right] \tag{11.20}$$

(see Appendix B). In this case it is not sufficient to have the transient 90% completed; we must wait somewhat longer. A good compromise for the recovery time $T_R = t - T_s$ is

$$T_R \simeq 4R_cC \tag{11.21}$$

If we want T_R to be smaller than the pulse duration T_s, we must require that

$$4R_cC < 0.69RC \quad \text{or} \quad R > 5.8R_c \tag{11.22}$$

We cannot make the resistance R too large, however; otherwise the transistor is not saturated. In analogy with Eq. (11.2), this requires that

$$\beta_F > \frac{R_c + R}{R_c} \quad \text{or} \quad R < \beta_F R_c \tag{11.23}$$

so that the requirements for R are

$$5.8R_c < R < \beta_F R_c \tag{11.23a}$$

Therefore $R \simeq 10R_c$ is quite sufficient for high-β transistors. These considerations are also important for astable multivibrators (Section 11.3a).

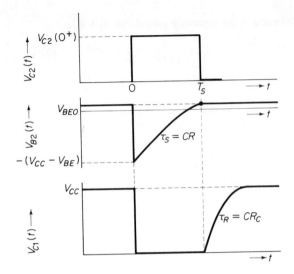

Fig. 11.5. Wave form of V_{C2}, V_{B2} and V_{C1} for circuit of Fig. 11.4.

Figure 11.5 shows the wave form of $V_{C2}(t)$, $V_{B2}(t)$, and $V_{C1}(t)$.

EXAMPLE 1: If $R = 10,000 \ \Omega$, $R_c = 1000 \ \Omega$, and $C = 10^3$ pF, find T_s and T_R and investigate whether the condition $T_R < T_s$ is satisfied.

ANSWER:

$$T_s = 0.69RC = 0.69 \times 10^4 \times 10^{-9} = 6.9 \ \mu s$$

$$T_R = 4R_cC = 4 \times 10^3 \times 10^{-9} = 4.0 \ \mu s \quad \text{or} \quad T_R < T_s$$

Up to now we assumed that the base of the transistor Q_2 could withstand the back voltage $V_{B2}(0^+) = V_{BE0} - V_{CC}$. Often this is not the case, but the emitter-base junction breaks down at the voltage $-BV_{EB0}$ given in device manuals. This breakdown affects the wave form and the duration of the pulse in the output. It is not difficult to take this into account, but it is beyond the scope of this book to do so (Compare Problem 11.4).

11.2b. MOSFET Monostable Multivibrator

Figure 11.6 shows a monostable multivibrator involving *p*-channel enhancement mode MOSFETs. Device Q_2 is normally on and operating in the ohmic regime, so that Q_1 is off. By triggering Q_1 one can turn Q_1 temporarily on, until the stable configuration "Q_2 on and Q_1 off" is restored.

Let Q_1 be triggered at $t = 0$. Just before triggering, that is, at $t = 0^-$,

$$V_{G2}(0^-) = V_{DD}; \qquad V_{D1}(0^-) = V_{DD}; \qquad V_c(0^-) = 0 \qquad (11.24)$$

where $V_c(t)$ is the voltage across the capacitor C. After triggering and com-

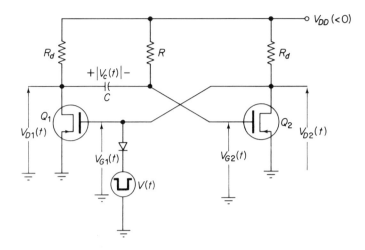

Fig. 11.6. MOSFET monostable circuit with trigger circuit.

pleted switching, that is, at $t = 0^+$,

$$V_{D1}(0^+) = V_{G2}(0^+) = V_{\text{on}} > V_T \qquad (11.24a)$$

and $V_{D1}(t) = V_{\text{on}}$ is fixed from then on. Therefore the capacitor charges through the resistor R and $V_{G2}(t)$ would decrease to the final value $V_{G2}(\infty) = V_{DD}$ if nothing intervened. Hence the time constant of the transient is $\tau_s = CR$ and (see Appendix B)

$$V_{G2}(t) = V_{G2}(0^+)\exp\left(-\frac{t}{CR}\right) + V_{G2}(\infty)\left[1 - \exp\left(-\frac{t}{RC}\right)\right]$$

$$= V_{DD} - (V_{DD} - V_{\text{on}})\exp\left(-\frac{t}{RC}\right) \qquad (11.25)$$

This part of the transient stops when $V_{G2}(t)$ reaches the value V_T. Let this be the case at $t = T_s$; then

$$\exp\left(-\frac{T_s}{RC}\right) = \frac{V_{\text{on}} - V_{DD}}{V_T - V_{DD}}; \qquad T_s = RC\ln\left(\frac{V_{\text{on}} - V_{DD}}{V_T - V_{DD}}\right) \qquad (11.26)$$

This is a relatively short time interval (see Example 2).

Just before switching, that is, at T_s^-, we have $V_{D1}(T_s^-) = V_{\text{on}}$ and $V_{G2}(T_s^-) = V_T$, so that $V_C(T_s^-) = V_{\text{on}} - V_T$. Just after switching, that is, at $t = T_s^+$, we thus have $V_C(T_s^+) = V_{\text{on}} - V_T$. The original situation has therefore not been fully restored, for the charge on the capacitor C needs to be removed. Since the discharging of C occurs through $R + R_d$, the recovery part of the transient has a time constant

$$\tau_R = C(R + R_d) \qquad (11.27)$$

To speed up the recovery, we must require that τ_R not be much larger than τ_s, or $R_d < R$. The full recovery time T_R is then a few times τ_R. This is basically not difficult, but usually T_R is considerably larger than T_s. Figure 11.7 shows the wave form of $V_{D2}(t)$.

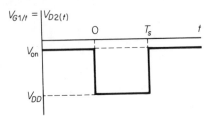

Fig. 11.7. Wave form of V_{D2} for circuit of Fig. 11.6.

EXAMPLE 2: If $V_{DD} = -20.0$ V, $V_T = -4.0$ V, $R = 10,000$ Ω, $R_d = 5000$ Ω, $C = 1000$ pF and $V_{on} = -1.0$ V, find τ_s, τ_R, T_s, and T_R.

ANSWER:

$$\tau_s = CR = 10^{-9} \times 10^4 = 10^{-5} \text{ s}$$

$$\tau_R = C(R + R_d) = 10^{-9} \times 1.5 \times 10^4 = 1.5 \times 10^{-5} \text{ s}$$

$$T_s = \tau_s \ln\left(\frac{V_{on} - V_{DD}}{V_T - V_{DD}}\right) = 10^{-5} \ln\left(\frac{19}{16}\right) = 1.7 \times 10^{-6} \text{ s}$$

$$T_R \simeq 4\tau_r = 6 \times 10^{-5} \text{ s} \gg T_s$$

11.3. ASTABLE MULTIVIBRATOR

An astable multivibrator is a circuit that generates a periodical non-sinusoidal signal and does not need triggering. It can be made by a modification of the monostable circuits. We discuss a few examples in greater detail.

11.3a. Astable Transistor Multivibrator

Figure 11.8 shows an astable transistor multivibrator circuit. It contains two equal capacitors C_1 and C_2 to connect the one collector with the other base and equal resistors R_1 and R_2 to feed the two bases, so that

$$C_1 R_1 = C_2 R_2 = CR = \tau \tag{11.28}$$

Since the theory of this circuit is identical to the one given for the monostable circuit in Section 11.2a, we can apply the results of that section directly. We must require that the transistor that is on be saturated and that the recovery time T_R of the circuit be shorter than the duration of the pulse T_s. According to Eq. (11.23a), this means that

$$5.8R_c < R < \beta_F R_c \tag{11.29}$$

which is easily met in practice. Moreover, the total period T of the wave form is $2T_s$, since T_s is the time that one device is on, and the full period

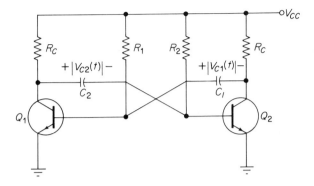

Fig. 11.8. Astable transistor circuit.

includes an "on" period for each device. Therefore

$$T = 2T_s \simeq 2CR \ln 2 = 1.38CR \qquad (11.30)$$

EXAMPLE 1: If $R = 10,000 \, \Omega$, $R_c = 1000 \, \Omega$, $\beta_F = 50$, and $C = 1000 \, \text{pF}$, investigate whether the condition (11.29) is satisfied and calculate the full period T of the wave form.

ANSWER: Equation (11.29) becomes $5.8 \times 10^3 < R < 50 \times 10^3$, and this is true. Since $\tau = CR = 10^{-9} \times 10^4 = 10^{-5} \, \text{s}$, $T = 1.38\tau = 13.8 \, \mu\text{s}$.

11.3b. Astable JFET Multivibrator

The circuit of Fig. 11.6 has the disadvantage of a relatively short pulse time and a relatively long recovery time. If such a monostable circuit is modified to make it an astable circuit, the wave form is not so easily evaluated. Therefore we shall discuss a JFET circuit that does not have this defect.

Figure 11.9 shows the circuit. We assume that $C_1 = C_2 = C$ and that $R_{g1} = R_{g2} = R_g$. Let V_P be the pinch-off voltage of the JFETs. We shall assume that at $t = 0$ the circuit switches from Q_1 off, Q_2 on to Q_1 on, Q_2 off.

Just before switching, that is, at $t = 0^-$, we have

$$V_{D1}(0^-) = V_{DD}; \qquad V_{G1}(0^-) = V_P; \qquad V_{G2}(0^-) = V_G' \simeq 0.70 \, \text{V} \qquad (11.31)$$

since the device that is on is drawing a gate current V_{DD}/R_{g2}. Just after switching, that is, at $t = 0^+$, we have instead

$$V_{D1}(0^+) = V_{on}; \qquad V_{G1}(0^+) = V_G' \simeq 0.70 \, \text{V} \qquad (11.32)$$

and hence, since V_{D1} shifts by an amount $V_{DD} - V_{on}$ during switching,

$$V_{G2}(0^+) = V_G' - V_{DD} + V_{on} \qquad (11.33)$$

Now the capacitor C is charged through the resistance R_g, since $V_{D1}(t) = V_{on}$ during this part of the cycle. Hence the time constant is

$$\tau_s = C_2 R_{g2} = CR_g \qquad (11.34)$$

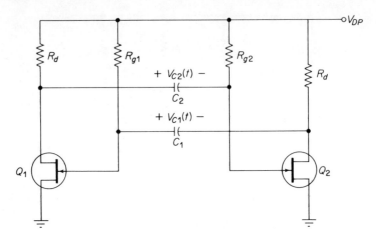

Fig. 11.9. Astable JFET circuit.

Since $V_{G2}(t)$ rises to the final value $V_{G2}(\infty) = V_{DD}$ if undisturbed, we have (see Appendix B)

$$V_{G2}(t) = V_{G2}(0^+)\exp\left(-\frac{t}{\tau_s}\right) + V_{G2}(\infty)\left[1 - \exp\left(-\frac{t}{\tau_s}\right)\right]$$

$$= V_{DD} - (2V_{DD} - V_G' - V_{on})\exp\left(-\frac{t}{\tau_s}\right) \qquad (11.35)$$

The FET Q_2 switches on when $V_{G2}(t) = V_P$. Let this occur at $t = T_s$; then

$$\exp\left(-\frac{T_s}{\tau_s}\right) = \frac{V_{DD} - V_P}{2V_{DD} - V_G' - V_{on}} \qquad \text{or}$$

$$T_s = CR_g \ln\left(\frac{2V_{DD} - V_G' - V_{on}}{V_{DD} - V_P}\right) \simeq CR_g \ln 2 \qquad (11.36)$$

so that the full period T of the wave form is

$$T = 2T_s \simeq 1.38CR_g \qquad (11.36a)$$

When Q_2 switches on, $V_{G2}(t)$ suddenly rises from V_P to V_G'. Therefore $V_{D1}(t)$, which was equal to V_{on} at $t = T_s^-$, becomes equal to $V_{on} + V_G' - V_P$ at $t = T_s^+$, because the voltage on the capacitor C cannot change abruptly. Now the capacitance C charges through R_d, since $V_{G2}(t) = V_G'$ during this part of the wave form. The time constant involved is thus

$$\tau_R = CR_d \qquad (11.37)$$

and the final value of $V_{D1}(t)$ is $V_{D1}(\infty) = V_{DD}$. It takes a time T_R, the so-called recovery time, for $V_{D1}(t)$ to reach this value, where

$$T_R \simeq 4\tau_R = 4CR_d \qquad (11.38)$$

We must now require that

$$T_R < T_s \quad \text{or} \quad 4CR_d < CR_g \ln 2 \quad \text{or} \quad R_g > \frac{4}{\ln 2} R_d = 5.8 R_d \qquad (11.39)$$

For example, $R_d = 5 \text{ k}\Omega$ and $R_g = 50 \text{ k}\Omega$ would be sufficient.
Figure 11.10 shows the generated wave form for $V_{D1}(t)$ and $V_{G2}(t)$.

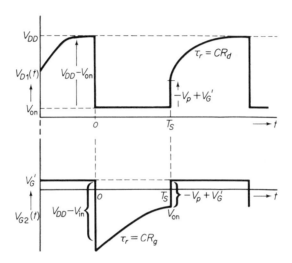

Fig. 11.10. Wave form of V_{D1} and V_{G2} for circuit of Fig. 11.9.

EXAMPLE 2: Find the full period of the wave form if $R_d = 5 \text{ k}\Omega$, $R_g = 50 \text{ k}\Omega$, $C = 1000 \text{ pF}$, $V_{DD} = 20 \text{ V}$, $V_P = -3 \text{ V}$, $V'_G = 0.70 \text{ V}$, and $V_{on} = 1.0 \text{ V}$.

ANSWER:

$$T = 2T_1 = 2CR_g \ln\left(\frac{2V_{DD} - V'_G - V_{on}}{V_{DD} - V_P}\right)$$

$$= 2 \times 10^{-9} \times 5 \times 10^4 \ln\left(\frac{38.3}{23.0}\right) = 0.51 \times 10^{-4} \text{ s}$$

so that the approximation $T = 2CR_g \ln 2$ is not too accurate in this case.

PROBLEMS

11.1. If Q_1 is off and triggered on at $t = 0$, find the output wave form in Fig. 11.2(a) if $V_{DD} = 30 \text{ V}$, $V_{GG} = -30 \text{ V}$, $V_P = -3 \text{ V}$, $R_{on} = 300 \text{ }\Omega$, $R_d = 5000 \text{ }\Omega$, $R_1 = 10,000 \text{ }\Omega$, and $R_2 = 15,000 \text{ }\Omega$.

11.2. Repeat Problem 11.1 for the circuit of Fig. 11.3: $V_{DD} = -20 \text{ V}$, $V_T = -4 \text{ V}$, and $K_1/K_2 = 4.0$.

11.3. Repeat Problem 11.1 for the circuit of Fig. 11.1(a) if $V_{CC} = 4$ V, $R_1 = 1000$ Ω, $R_c = 500$ Ω, $V_{BE0} = 0.70$ V, $V_{CEsat} = 0.10$ V, and $\beta_F = 50$.

11.4. In the circuit of Fig. 11.4, $R = 10{,}000$ Ω, $R_1 = 2000$ Ω, $R_c = 1000$ Ω, $V_{CC} = 4.0$ V, $V_{BE} = 0.70$ V, and $V_{CEsat} \simeq 0.00$ V. (a) Choose the capacitance C so that the duration T_s of the output pulse is 1.0 μs. Use the accurate expression for T_s. (b) Find the accurate height of the output pulse if taken from the drain of Q_2. (c) Repeat if the output is taken from the drain of Q_1.

11.5. Repeat Problem 11.4 for the case $V_{CC} = 6.0$ V, $BV_{BE0} = 3.5$ V, $V_{BE} = 0.70$ V, $V_{CE\,sat} = 0.00$ V, $\beta_F = 50$, $R = 10{,}000$ Ω, $R_1 = 2000$ Ω, and $R_c = 1000$ Ω. Here $-BV_{BE0} = -V_{B0}$ is the voltage at which the base-emitter junction breaks down.

11.6. In the circuit of Fig. 11.6, $V_{DD} = -20$ V, $V_T = -4.0$ V, $R = 10{,}000$ Ω, $R_d = 5000$ Ω, and $V_{on} = -1.0$ V. (a) Find the capacitance C so that the length of the pulse is 10^{-6} s. (b) Find the height of the output pulse if taken from the drain of Q_2. (c) Repeat if the output pulse is taken from the drain of Q_1.

11.7. In the circuit of Fig. 11.8, $V_{CC} = 4.0$ V, $V_{BE} = 0.70$ V, $V_{CE\,sat} = 0.00$ V, $\beta_F = 50$, $R_c = 1000$ Ω, and $R_1 = R_2 = R = 10{,}000$ Ω. Find the capacitance C so that the period of the generated square wave is 10^{-4} s. Use the accurate expression for T_{s1}.

11.8. In the circuit of Fig. 11.9, $V_{DD} = 20$ V, $V_P = -3.0$ V, $V_G' = 0.70$ V, $V_{on} = 1.0$ V, $R_d = 5$ kΩ, and $R_g = 50$ kΩ. Find the capacitance C needed to make a square wave with a period of 10^{-4} s.

12

Oscillators

Oscillators are usually made by feeding so much signal back from the output of an amplifier to the input that the output voltage can be sustained. Oscillators can also be made by connecting a negative conductance in parallel with a parallel-tuned circuit or a negative resistance in series with a series-tuned circuit. We shall show that the two methods of approach are identical.

12.1. OSCILLATORS AND FEEDBACK

12.1a. Nyquist and Barkhausen Conditions

Figure 12.1 shows an amplifier giving an amplification μ with a feedback network feeding back a voltage βv_0 to the input. We shall now investigate how the output voltage v_0 is modified by the feedback.

Let v_s be the voltage applied to the feedback amplifier; then the input voltage v_i is

$$v_i = v_s + \beta v_0 \tag{12.1}$$

However,

$$v_0 = \mu v_i \tag{12.2}$$

so that

$$v_i = v_s + \mu \beta v_i \quad \text{or} \quad v_i = \frac{v_s}{1 - \mu \beta} \tag{12.3}$$

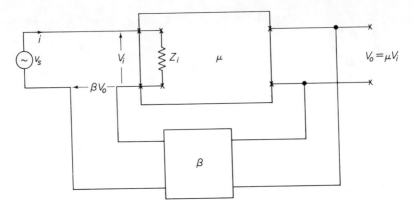

Fig. 12.1. Feedback circuit (voltage feedback).

and hence

$$v_0 = \mu v_i = \frac{\mu}{1 - \mu\beta} v_s \tag{12.4}$$

If $\mu\beta$ is real and negative and sufficiently large (negative feedback),

$$v_0 \simeq \frac{\mu}{-\mu\beta} v_s = -\frac{v_s}{\beta} \tag{12.4a}$$

which is independent of the amplification μ. *Negative feedback can thus be used to make very stable amplifiers.*

On the other hand, if $\mu\beta$ is real and positive and $\mu\beta > 1$, oscillations usually occur. The general conditions for stability and instability are quite complicated, but the following condition, known as the *Nyquist condition*, applies in most simple cases: *If the factor $\mu\beta$ is plotted in the complex plane with the frequency ω as a parameter, one obtains a closed curve; the circuit is stable if the curve does not encircle the point $(1, j0)$ and is unstable if the curve*

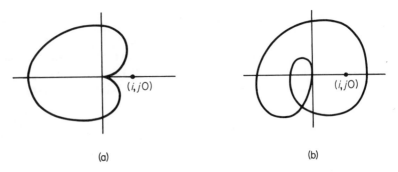

(a)   (b)

Fig. 12.2. (a) Nyquist plot of stable circuit. (b) Nyquist plot of unstable circuit.

encircles the point $(1, j0)$ *once.* The plot of $\mu\beta$ in the complex plane is called the *Nyquist plot*. Figure 12.2(a) shows the Nyquist plot of a stable circuit and Fig. 12.2(b) the Nyquist plot of an unstable one.

We shall now investigate the input impedance Z_{in} seen by the signal source v_s of Fig. 12.1. If Z_i is the input impedance of the amplifier, then the current i flowing into the circuit is

$$i_i = \frac{v_s + \beta v_0}{Z_i} = \frac{v_s}{Z_i}\left(1 + \frac{\mu\beta}{1 - \mu\beta}\right) = \frac{v_s}{Z_i(1 - \mu\beta)} \qquad (12.5)$$

as is found by substituting for v_0. Here we have assumed that the signal βv_0 is developed across an impedance that is small in comparison with Z_i. The input impedance Z_{in} "seen" by the signal v_s is therefore

$$Z_{in} = \frac{v_s}{i_i} = Z_i(1 - \mu\beta) \qquad (12.6)$$

The input impedance is thus zero for $\mu\beta = 1$. This important result is in agreement with results to be developed in Section 12.2.

If $v_s = 0$ and there is a frequency ω for which $\mu\beta$ is real and positive, then there are three possibilities:

1. $\mu\beta < 1$. If oscillations are present at $t = 0$, their amplitude will decay exponentially.
2. $\mu\beta > 1$. If oscillations are present at $t = 0$, their amplitude will increase exponentially.
3. $\mu\beta = 1$. Oscillations, once present, will be maintained with a stable amplitude.

Now the amplification μ at a particular frequency ω for any amplifier depends on the input amplitude v_i. One generally finds for sufficiently large values of v_i that μ decreases with increasing amplitude v_i. Therefore, if we have a feedback amplifier with $\mu\beta > 1$ at a particular frequency ω for small amplitudes v_i, then oscillations of frequency ω with growing amplitude will occur. However, $\mu\beta$ will ultimately decrease with an increasing value of v_i and reach the value $\mu\beta = 1$. At this point the amplitude stops growing and a stable situation develops. We thus have the following rule: *In an oscillator of frequency ω a stable amplitude will develop for which $\mu\beta = 1$.* This is called the *Nyquist condition* for oscillators.

A related condition, the *Barkhausen condition*, can be formulated as follows: *If in an oscillator the input lead is broken and an input voltage v_i is applied to the input, then the same voltage must appear at the other side of the break, both in amplitude and phase, if stable oscillations are to occur.*

To show that this corresponds to the Nyquist condition, we turn to Fig. 12.3. If v_i is the voltage applied to the input, then the output voltage

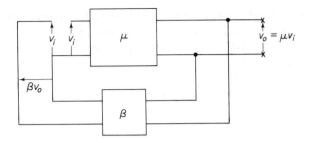

Fig. 12.3. Circuit for demonstrating the equivalence of the Barkhausen and the Nyquist condition for oscillation.

$v_0 = \mu v_i$. If the part β of the output voltage is applied to the input, then according to the Barkhausen condition,

$$v_i = \beta v_0 = \mu\beta v_i \quad \text{or} \quad \mu\beta = 1 \tag{12.7}$$

which is nothing but the Nyquist condition.

12.1b. Applications of the Barkhausen Condition

As a first example we shall consider the Wien-bridge oscillator of Fig. 12.4(a), which consists of a two-stage amplifier that is fed back to the input by means of an RC-network with $R_1 = R_2 = R$ and $C_1 = C_2 = C$. The output resistance R_0 of the amplifier is small in comparison with R_1, the C_bs are blocking capacitors of negligible impedance, and the resistances R_g are so large that

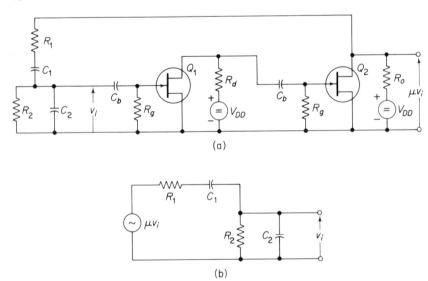

(a)

(b)

Fig. 12.4. (a) Wien-bridge oscillator. (b) Equivalent circuit used to derive the condition for oscillation.

their effects can be neglected. The Barkhausen condition can now be illustrated with the help of Fig. 12.4(b). Since the two impedances can be expressed as

$$R_1 + \frac{1}{j\omega C_1} = \frac{1 + j\omega C_1 R_1}{j\omega C_1} \tag{12.8}$$

and

$$\frac{R_2[1/(j\omega C_2)]}{R_2 + 1/(j\omega C_2)} = \frac{R_2}{1 + j\omega C_2 R_2} \tag{12.9}$$

the Barkhausen condition may be written as

$$\begin{aligned}
v_i &= \mu v_i \frac{R_2/(1 + j\omega C_2 R_2)}{(1 + j\omega C_1 R_1)/(j\omega C_1) + R_2/(1 + j\omega C_2 R_2)} \\
&= \mu v_i \frac{j\omega C_1 R_2}{(1 + j\omega C_1 R_1)(1 + j\omega C_2 R_2) + j\omega C_1 R_2} \\
&= \mu v_i \frac{j\omega CR}{1 + 3j\omega CR - \omega^2 C^2 R^2}
\end{aligned} \tag{12.10}$$

The first step is obtained by multiplying the denominator and numerator by $j\omega C_1(1 + j\omega C_2 R_2)$, and the last step is obtained by putting $C_1 = C_2 = C$ and $R_1 = R_2 = R$. Dividing by v_i and multiplying both sides by $1 + 3j\omega CR - \omega^2 C^2 R^2$ yields

$$1 + 3j\omega CR - \omega^2 C^2 R^2 = j\omega CR\mu \tag{12.10a}$$

or, since the real parts must be equal and the imaginary parts equal,

$$1 - \omega^2 C^2 R^2 = 0, \quad \text{or} \quad \omega CR = 1, \quad \text{and} \quad \mu = 3 \tag{12.10b}$$

The first condition thus yields the frequency of oscillation; this condition indicates that ω changes by a factor of 10 when C changes by a factor of 10. The second condition gives the amplitude of the oscillation. To make the amplification start, the low-amplitude value of μ must be somewhat larger than 3. The amplitude v_i will then increase until the stability condition $\mu(v_i) = 3$ is reached.

EXAMPLE 1: If $R = 1000 \, \Omega$, for what value of C is the frequency of oscillation equal to 1 MHz?

ANSWER:

$$2\pi f CR = 1; \quad C = \frac{1}{2\pi f R} = \frac{1}{2\pi \times 10^6 \times 10^3} = \frac{10^{-9}}{2\pi} = 160 \text{ pF}$$

EXAMPLE 2: If $\mu = 6/(1 + 0.25v_i^2)$, find the value of v_i for which a stable amplitude is obtained.

ANSWER: $3 = 6/(1 + 0.25v_i^2)$, $1 + 0.25v_i^2 = 2$, $0.25v_i^2 = 1$, and $v_i = 2$ V.

As a second example we shall consider the circuit of Fig. 12.5(a). This can also be discussed with the help of the Barkhausen condition. The condition is illustrated with the help of Fig. 12.5(b), where we have neglected the

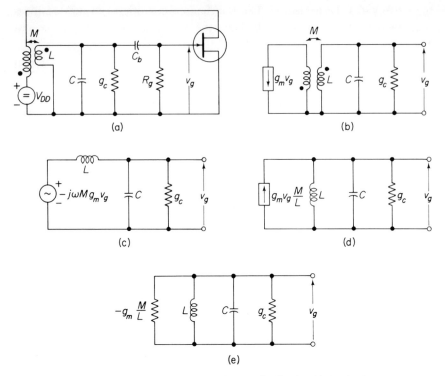

Fig. 12.5. (a) Oscillator with inductive feedback. (b) Equivalent circuit of (a). (c) Modified equivalent circuit of (b). (d) Modified equivalent circuit of (c). (e) Equivalent circuit of oscillator showing negative conductance.

effect of the resistance R_g in comparison with the tuned circuit conductance g_c. This circuit, in turn, can be represented by the equivalent circuit of Fig. 12.5(c), which finally corresponds to Fig. 12.5(d). The Barkhausen condition is then

$$\frac{(g_m M/L)v_g}{1/(j\omega L) + j\omega C + g_c} = v_g \tag{12.11}$$

Multiplying both sides by $1/(j\omega L) + j\omega C + g_c$ and dividing by v_g yields

$$g_m \frac{M}{L} = \frac{1}{j\omega L} + j\omega C + g_c \tag{12.12}$$

Equating the real and imaginary parts of Eq. (12.12) gives

$$\frac{1}{j\omega L} + j\omega C = 0 \quad \text{or} \quad \omega^2 LC = 1 \quad \text{and} \quad g_m \frac{M}{L} = g_c \tag{12.12a}$$

The first equation gives the frequency of oscillation. We see that the frequency changes only by a factor of 3 when C changes by a factor of 9, whereas in the Wien-bridge oscillator the frequency changed by a factor of 9 under this condition. The second condition gives the *amplitude* of oscillation. In this circuit g_m decreases with increasing amplitude v_g. If $g_m M/L > g_c$ for small amplitudes, v_g will increase with time until a stable amplitude is reached for which

$$g_m(v_g)\frac{M}{L} = g_c \tag{12.12b}$$

Equation (12.12) may also be written as ·

$$\frac{1}{j\omega L} + j\omega C + g_c - g_m\frac{M}{L} = 0 \tag{12.13}$$

This is easily understood with the help of Fig. 12.5(d). The current generator $(g_m M/L)v_g$ of Fig. 12.5(d) that establishes v_g corresponds to a feedback conductance $-g_m M/L$ in parallel with the tuned circuit [see Fig. 12.5(e)]. We shall derive Eq. (12.13) in the next section from another point of view.

EXAMPLE 3: Find the value of the capacitance C for an oscillator according to Fig. 12.5(a) which generates 1 MHz for an inductance of 10^{-4} H.

ANSWER:

$$C = \frac{1}{\omega^2 L} = \frac{1}{4\pi^2 \times 10^{12} \times 10^{-4}} \simeq \frac{10^{-9}}{4} = 250 \text{ pF}$$

EXAMPLE 4: The FET used in the oscillator has a transconductance $g_m = 10^{-3}$ mho for small signals, whereas the tuned circuit conductance is 10^{-4} mho. Find the value of M/L so that $g_m M/L = 4g_c$ for small signals.

ANSWER:

$$\frac{M}{L} = \frac{4g_c}{g_m} = \frac{4 \times 10^{-4}}{10^{-3}} = 0.4$$

EXAMPLE 5: In an oscillator with variable C the minimum circuit capacitance is 30 pF and the maximum circuit capacitance is 270 pF, so that the frequency of oscillation can change over a factor of 3. Find the value of L so that the frequency of oscillation can change from 1 to 3 MHz.

ANSWER: The lowest frequency is obtained for the largest value of C. Therefore

$$L = \frac{1}{\omega^2 C} = \frac{1}{4\pi^2 \times 10^{12} \times 270 \times 10^{-12}} = \frac{1}{4\pi^2 \times 270} = 0.94 \times 10^{-4} \text{ H}$$

12.2. ADMITTANCE AND IMPEDANCE CONSIDERATIONS IN OSCILLATORS

There are cases in which the feedback consideration of Section 12.1 does not apply. Certain solid-state devices, such as tunnel diodes, have a negative conductance. If such a device is connected in parallel with a tuned circuit, oscillations can occur, even though no feedback exists. On the other hand, feedback circuits that produce oscillations can also be represented by an input admittance Y_{in} or an input impedance Z_{in}, each with a negative real part. It is therefore necessary to consider oscillators also from an impedance or admittance point of view.

12.2a. Transient Response of Tuned Circuits

Figure 12.6(a) shows an L-C-g parallel circuit with g either positive or negative. We shall now investigate the transient response of this circuit by finding the general solution of the homogeneous integro-differential equation

$$C\frac{dv}{dt} + gv + \frac{1}{L}\int v\, dt = 0 \qquad (12.14)$$

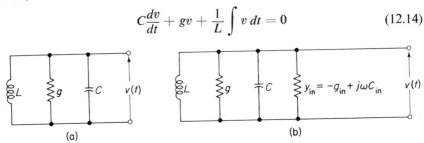

Fig. 12.6. (a) Pulse response in tuned parallel circuit. (b) Equivalent circuit of oscillator with parallel tuned circuit.

This can be done by substituting $v = \exp(st)$ and solving for s. When this has been accomplished, the general solution can be written in the form

$$v = A\exp\left(-\frac{1}{2}\frac{g}{C}t\right)\cos\left\{\omega_0 t\left[1 - \left(\frac{g}{2\omega_0 C}\right)^2\right]^{1/2} + \varphi\right\} \qquad (12.14a)$$

for $|2\omega_0 C/g| > 1$, whereas for $|2\omega_0 C/g| < 1$

$$v = A\exp\left(-\frac{1}{2}\frac{g}{C}t\right)\cosh\left\{t\left[\left(\frac{g}{2C}\right)^2 - \omega_0^2\right]^{1/2} + \varphi\right\} \qquad (12.14b)$$

Here the frequency ω_0 is defined by

$$\omega_0^2 LC = 1 \qquad (12.14c)$$

We thus have the following possibilities:

1. $g > 2\omega_0 C$. The solution dies out aperiodically.
2. $0 < g < 2\omega_0 C$. The solution is a damped oscillation.
3. $g = 0$. The solution is an undamped oscillation of frequency ω_0.
4. $-2\omega_0 C < g < 0$. The solution is an oscillation with an exponentially growing amplitude.
5. $g < -2\omega_0 C$. The solution is nonoscillatory and grows exponentially.

Since $g = 0$ and $\omega_0^2 LC = 1$ for an undamped oscillation, we may write for $\omega = \omega_0$

$$Y_{par} = g + \frac{1}{j\omega_0 L} + j\omega_0 C = 0 \qquad (12.15)$$

Figure 12.6(b) shows an actual L-g_c-C parallel circuit connected in parallel to an admittance $Y_{in} = -g_{in} + j\omega C_{in}$ with negative real part, caused by feedback. Applying the previous discussion, the condition for oscillation is

$$Y_{par} = g_c + \frac{1}{j\omega_0 L} + j\omega_0 C - g_{in} + j\omega_0 C_{in} = 0 \qquad (12.16)$$

Equation (12.16) implies that both the real and imaginary parts of Y_{par} must be zero, which yields

$$\frac{1}{j\omega_0 L} + j\omega_0(C + C_{in}) = 0 \quad \text{or} \quad \omega_0 = [L(C + C_{in})]^{-1/2}; \qquad g_c - g_{in} = 0$$
$$(12.16a)$$

The first condition gives the frequency of oscillation and the second condition gives the amplitude v of the oscillations. To make the oscillations start, $g_c - g_{in}(v)$ must be negative for small values of v. The oscillations will then start and grow in amplitude. Because of the nonlinearity of the device, $g_{in}(v)$ will ultimately decrease with increasing v. The amplitude v of the oscillations will thus grow until $g_c - g_{in}(v) = 0$. When this happens, a stable amplitude has been attained.

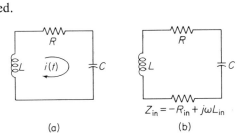

(a)   (b)

Fig. 12.7. (a) Pulse response in tuned series circuit. (b) Equivalent circuit of oscillator with series tuned circuit.

We shall now apply similar considerations to an L-R-C series circuit [Fig. 12.7(a)], where R may again be positive or negative, and investigate the transient response of the network by solving the homogeneous integro-differential equation

$$L\frac{di}{dt} + Ri + \frac{1}{C}\int i\, dt = 0 \qquad (12.17)$$

Going through the same considerations as before, we find that a stable oscillation of frequency $\omega_0 = (LC)^{-1/2}$ is obtained if $R = 0$. This corresponds to

$$Z_{loop} = j\omega_0 L + \frac{1}{j\omega_0 C} + R = 0 \qquad (12.18)$$

Figure 12.7(b) shows an actual L-R_c-C series circuit applied to an impedance $Z_{in} = -R_{in} + j\omega L_{in}$ with negative real part, caused by feedback. Equation (12.18) then yields

$$Z_{loop} = j\omega_0 L + \frac{1}{j\omega_0 C} + R_c - R_{in} + j\omega_0 L_{in} = 0 \qquad (12.19)$$

Equating the real part equal to zero and the imaginary part equal to zero, this yields

$$j\omega_0 L + \frac{1}{j\omega_0 C} + j\omega_0 L_{in} = 0 \quad \text{or} \quad \omega_0 = [(L + L_{in})C]^{-1/2};$$

$$R_c - R_{in}(i) = 0 \qquad (12.20)$$

The first condition again gives the frequency of the oscillations and the second condition gives the (current) amplitude. To make the oscillations start, one must require that $R_c - R_{in}(i) < 0$ for small amplitudes. The amplitude i of the oscillations will then grow, and this ultimately causes a decrease in $R_n(i)$, so that finally a stable amplitude will be reached for which $R_c - R_{in}(i) = 0$.

12.2b. Applications

As a first example we shall discuss the circuit of Fig. 12.5(a) from an admittance point of view. Since a parallel-tuned circuit is involved, the condition for oscillation is $Y_{par} = 0$. To apply this condition, the admittance Y_{in} caused by the feedback must be evaluated. To that end, C and g_c in Fig. 12.5(a) are replaced by an EMF v_g and the FET is replaced by a current generator $g_m v_g$; this gives the circuit of Fig. 12.8(a). It is easily seen by inspection that the current generator $g_m v_g$ gives rise to a current $i = (-g_m v_g j\omega M)/(j\omega L)$.

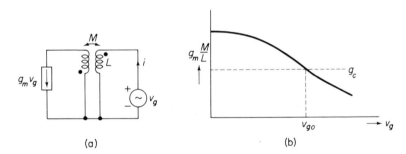

(a)   (b)

Fig. 12.8. (a) Circuit of Fig. 12.5(a) as a negative conductance circuit. (b) Graphical determination of amplitude of oscillation.

Therefore Y_{in} is

$$Y_{in} = \frac{i}{v_g} = -g_m \frac{M}{L} \qquad (12.21)$$

and from Eq. (12.16) the condition for oscillation can be written as

$$g_c + j\omega_0 C + \frac{1}{j\omega_0 L} - g_m \frac{M}{L} = 0 \qquad (12.22)$$

which corresponds exactly to Eq. (12.13).

To understand the stable operating point, we have to evaluate how the transconductance g_m depends on the gate voltage v_g. To that end we return to Fig. 12.5(a) and observe that the gate of the FET begins to rectify if the amplitude of the oscillations increases. This effectively clamps the maximum gate voltage $V_{g\,max}$ to the turn-on voltage of the gate diode, which is about 0.70 V. Because of the gate current that flows, the dc bias of the gate shifts further negative until the FET passes current only part of the time. The current $I(t)$ flowing through the FET then has a fixed peak value and the width of the current pulse decreases with increasing amplitude. If we now write

$$I(t) = I_0 + i_1 \cos \omega_0 t + i_2 \cos 2\omega_0 t + \cdots \qquad (12.23)$$

we have that

$$i_1 = \frac{1}{\pi} \int_{-\pi}^{\pi} I(t) \cos \omega_0 t \, d(\omega_0 t) \qquad (12.23a)$$

decreases with increasing amplitude v_g. Since the transconductance g_m must be defined as

$$g_m = \frac{i_1}{v_g} \qquad (12.24)$$

we see that g_m decreases with increasing amplitude v_g.

If $g_m M/L$ is plotted versus v_g, we thus obtain the curve shown in Fig. 12.8(b). Also plotting the curve for g_c, we see that a stable amplitude v_{g0} is obtained when

$$g_m(v_{g0})\frac{M}{L} = g_c \qquad (12.24a)$$

This amplitude is stable, for if for some reason v_g would increase, $-g_m M/L + g_c$ would become positive, and this would tend to decrease the amplitude. Or if v_g would decrease, $-g_m M/L + g_c$ would become negative, and this would tend to increase the amplitude.

As a second example we shall consider the circuit of Fig. 12.9(a), where jX_1 and jX_2 are reactances of the same sign. If both are capacitive reactances, jX_2 must be shunted by a choke or resistor to allow passage of emitter cur-

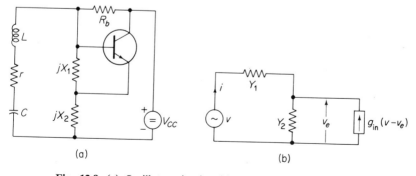

(a) (b)

Fig. 12.9. (a) Oscillator circuit with very stable frequency. (b) Equivalent circuit used to derive the condition for oscillation.

rent; if they are inductive reactances, a blocking capacitor must be put in series with jX ; otherwise the base is short-circuited to the emitter.

Since this is a series-tuned circuit, the condition for oscillation is $Z_{\text{loop}} = 0$. To evaluate the input impedance Z_{in} provided by the feedback circuit, we replace the series-tuned circuit by an EMF v, jX_1 by $Y_1 = 1/(jX_1)$, and jX_2 by $Y_2 = 1/(jX_2)$ and obtain the circuit of Fig. 12.9(b). If $|X_1| \ll r_\pi$, we obtain for the current i

$$i = (v - v_e)Y_1; \qquad v_e Y_2 = g_m(v - v_e) + Y_1(v - v_e) \qquad (12.25)$$

Solving for v_e yields

$$v_e = \frac{g_m + Y_1}{g_m + Y_1 + Y_2} v; \qquad v - v_e = \frac{Y_2}{g_m + Y_1 + Y_2} v \qquad (12.26)$$

so that

$$i = \frac{Y_1 Y_2}{g_m + Y_1 + Y_2} v \qquad (12.27)$$

Consequently

$$Z_{\text{in}} = \frac{v}{i} = \frac{g_m + Y_1 + Y_2}{Y_1 Y_2} = \frac{g_m}{Y_1 Y_2} + \frac{1}{Y_1} + \frac{1}{Y_2}$$

$$= j(X_1 + X_2) - g_m X_1 X_2 \qquad (12.28)$$

From Eq. (12.19) the condition for oscillation is therefore

$$\frac{1}{j\omega_0 C} + r + j\omega_0 L + j(X_1 + X_2) - g_m X_1 X_2 = 0 \qquad (12.29)$$

so that

$$\frac{1}{j\omega_0 C} + j\omega_0 L + j(X_1 + X_2) = 0; \qquad r - g_m X_1 X_2 = 0 \qquad (12.29a)$$

The first equation determines the frequency of oscillation and the second determines the amplitude, because g_m again decreases with increasing amplitude. The merit of the circuit is that X_1 and X_2 can be quite small because g_m is large so that $j(X_1 + X_2)$ is small. Therefore the frequency of the oscillation is hardly affected by the properties of the transistor. Consequently, if the series-tuned circuit is kept at a fixed temperature, an extremely stable frequency is generated. At lower frequencies X_1 and X_2 are usually chosen inductive and at higher frequencies they are capacitive.

EXAMPLE: An oscillator at 1 MHz has $C = 100$ pF and $g_m = 10^{-1}$ mho and the tuned circuit has a Q-factor $Q_0 = 100$. If $X_1 = X_2 = X$, find the value of X needed for oscillations to occur.

ANSWER:

$$\frac{1}{Q_0} = \omega_0 C r; \qquad r = \frac{1}{\omega_0 C Q_0} = \frac{1}{2\pi \times 10^6 \times 10^{-10} \times 100} = 16\,\Omega$$

$$|X_1| = |X_2| = |X| = \left(\frac{r}{g_m}\right)^{1/2} = \left(\frac{16}{0.1}\right)^{1/2} = 12.6\,\Omega$$

so that the condition $|X_1| \ll r_\pi$ is well satisfied. If the reactances are capacitive, we have

$$X_1 = X_2 = -\frac{1}{\omega_0 C}; \qquad C = \frac{1}{\omega_0 |X|} = \frac{1}{2\pi \times 10^6 \times 12.6} = 12{,}600 \text{ pF}$$

PROBLEMS

12.1. Figure 12.10 shows a tuned drain oscillator. Apply the Barkhausen condition and show that the condition for oscillation is

$$Y_{\text{par}} = \frac{1}{R} + \frac{1}{j\omega_0 L} + j\omega_0 C - g_m \frac{M}{L} = 0$$

You may ignore the effects of R_g and C_g.

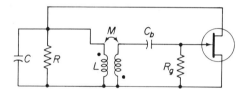

Fig. 12.10. Tuned drain oscillator.

12.2. Figure 12.11 shows the Colpitts oscillator. Apply the Barkhausen condition and show that the condition for oscillation yields

$$\omega_0^2 L \frac{C_1 C_2}{C_1 + C_2} = 1; \qquad g_m = \omega_0^2 C_1 C_2 r$$

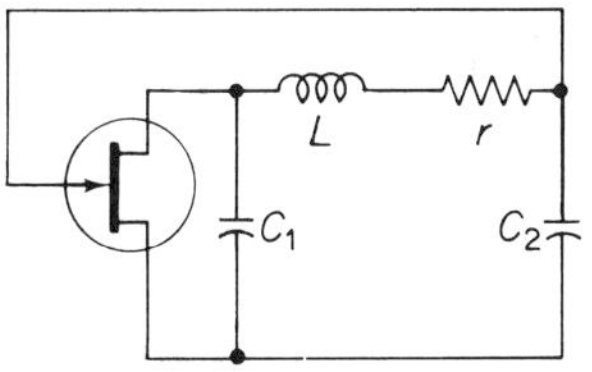

Fig. 12.11. Colpitts oscillator.

12.3. Figure 12.12 shows the Hartley oscillator. Find the condition for oscillation by applying the Barkhausen condition and show that

$$\omega_0^2 (L_1 + L_2) C = 1; \qquad g_m = \frac{r_1 + r_2}{\omega_0^2 L_1 L_2}$$

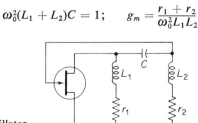

Fig. 12.12. Hartley oscillator.

12.4. Figure 12.13 has $\omega^2 C_{dg} L_g \ll 1$ for all frequencies of practical interest. If the device has a transconductance g_m, show that the condition for oscillation is

$$Y_{\text{par}} = \frac{1}{j\omega_0 L} + \frac{1}{R} + j\omega_0(C + C_{dg}) - g_m \omega_0^2 L_g C_{ag} = 0$$

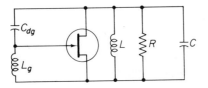

Fig. 12.13.

12.5. The phase-shift oscillator of Fig. 12.14 has a three-stage CR phase shift network. If the amplifier has the output voltage v_0 180 degrees out of phase with respect to the input voltage v_i, the output impedance of the amplifier is very low, and the input impedance very high, show that the condition for oscillation is

$$|g_v| = 29; \qquad \omega_0 CR = \frac{1}{(6)^{1/2}}$$

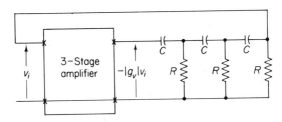

Fig. 12.14. Phase-shift oscillator.

12.6. A tunnel diode oscillator has a circuit as shown in Fig. 12.15. At the operating point the device gives a negative conductance

$$-\frac{g_{d0}}{1 + (v/v_1)^2}$$

where v is the amplitude of the oscillation. Here $g_{d0} = 10^{-2}$ mho and $v_1 = 0.05$ V. (a) Write the condition for oscillation. (b) Find the frequency of oscillation. (c) Find the stable amplitude of oscillation.

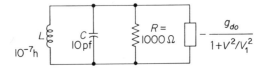

Fig. 12.15. Tunnel diode oscillator.

12.7. The circuit of Fig. 12.16 shows a bridge-type oscillator that performs better than the corresponding bridge circuit discussed in the text. If $g_v = 30$, find the value of R_2 for which the circuit just oscillates at the frequency $\omega_0 = 1/(CR)$. C_b is a blocking capacitor of negligible impedance.

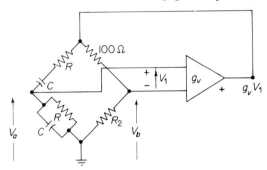

Fig. 12.16. Bridge-type oscillator.

13

Noise

The term *noise* refers to random fluctuating signals generated in electron devices and electric circuits. In conductors the fluctuations are caused by the random thermal agitation of the carriers in the conductor; this is called *thermal noise*. In thermionic vacuum diodes the fluctuating signals are caused by the fact that the emission of electrons by the cathode constitutes a series of independent events occurring at random instants; this is called *shot noise*. Shot noise also occurs in *p-n* junction diodes and in transistors, because the transmission of carriers across junctions constitutes a series of independent events occurring at random instants. Most electronic devices also show flicker noise; this is a noise form that is most pronounced at low frequencies and that is caused by complicated, partly unknown processes.

Noise considerations are important in the design of small-signal amplifiers, because the noise sources operating in the amplifier set a lower limit to the signals that can be processed.

If one measures noise in sensitive amplifiers or equipment, one finds that the observed results are independent of *when* the measurements are performed. Such noise processes are said to be *stationary*; practically all noise sources encountered by the electrical engineer fall into this catagory.

In such measurements one usually measures the mean square value of the noise by averaging over a sufficiently long time interval. This is called a *time average*. When one performs calculations, however, it is much more convenient to average at a given instant over an *ensemble*, that is, an assembly of a large number of identical systems subjected to independent fluctuations. This is called an *ensemble average*. Fortunately for practically all noise sources encountered by the electrical engineer the two types of average are identical (such noise sources are said to be *ergodic*).

We shall denote an average by a bar over the quantity that is being averaged. If $X(t)$ is a noise signal, then $\overline{X(t)}$ denotes its average value and $\overline{X^2(t)}$ its mean square value.

13.1. FOURIER ANALYSIS; SPECTRAL INTENSITY

13.1a. Fundamental Theorem of Fourier Analysis

Let $X(t)$ be a noise signal with zero average value $[\overline{X(t)} = 0]$, and let a Fourier analysis be made of $X(t)$. The fundamental theorem of Fourier analysis then states that

1. It is possible to introduce a spectral intensity $S_x(f)$, so that $[S_x(f)\,\Delta f]^{1/2}$ represents the rms Fourier component for the frequency interval Δf around f.
2. Fourier components in different frequency intervals are independent of each other.
3. Alternating-current circuit analysis can be applied to these rms Fourier components.

This means that if $X(t)$ is a noise voltage, then the noise can be represented by an rms noise EMF $[S_x(f)\,\Delta f]^{1/2}$, whereas if $X(t)$ is a noise current, then the noise can be represented by an rms noise current generator $[S_x(f)\,\Delta f]^{1/2}$. Noise calculations thus amount to calculating the spectral intensity $S_x(f)$ of $X(t)$.

We cannot give a rigorous proof of this theorem here. The following discussion is offered for those readers who would like to see its plausibility.

Let us consider a noise signal $X(t)$ for the time interval $0 \leq t \leq T$. To satisfy the requirement for mathematical rigor, we redefine the signal at the boundaries by putting

$$X(0) = X(T) = \lim_{h \to 0} \tfrac{1}{2}[X(h) + X(T - h)] \tag{13.1}$$

where $h > 0$. If we now make a Fourier analysis of $X(t)$ for the time interval $0 \leq t \leq T$, it is allowed to equate the sum of the Fourier series to $X(t)$:

$$X(t) = \sum_{n=-\infty}^{\infty} a_n \exp(j\omega_n t) \qquad \text{for } 0 \leq t \leq T \tag{13.2}$$

where $\omega_n = 2\pi n/T$ and $n = 0, \pm 1, \pm 2, \ldots$, and

$$a_n = \frac{1}{T} \int_0^T X(t) \exp(-j\omega_n t)\, dt \tag{13.2a}$$

is the Fourier coefficient of frequency $f_n = n/T$. It is easily seen that if a_n^* denotes the complex conjugate of a_n, then $a_{-n} = a_n^*$.

It may now be shown that the limit

$$\lim_{T \to \infty} 2\overline{T a_n a_m^*} = 0 \qquad \text{for n} \neq m \tag{13.3}$$

whereas the limit

$$\lim_{T \to \infty} 2\overline{T a_n a_n^*}$$

exists and is nonzero. It is called the *spectral intensity* of $X(t)$ and is denoted by $S_x(f_n)$:

$$S_x(f_n) = \lim_{T \to \infty} 2\overline{T a_n a_n^*} \tag{13.4}$$

Thus for large T we may write

$$\overline{2a_n a_m^*} \simeq 0 \qquad (m \neq n); \qquad \overline{2a_n a_n^*} \simeq \frac{S_x(f_n)}{T} = S_x(f_n)\,\Delta f_n \tag{13.5}$$

where $\Delta f_n = 1/T$. This means that the Fourier coefficients can be treated as independent ac quantities. Furthermore, it is easily shown that $[S_x(f_n)\,\Delta f_n]^{1/2}$ is the rms value of the Fourier component $a_n \exp(j\omega_n t) + a_{-n} \exp(-j\omega_n t)$ of frequency f_n. If we now omit the subscript n, we have arrived at the statements of the theorem.

13.1b. Applications of the Theorem

We shall now state the following theorems:

1. If a noise signal has a spectral intensity $S_x(f)$, then

$$\overline{X^2} = \int_0^\infty S_x(f)\, df \tag{13.6}$$

2. If a noise signal $X(t)$ has a spectral intensity $S_x(f)$ and the signal is applied (Fig. 13.1) to an amplifier having a voltage gain $g(f)$, then the output noise signal $Y(t)$ has a mean square value

$$\overline{Y^2} = \int_0^\infty S_x(f)\,|g(f)|^2\, df \tag{13.7}$$

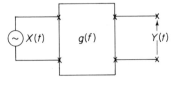

Fig. 13.1. Noise signal $X(t)$, applied to amplifier, gives output $Y(t)$.

For those readers who are interested in finding out how the fundamental theorem of Section 13.1a is applied, we offer the following proofs.

To prove Eq. (13.6), we observe that

$$X(t) = \sum_{n=-\infty}^{\infty} a_n \exp(j\omega_n t)$$

Therefore

$$\overline{X^2} = \sum_{n=-\infty}^{\infty} \overline{a_n a_n^*} = \sum_{n=0}^{\infty} \overline{2a_n a_n^*} = \sum_{n=0}^{\infty} S_x(f_n) \Delta f_n = \int_0^{\infty} S_x(f_n) \, df_n$$

since $\overline{a_{-n} a_{-n}^*} = \overline{a_n a_n^*}$ and $\overline{a_n a_m^*} = 0$ for $m \neq n$. Dropping the subscript n gives Eq. (13.6).

To prove Eq. (13.7), we put

$$X(t) = \sum_{n=-\infty}^{\infty} a_n \exp(j\omega_n t); \qquad Y(t) = \sum_{n=-\infty}^{\infty} b_n \exp(j\omega_n t)$$

However, according to ac circuit analysis,

$$b_n = a_n g(f_n)$$

so that

$$S_y(f) = \lim_{T \to \infty} 2T \overline{b_n b_n^*} = \lim_{T \to \infty} 2T \overline{a_n a_n^*} \, |g(f_n)|^2 = S_x(f_n) |g(f_n)|^2$$

and

$$\overline{Y^2} = \int_0^{\infty} S_y(f_n) \, df_n = \int_0^{\infty} S_x(f_n) |g(f_n)|^2 \, df_n$$

By again dropping the subscript n, we obtain Eq. (13.7).

Equation (13.7) gives a very convenient method for measuring the spectral $S_x(f)$ as follows. Let the amplifier be a tuned amplifier with a relatively narrow passband of width B (difference between half-power frequencies) around a center frequency f_0, with $|g(f_0)| = g_0$; then $S_x(f) \simeq S_x(f_0)$ for all the frequencies of the passband, and consequently

$$\overline{Y^2} = S_x(f_0) \int_0^{\infty} |g(f)|^2 \, df = S_x(f_0) g_0^2 B_{\text{eff}} \tag{13.8}$$

where the effective bandwidth B_{eff} is defined by

$$B_{\text{eff}} = \int_0^{\infty} \frac{|g(f)|^2}{g_0^2} \, df \tag{13.9}$$

One can now determine $\overline{Y^2}$ with the help of a quadratic detector and measure g_0 and B_{eff}; $S_x(f_0)$ can thus be evaluated.

We cannot discuss here the various methods of calculating the spectral intensities of individual noise processes. It is sufficient for our purpose that a spectral intensity $S_x(f)$ can be defined, calculated, and measured.

Most fundamental noise processes have a spectral intensity that is constant for a wide frequency range. Such noise processes are said to be *white*. For example, the thermal noise of a resistor R has a constant spectral intensity up to infrared frequencies, whereas shot noise of a saturated thermionic diode has a constant spectral intensity up to a few hundred megahertz. The only notable exception is flicker noise, which has a spectral intensity of the form constant $/f^\alpha$, with $\alpha \simeq 1$. This peculiar spectrum comes about because of the mechanisms by which it is generated. In FETs and transistors the noise seems to be associated with wave-mechanical tunneling of carriers to empty trapping sites in surface oxide layers where they are temporarily stored.

13.1c. Thermal Noise and Shot Noise

According to *Nyquist's theorem*, the thermal noise of a resistance R at the absolute temperature T can be represented by a noise EMF $[S_v(f)\,\Delta f]^{1/2}$ in series with R [Fig. 13.2(a)] or by a noise current generator $[S_i(f)\,\Delta f]^{1/2}$ in parallel with R [Fig. 13.2(b)], where

$$S_v(f) = 4kTR; \qquad S_i(f) = 4kTg \qquad (13.10)$$

Here k is Boltzmann's constant $(= 1.38 \times 10^{-23}\ \text{J/deg})$ and $g = 1/R$.

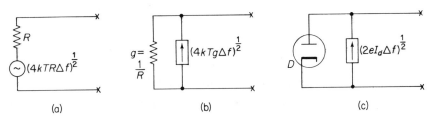

(a)   (b)   (c)

Fig. 13.2. (a) Nyquist's theorem. (b) Alternate equivalent formulation of Nyquist's theorem. (c) Saturated thermionic diode as a noise current source.

According to *Schottky's theorem*, the shot noise of a saturated thermionic diode carrying a dc current I_d can be represented by a current generator $[S_i(f)\,\Delta f]^{1/2}$ in parallel with the diode [Fig. 13.2(c)], where

$$S_i(f) = 2eI_d \qquad (13.11)$$

and e is the electron charge.

We shall not prove these formulas, but we shall apply them to particular cases. As a first application we evaluate the mean square value of the noise voltage $V_C(t)$ developed across the capacitance C of an RC parallel circuit kept at the absolute temperature T. This application is important, since it represents the input noise of a wide-band amplifier.

According to Nyquist's theorem, the noise can be described by a noise EMF $[S_v(f)\,\Delta f]^{1/2}$ in series with R. Consequently the spectral intensity

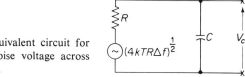

Fig. 13.3. Equivalent circuit for calculating noise voltage across an RC circuit.

$S_{v_c}(f)$ of $V_c(t)$ is (Fig. 13.3)

$$S_{v_c}(f) = S_v(f)\frac{1/(\omega C)^2}{R^2 + 1/(\omega C)^2} = \frac{4kTR}{1 + \omega^2 C^2 R^2}$$

Consequently

$$\overline{V_c^2}(t) = \int_0^\infty S_{v_c}(f)\,df = 4kTR\int_0^\infty \frac{df}{1 + \omega^2 C^2 R^2}$$

$$= \frac{4kTR}{2\pi CR}\int_0^\infty \frac{dx}{1 + x^2} = \frac{4kTR}{2\pi CR}\frac{\pi}{2} = \frac{kT}{C} \qquad (13.12)$$

where $x = \omega CR$ is introduced as a new variable.

EXAMPLE 1: Find $(\overline{V_c^2})^{1/2}$ at $T = 300°K$ if $C = 30$ pF.

ANSWER:

$$(\overline{V_c^2})^{1/2} = \left(\frac{1.38 \times 10^{-23} \times 3 \times 10^2}{3 \times 10^{-11}}\right)^{1/2} = 11.8 \; \mu V$$

As a second example we shall apply Schottky's theorem to a *p-n* junction diode. We know that the current

$$I = I_{rs}\left[\exp\left(\frac{eV}{kT}\right) - 1\right] \qquad (13.13)$$

flowing in a *p-n* junction diode can be considered as being caused by two independent currents, $I_{rs} \exp(eV/kT)$ and $-I_{rs}$. Both currents fluctuate independently and each shows full shot noise; therefore

$$S_i(f) = 2eI_{rs}\exp\left(\frac{eV}{kT}\right) + 2eI_{rs} = 2e(I + 2I_{rs}) \qquad (13.14)$$

For normal operation $I \gg I_{rs}$ and the differential conductance $g_0 = eI/kT$, so that Eq. (13.14) may be approximated as

$$S_i(f) \simeq 2eI = 2kT\frac{eI}{kT} = 2kTg_0 \qquad (13.15)$$

corresponding to *half thermal noise* of the differential conductance g_0. This is important for understanding transistor noise.

13.1d. Application to FETs and Transistors

The basic noise source in an FET is the thermal noise of the conducting channel. That is, if the FET is kept at the temperature T and a section Δx

of the channel has a resistance ΔR, then the noise EMF in series with ΔR is $(4kT\,\Delta R\,\Delta f)^{1/2}$. When the drain is HF short-circuited to the source, the spectral intensity $S_i(f)$ of the fluctuating drain current $I(t)$ is obtained by adding the contribution of all sections Δx to $S_i(f)$ quadratically. After some calculations one obtains

$$S_i(f) = \tfrac{2}{3}\cdot 4kTg_{m0} \tag{13.16}$$

for a device operating in pinch off. Here g_{mo} is the transconductance of the device at pinch off.

We can also represent the noise by an EMF $V(t)$ in series with the gate. Since $I(t) = g_{mo}V(t)$, we have

$$S_i(f) = g_{mo}^2 S_v(f) \quad \text{or} \quad S_v(f) = \frac{2}{3}\frac{4kT}{g_{mo}} \tag{13.17}$$

We can now equate $S_v(f)$ to the thermal noise of a resistance R_n at the temperature T_0 (fixed reference temperature of 290°K)

$$S_v(f) = 4kT_0 R_n = \frac{2}{3}\frac{4kT_0}{g_{mo}} \quad \text{or} \quad R_n = \frac{2}{3g_{mo}} \tag{13.18}$$

for $T = T_0$. The quantity R_n is called the *equivalent noise resistance* of the FET and is expressed in ohms. The noise equivalent circuit of the FET is thus as shown in Fig. 13.4.

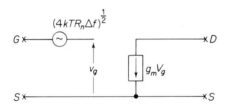

Fig. 13.4. Equivalent noise circuit of an FET.

EXAMPLE 2: If $g_{mo} = 10{,}000\ \mu$mhos, find R_n.

ANSWER:

$$R_n = \frac{2}{3g_{mo}} = \frac{2}{3 \times 10^{-2}} = 67\ \Omega$$

The basic noise sources in a transistor are the shot noises of the base current I_B and the collector current I_C. Since the carriers injected into the base by the emitter end up *either* in the base *or* in the collector, one would expect the two currents to fluctuate independently and show shot noise, so that

$$S_{I_B}(f) = 2eI_B; \qquad S_{I_C}(f) = 2eI_C \tag{13.19}$$

We thus obtain the equivalent circuit of Fig. 13.5(a).

The expression for $S_{I_B}(f)$ may be written as

$$S_{I_B}(f) = 2kT\frac{eI_B}{kT} = \frac{2kT}{r_\pi} \tag{13.20}$$

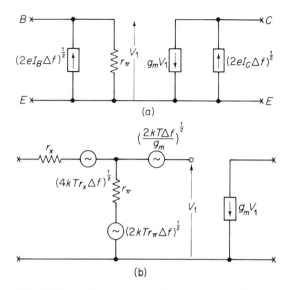

Fig. 13.5. (a) Equivalent noise circuit of transistor. (b) Alternate equivalent noise circuit.

where $r_\pi = kT/eI_B$. Since the noise current generator $[S_{I_B}(f)\,\Delta f]^{1/2}$ is in parallel with r_π, the noise can also be represented by an EMF

$$[S_{I_B}(f)\Delta f]^{1/2}r_\pi = (2kTr_\pi\,\Delta f)^{1/2} \qquad (13.21)$$

in series with r_π. This corresponds to half thermal noise of r_π.

The noise current generator $[S_{I_C}(f)\,\Delta f]^{1/2}$ in parallel with the collector junction can be represented by a noise EMF $[S_{I_C}(f)\,\Delta f]^{1/2}/g_m$ in series with the base. However, since the transconductance $g_m = eI_C/kT$, we may write

$$\left[\frac{S_{I_C}(f)\Delta f}{g_m^2}\right]^{1/2} = \left[\frac{2kT(eI_C/kT)\,\Delta f}{g_m^2}\right]^{1/2} = \left(\frac{2kT\,\Delta f}{g_m}\right)^{1/2} \qquad (13.22)$$

If we now bear in mind that r_x shows thermal noise, we obtain the equivalent circuit of Fig. 13.5(b).

13.2. NOISE FIGURE

The noise figure F is a measure for the noisiness of amplifiers and amplifier stages. We shall now define and calculate it for several cases.

13.2a. Definition of the Noise Figure F

Let an amplifier be connected to a signal source of internal resistance R_s at the reference temperature T_0. Part of the output noise power then comes from the thermal noise of the resistance R_s. We shall now define the noise figure F as the following ratio:

$$F = \frac{\text{total output noise power}}{\text{part caused by thermal noise of } R_s} \qquad (13.23)$$

To prepare for understanding the method of measurement of F, we refer all noise sources of the amplifier back to R_s and represent them by an equivalent current generator $(\overline{i_{eq}^2})^{1/2}$ in parallel with R_s for the frequency interval Δf (Fig. 13.6). If we now compare $(\overline{i_{eq}^2})^{1/2}$ with the thermal noise $(4kT_0 \,\Delta f/R_s)^{1/2}$ of the resistance R_s, we see that the definition (13.23) corresponds to

$$F = \frac{\overline{i_{eq}^2}}{4kT_0 \,\Delta f/R_s} = \frac{\overline{i_{eq}^2} R_s}{4kT_0 \,\Delta f} \qquad (13.24)$$

If we can determine $\overline{i_{eq}^2}$, we can evaluate F.

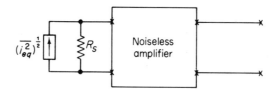

Fig. 13.6. Noise of an amplifier represented by an equivalent current generator $(\overline{i_{eq}^2})^{1/2}$ at input.

To determine $\overline{i_{eq}^2}$, we connect a saturated thermionic diode in parallel with R_s and adjust the saturated diode current I_d so that the noise output power of the amplifier is doubled. Let this occur at $I_d = I_{eq}$; then

$$\overline{i_{eq}^2} = 2eI_{eq} \,\Delta f \qquad (13.24a)$$

or

$$F = \frac{e}{2kT_0} I_{eq} R_s \qquad (\simeq 20 I_{eq} R_s \quad \text{for} \quad T = 290°K) \qquad (13.25)$$

EXAMPLE 1: If the source resistance R_s is 1000 Ω and the measured value of $I_{eq} = 90 \,\mu A$, find the noise figure F.

ANSWER: $F = 20 \times 90 \times 10^{-6} \times 10^3 = 1.8$.

The above definition holds for complete amplifiers as well as for amplifier stages. In a complete amplifier the noise almost always comes mainly from the first stage. It is therefore sufficient to calculate the noise figure of a single FET or transistor amplifier stage. This is done in Sections 13.2b and c.

13.2b. Noise Figure of the FET Amplifier Stage

To evaluate the noise figure of an FET amplifier stage, we add a source resistance R_s with a thermal noise EMF $(4kT_0 R_s \,\Delta f)^{1/2}$ in series with R_s.

We then obtain the equivalent circuit of Fig. 13.7. Consequently

$$F \cdot 4kT_0R_s\,\Delta f = 4kT_0R_s\,\Delta f + 4kT_0R_n\,\Delta f \quad \text{or} \quad F = 1 + \frac{R_n}{R_s} \qquad (13.26)$$

The noise figure thus comes closer to unity if R_s is increased. Usually, however, one has to meet requirements about the upper cutoff frequency of the amplifier stage, and this limits the value of R_s that can be used.

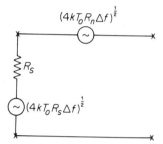

Fig. 13.7. Equivalent noise circuit of an FET amplifier.

EXAMPLE 2: If $R_n = 200\ \Omega$ and $R_s = 1000\ \Omega$, find F.
ANSWER: $F = 1 + (200/1000) = 1.20$.

13.2c. Noise Figure of a Transistor Amplifier Stage

To evaluate the noise figure of a transistor amplifier stage, we put a source resistance R_s with a thermal noise EMF $(4kT_0R_s\,\Delta f)^{1/2}$ in series with R_s. We thus obtain the equivalent circuit of Fig. 13.8. Consequently the mean

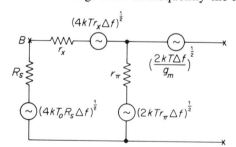

Fig. 13.8. Equivalent noise circuit of a transistor amplifier.

square of the equivalent voltage v_b at the base for $T = T_0$ is

$$\overline{v_b^2} = 4kT_0R_s\,\Delta f\left(\frac{r_\pi}{R_s + r_x + r_\pi}\right)^2 + 4kT_0r_x\,\Delta f\left(\frac{r_\pi}{R_s + r_x + r_\pi}\right)^2$$

$$+ 2kT_0r_\pi\,\Delta f\left(\frac{R_s + r_x}{R_s + r_x + r_\pi}\right)^2 + \frac{2kT_0\,\Delta f}{g_m} \qquad (13.27)$$

Since the first term comes from the source resistance R_s, the noise figure F is

$$F = \frac{\overline{v_b^2}}{4kT_0R_s\,\Delta f\,r_\pi^2/(R_s + r_x + r_\pi)^2} = 1 + \frac{r_x}{R_s} + \frac{1}{2}\frac{(R_s + r_x)^2}{r_\pi R_s}$$

$$+ \frac{1}{2g_mR_s}\frac{(R_s + r_x + r_\pi)^2}{r_\pi^2}$$

which may be approximated as

$$F \simeq 1 + \frac{r_x}{R_s} + \frac{R_s}{2r_\pi} + \frac{1}{2g_m R_s} = 1 + \frac{1 + 2g_m r_x}{2g_m R_s} + \frac{R_s}{2r_\pi} \qquad (13.28)$$

if we replace $(R_s + r_x)^2$ by R_s^2 and $(R_s + r_x + r_\pi)^2$ by r_π^2, which is allowed if $R_s \gg r_x$ and $r_\pi \gg R_s + r_x$.

Considered as a function of R_s, this has a minimum value if

$$\frac{\partial F}{\partial R_s} = -\frac{(1 + 2g_m r_x)}{2g_m R_s^2} + \frac{1}{2r_\pi} = 0$$

or

$$R_s = (R_s)_{\text{opt}} = \frac{[\beta_F(1 + 2g_m r_x)]^{1/2}}{g_m} \qquad (13.29)$$

Substituting into Eq. (13.28), the minimum noise figure becomes

$$F = F_{\text{min}} = 1 + \left(\frac{1 + 2g_m r_x}{\beta_F}\right)^{1/2} \qquad (13.30)$$

EXAMPLE 3: Find $(R_s)_{\text{opt}}$ and F_{min} if $g_m = 40 \times 10^{-3}$ mho, $r_x = 25\ \Omega$, and $\beta_F = 100$.

ANSWER:

$$(R_s)_{\text{opt}} = \frac{[100(1 + 2)]^{1/2}}{40 \times 10^{-3}} = 430\ \Omega$$

$$F_{\text{min}} = 1 + \left(\frac{1 + 2}{100}\right)^{1/2} = 1.17$$

Since $r_\pi = 100/(40 \times 10^{-3}) = 2500\ \Omega$, the conditions $r_x \ll R_s$ and $R_s + r_x \ll r_\pi$ are reasonably well satisfied.

PROBLEMS

13.1. An amplifier is operated at a source impedance R_s of 300 Ω and a thermionic saturated diode is connected in parallel with R_s. If the saturated diode operates properly for $I_D \leq 2.0$ mA and the noise figure is measured by doubling the noise output power of the amplifier, find the maximum noise figure that can be measured with this arrangement.

13.2. An FET amplifier is provided with a source impedance R_s of 1000 Ω. A switch alternately short-circuits R_s. The measured output power now gives readings M_s when R_s is short-circuited and M_0 when the switch is open. Find the noise resistance of the FET amplifier. Apply it to the case $M_0 = 3M_s$.

13.3. A tunnel diode has a negative conductance $-g_d$ of -0.0070 mho at a diode current I_d of 1.0 mA. An amplifier is made by connecting a source conductance of 0.010 mho in parallel with the tunnel diode (Fig. 13.9). The diode current has shot noise. (a) Calculate the available gain of this arrangement. (b) Calculate the noise figure of the circuit. *Hint:* Define the available gain in this case as

$$G_{av} = \frac{\text{available power of source} + \text{negative resistance}}{\text{available power of source}}$$

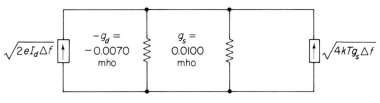

Fig. 13.9.

13.4. In a transistor $g_m r_x$ has a value of unity. Find the value of β_F needed to make $F_{min} = 1.20$.

14

Multistage
Amplifiers

In Chapters 4, 6, and 8 we discussed the interstage networks of cascaded amplifiers and the gain and cutoff frequencies of an individual amplifier stage in the cascaded chain. In this chapter we shall consider the chain as a whole and look for the overall response of the system.

14.1. CASCADE LOW-FREQUENCY AMPLIFIER STAGES

We shall first investigate the degradation in frequency response that occurs when a number of stages are connected in cascade. Let an individual stage have a voltage gain $g_v(f)$, a midband gain g_{v0}, a lower cutoff frequency f_2, and an upper cutoff frequency f_1, where f_1 and f_2 are defined by

$$|g_v(f)| = 2^{-1/2}|g_{v0}| \qquad \text{for } f = f_1 \quad \text{and} \quad f = f_2 \qquad (14.1)$$

If we now have n such stages in cascade and the voltage gain of the n-stage amplifier is denoted by $g_{vn}(f)$ and the midband gain is denoted by g_{v0n}, then for identical interstage networks

$$g_{vn}(f) = [g_v(f)]^n; \qquad g_{v0n} = (g_{v0})^n \qquad (14.2)$$

In analogy with the definition (14.1) we can now define the upper and lower cutoff frequencies f_{1n} and f_{2n} of the n-stage amplifier by the definition

$$|g_{vn}(f)| = 2^{-1/2}|g_{v0n}| \qquad \text{for } f = f_{1n} \quad \text{and} \quad f = f_{2n} \qquad (14.3)$$

Substituting (14.2), we see that for identical interstage networks this corresponds to

$$|g_v(f)| = 2^{-1/2n}|g_{v0}| \quad \text{for } f = f_{1n} \text{ and } f = f_{2n} \qquad (14.4)$$

If we bear in mind that $g_v(f)$ decreases *gradually* for increasing and for decreasing frequencies when compared with the midband region, we see from comparison of Eqs. (14.4) and (14.1) that

$$f_{1n} < f_1 \quad \text{and} \quad f_{2n} > f_2 \qquad (14.4a)$$

This is what is meant by *degradation* in the frequency response. It would not occur if an individual stage had the following high-frequency response:

$$|g_v(f)| = |g_{v0}| \quad \text{for } f < f_1; \quad |g_v(f)| = 0 \quad \text{for } f > f_1 \qquad (14.5)$$

The upper cutoff frequency of a single stage is then f_1. The response of an n-stage amplifier would then be

$$|g_{vn}(f)| = |g_{v0}|^n \quad \text{for } f < f_1; \quad |g_{vn}(f)| = 0 \quad \text{for } f > f_1 \qquad (14.5a)$$

and the upper cutoff frequency of the n-stage amplifier would still be equal to f_1. We shall see in Section 14.1b how this can be approximated. The lower cutoff frequency f_2 of n-stage amplifiers cannot be improved in this manner.

14.1a. Cascaded *RC* Amplifiers

We saw in Chapter 4, Section 4.3a, that in a single amplifier stage the high-frequency gain was given by

$$g_v(f) = \frac{g_{v0}}{1 + jf/f_1} \qquad (14.6)$$

The midband gain g_{v0} followed from

$$g_{v0} = -g_m R_{par} \qquad (14.6a)$$

whereas the upper cutoff frequency f_1 was given by

$$2\pi f_1 C_{par} R_{par} = 1 \qquad (14.6b)$$

Here g_m is the transconductance of the amplifying device and C_{par} and R_{par} are the parallel resistance and capacitance of the interstage networks.

If we now condiser n identical stages in cascade and apply definition (14.4), we obtain

$$\left|1 + j\frac{f_{1n}}{f_1}\right| = \left(1 + \frac{f_{1n}^2}{f_1^2}\right)^{1/2} = 2^{1/2n} \quad \text{or} \quad 1 + \frac{f_{1n}^2}{f_1^2} = 2^{1/n}$$

so that

$$f_{1n} = f_1(2^{1/n} - 1)^{1/2} \qquad (14.7)$$

For large n the numerical calculation can be simplified by observing that

$$2^{1/n} = \exp\left(\frac{\ln 2}{n}\right) \simeq 1 + \frac{\ln 2}{n} + \frac{1}{2}\left(\frac{\ln 2}{n}\right)^2$$

so that

$$f_{1n} \simeq f_1 \left(\frac{\ln 2}{n}\right)^{1/2} \left(1 + \frac{1}{2}\frac{\ln 2}{n}\right)^{1/2} \simeq f_1 \left(\frac{\ln 2}{n}\right)^{1/2} \left(1 + \frac{1}{4}\frac{\ln 2}{n}\right) \qquad (14.7a)$$

since $(1 + x)^{1/2} \simeq 1 + \frac{1}{2}x$ for small x. In practice this turns out to be a surprisingly good approximation, even for relatively small values of n.

EXAMPLE 1: What must be the individual gain per stage so that a 10-stage amplifier will have a midband gain of 10^5?

ANSWER: $10^5 = |g_{vo}|^{10}$; $|g_{vo}| = 10^{1/2} = 3.16$.

EXAMPLE 2: If the upper cutoff frequency of a single stage is 100 MHz, what is the upper cutoff frequency of a 10-stage amplifier made up of such stages?

ANSWER:

$$f_{1n} = f_1(2^{1/10} - 1)^{1/2} \simeq f_1 \left(\frac{\ln 2}{10}\right)^{1/2} \left(1 + \frac{\ln 2}{40}\right) = 0.268 f_1 = 26.8 \text{ MHz}$$

We now turn to the lower cutoff frequency f_{2n}. Since in this case the voltage gain of a single stage may be written as

$$g_v(f) = \frac{g_{vo}}{1 - jf_2/f} \qquad (14.8)$$

definition (14.4) yields

$$\left|1 - j\frac{f_2}{f_{2n}}\right| = \left[1 + \left(\frac{f_2}{f_{2n}}\right)^2\right]^{1/2} = 2^{1/2n} \quad \text{or} \quad 1 + \left(\frac{f_2}{f_{2n}}\right)^2 = 2^{1/n}$$

so that

$$f_{2n} = \frac{f_2}{(2^{1/n} - 1)^{1/2}} \qquad (14.9)$$

Therefore f_{2n} increases with increasing n, as stated previously.

14.1b Improvement in the High-Frequency Response of *n*-Stage Amplifiers

We shall now investigate how conditions (14.5) and (14.5a) can be approximated.

We saw in Chapter 6, Section 6.7b, that for an optimally flat series-peaked stage ($m = \frac{1}{2}$) the frequency response of a single interstage network in given by

$$|g_v(f)| = \frac{g_m R_L}{[1 + (f/f_1')^4]^{1/2}} \qquad (14.10)$$

where $m = L/(R_L^2 C_{par}) = \frac{1}{2}$, L the peaking inductance,

$$f_1' = f_1\sqrt{2} \quad \text{and} \quad 2\pi f_1 C_{par} R_L = 1 \qquad (14.10a)$$

Here it is assumed that R_{par} can be approximated by the load resistance R_L of the interstage network.

If we now apply the same reasoning as in Section 14.1a, we obtain

$$\left[1 + \left(\frac{f'_{1n}}{f'_1}\right)^4\right]^{1/2} = 2^{1/2n} \quad \text{or} \quad f'_{1n} = f'_1(2^{1/n} - 1)^{1/4} \qquad (14.11)$$

Since f'_1 is larger than f_1 and $(2^{1/n} - 1)^{1/4}$ is considerably larger than $(2^{1/n} - 1)^{1/2}$ an appreciable improvement in frequency response is obtained.

EXAMPLE 3: If the upper cutoff frequency of a non-series-peaked interstage network is 100 MHz, find the upper cutoff frequency of a 10-stage amplifier in which each interstage network is series-peaked with $m = \frac{1}{2}$.

ANSWER: Combining Eq. (14.11) with the result of Example 2, we have

$$f'_{1n} = f'_1(2^{1/10} - 1)^{1/4} = (0.268)^{1/2}f'_1 = 0.517f_1\sqrt{2} = 73.1 \text{ MHz}$$

This is about a factor of 3 better than the non-series-peaked case.

Another approach, useful for an even number of cascaded stages, consists of combining two series-peaked stages, one with $m = 0$ and one with $m = 1$, into a pair and then using $n/2$ of these pairs in cascade. If $\tau_1 = C_{par}R_L$, the stage with $m = 0$ has

$$g_v(f) = -\frac{g_m R_L}{1 + j\omega\tau_1} \qquad (14.12)$$

and the one with $m = 1$ has

$$g'_v(f) = -\frac{g_m R_L}{1 - \omega^2\tau_1^2 + j\omega\tau_1} \qquad (14.13)$$

as follows immediately from Eq. (6.56). The pair thus has a voltage gain of

$$g_{v2}(f) = \frac{g_m^2 R_L^2}{(1 + j\omega\tau_1)(1 - \omega^2\tau_1^2 + j\omega\tau_1)} \qquad (14.14)$$

so that $g_{v02} = g_m^2 R_L^2$ and

$$\frac{|g_{v2}(f)|}{g_{v02}} = \frac{1}{(1 + \omega^2\tau_1^2)^{1/2}[(1 - \omega^2\tau_1^2)^2 + \omega^2\tau_1^2]^{1/2}} = \frac{1}{[1 + (f/f_1)^6]^{1/2}} \qquad (14.14a)$$

where again $2\pi f_1 C_{par}R_L = 2\pi f_1\tau_1 = 1$. Once more applying the condition

$$\frac{|g_{vn}(f)|}{(g_{v02})^{1/2n}} = \frac{1}{[1 + (f/f_1)^6]^{n/2}} = 2^{-1/2} \qquad (14.15)$$

yields

$$f_{1n} = f_1(2^{2/n} - 1)^{1/6} \qquad (14.15a)$$

This is called an *optimally flat pair.*

EXAMPLE 4: Find the upper cutoff frequency of a 10-stage amplifier in which $f_1 = 100$ MHz and the stages are arranged in five series-peaked pairs having $m = 0$ and $m = 1$, respectively.

ANSWER: $f = f_1(2^{1/5} - 1)^{1/6} = (0.149)^{1/6}f_1 = 0.728f_1 = 72.8$ MHz. This is about as good as in Example 3.

To expand this approach, we have to write the gain function in Laplace notation and determine its *poles*, i.e., the complex frequencies for which the gain function is infinite. This is done in the next section.

14.1c. Poles in the High-Frequency Response Function

We shall first consider a simple RC interstage network and write the gain function $g_v(f)$ in Laplace notation as $g_v(s)$ by putting $s = j\omega$. Again introducing $\tau_1 = C_{par}R_{par}$ as a parameter,

$$g_v(s) = -\frac{g_m R_{par}}{1 + s\tau_1} = -\frac{g_m}{2\pi C_{par}}\frac{\tau_1}{1 + s\tau_1} = -\frac{g_m}{2\pi C_{par}}\frac{1}{s + 1/\tau_1} \qquad (14.16)$$

This shows that $g_v(s)$, considered as a function of the complex variable s, becomes infinite for $s = s_1 = -1/\tau_1$. It is therefore said that $g_v(s)$ has a *pole* at $s = s_1$. We note that s_1 lies in the negative half of the complex s-plane.

Next we shall consider a series-peaked interstage network. According to Chapter 6, Eq. (6.56), the gain function may be written as

$$g_v(f) = -\frac{g_m R_L}{1 - m\omega^2\tau_1^2 + j\omega\tau_1}$$

where $\tau_1 = C_{par}R_L$, $m = L/(C_{par}R_L^2)$, and L is the peaking inductance; here R_{par} has been approximated by the load resistance R_L of the interstage network. In Laplace transform notation we thus have

$$g_v(s) = -\frac{g_m R_L}{1 + s\tau_1 + ms^2\tau_1^2} \qquad (14.17)$$

For $m \neq 0$ this function has two poles s_1 and s_2, which are the solutions of the equation

$$(s\tau_1)^2 + \frac{1}{m}s\tau_1 + \frac{1}{m} = 0 \qquad (14.17a)$$

so that

$$s_1, s_2 = \frac{1}{\tau_1}\left[-\frac{1}{2m} \pm \left(\frac{1}{4m^2} - \frac{1}{m}\right)^{1/2}\right] = \frac{1}{2m\tau_1}[-1 \pm (1 - 4m)^{1/2}] \qquad (14.17b)$$

The solutions s_1 and s_2 are real and negative for $m < \frac{1}{4}$, whereas s_1 and s_2 are each other's complex conjugate with negative real part for $m > \frac{1}{4}$. Note that in either case the poles s_1 and s_2 again lie in the negative half of the complex plane; this is a general rule for any arbitrary RCL-network.

In the particular case $m = 1/2$, used for the optimally flat series-peaked interstage circuit, we have

$$s_1, s_2 = \frac{1}{\tau_1}(-1 \pm j) = -\frac{\sqrt{2}}{\tau_1}\exp\left(\pm\frac{j\pi}{4}\right) \qquad (14.18)$$

In the case of an optimally flat pair, involving one stage with $m = 0$

and a second stage with $m = 1$, we have for the first stage a pole at $s = s_1 = -1/\tau_1$ and for the second stage

$$s_2, s_3 = \frac{1}{2\tau_1}(-1 \pm \sqrt{3}\,j) = -\frac{1}{\tau_1}\exp\left(\pm\frac{j\pi}{3}\right) \qquad (14.19)$$

Figure 14.1(a) shows the pole location for a simple RC-network, Fig. 14.1(b) for a single series-peaked stage with $m = 1/2$, and Fig. 14.1(c) the pole location for a series-peaked pair of stages, one with $m = 0$ and the other with $m = 1$. Note that in the second and third cases the poles are located symmetrically on a semicircle of radii $\sqrt{2}/\tau_1$ and $1/\tau_1$, respectively, with the origin at its center.

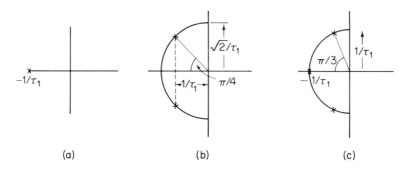

(a)   (b)   (c)

Fig. 14.1. (a) Pole location of an RC interstage circuit. (b) Pole location in a series peaked circuit with $m = \frac{1}{2}$. (c) Pole location in a pair, one with an RC interstage circuit, one with series peaked circuit with $m = 1$.

One can suggest the following improvement in the high-frequency response of multistage amplifiers. To that end, $2n$ poles are located symmetrically on a semicircle of radius τ_1, with the origin at its center, each by appropriate choice of the load resistances and the series peaking inductors. This is illustrated in Problem 5 at the end of the chapter for a pair of stages. Stagger-tuned amplifiers, discussed in Section 14.2, use an analogous method.

14.1d. Pulse Response of Cascaded RC-Amplifiers

To discuss the short-term pulse response of an n-stage amplifier with RC interstage networks, we shall first review the results obtained in Chapter 4 for a single stage. As shown in Chapter 4, the unit step function $u(t)$ gives the following response for a single amplifier stage:

$$v_0(t) = g_{v0}\left[1 - \exp\left(-\frac{t}{\tau_1}\right)\right] \qquad (14.20)$$

where

$$\tau_1 = \frac{1}{2\pi f_1} = C_{par}R_{par} \qquad (14.20a)$$

The *delay time* of this response is defined as the time τ_{D1} it takes the response to reach 50% of its final value. This yields

$$1 - \exp\left(-\frac{t}{\tau_1}\right) = \frac{1}{2}; \qquad \exp\left(-\frac{t}{\tau_1}\right) = \frac{1}{2}$$

$$t = \tau_{D1} = \tau_1 \ln 2 = 0.69\tau_1 \tag{14.21}$$

The *rise time* of this response is defined as the time τ_{R1} it takes the response to rise from 10 to 90% of the final response. Now

$$1 - \exp\left(-\frac{t}{\tau_1}\right) = 0.90 \text{ means} \qquad \exp\left(-\frac{t}{\tau_1}\right) = 0.10$$

$$t = \tau_1 \ln 10 = 2.30\tau_1$$

$$1 - \exp\left(-\frac{t}{\tau_1}\right) = 0.10 \text{ means} \qquad \frac{t}{\tau_1} \simeq 0.10 \quad \text{or} \quad t = 0.10\tau_1$$

Consequently

$$\tau_{R1} = (2.30 - 0.10)\tau_1 = 2.20\tau_1 \tag{14.22}$$

We could also have obtained Eq. (14.20) with the help of Laplace transform techniques. According to Eq. (14.16), the gain function of a simple RC interstage network in Laplace transform notation may be written as

$$g_v(s) = \frac{g_{v0}}{1 + s\tau_1} \tag{14.23}$$

Since the unit step function $u(t)$ has the Laplace transform $1/s$, the Laplace transform $V_0(s)$ of $v_0(t)$ is

$$V_0(s) = \frac{g_v(s)}{s} = \frac{g_{v0}}{s(1 + s\tau_1)} = g_{v0}\left(\frac{1}{s} - \frac{1}{s + 1/\tau_1}\right) \tag{14.24}$$

Since the Laplace transform of $\exp(-t/\tau_1) = 1/(s + 1/\tau_1)$, one obtains Eq. (14.20) by transforming back to the time domain.

The Laplace transform method is very useful for the calculation of the pulse response of n-stage amplifiers with identical stages. In analogy with Eq. (14.23) the gain function in Laplace notation may be written as

$$g_{vn}(s) = \frac{(g_{v0})^n}{(1 + s\tau_1)^n} \tag{14.25}$$

Since the unit step function $u(t)$ has the Laplace transform $1/s$, the Laplace transform of the output voltage $v_{0n}(t)$ is

$$V_{0n}(s) = \frac{g_{vn}(s)}{s} = \frac{(g_{v0})^n}{s(1 + s\tau_1)^n} \tag{14.26}$$

Transforming back to the time domain, one obtains

$$v_{0n}(t) = (g_{v0})^n\left[1 - \exp\left(-\frac{t}{\tau_1}\right)\sum_{r=0}^{n-1}\frac{(t/\tau_1)^r}{r!}\right] \tag{14.27}$$

as can be found in any textbook on circuit theory.

It is now much more difficult to find the delay time τ_D, defined again as the time it takes the response to rise to 50% of the final value, and the rise time τ_R, defined as the time it takes the response to rise from 10 to 90% of the final value. One can, of course, solve the problem graphically, and when one does so, one finds the following approximations for large n:

$$\tau_R \simeq \tau_{R1}\sqrt{n}\,; \qquad \tau_D = \tau_{D1}n \tag{14.27a}$$

In other words, *the delay times add linearly and the rise times quadratically.* If the n stages are not all identical, but have delay times $\tau_{D1}, \ldots, \tau_{Dn}$ and rise times $\tau_{R1}, \ldots, \tau_{Rn}$, one obtains approximately

$$\tau_R \simeq (\tau_{R1}^2 + \tau_{R2}^2 + \cdots + \tau_{Rn}^2)^{1/2}\,; \qquad \tau_D \simeq \tau_{D1} + \tau_{D2} + \cdots + \tau_{Dn} \tag{14.27b}$$

of which Eq. (14.27a) are special cases.

Next we shall discuss the *sag* of an n-stage amplifier. To that end we shall first review the results obtained in Chapter 4 for a single stage. As shown in Chapter 4, the unit step function $u(t)$ gives the following long-term response for a single amplifier stage:

$$V_0(t) = g_{v0} \exp\left(-\frac{t}{\tau_2}\right); \qquad \tau_2 = \frac{1}{2\pi f_2} = C_b R_{\text{loop}} \tag{14.28}$$

Here C_b is the capacitance of the blocking capacitor, and $R_{\text{loop}} = R_1 + R_2$, where R_1 and R_2 were defined in Chapter 4. Since $\exp(-t/\tau_2) \simeq 1 - t/\tau_2$ for small t, we defined the *sag* of the circuit as $100t/\tau_2$ and expressed it in percent per millisecond.

We could also have obtained Eq. (14.28) with the help of Laplace transform techniques. Putting $s = j\omega$ and $\tau_2 = 1/(2\pi f_2)$, the gain function (14.8) may be written as

$$g_v(s) = \frac{g_{v0}}{1 + 1/(s\tau_2)} = g_{v0}\frac{s}{s + 1/\tau_2} \tag{14.29}$$

Since the unit step function $u(t)$ has a Laplace transform $1/s$, the Laplace transform $V_0(s)$ of the output voltage $v_0(t)$ is

$$V_0(s) = \frac{g_{vs}}{s} = \frac{g_{v0}}{s + 1/\tau_2} \tag{14.30}$$

Transforming back to the time domain yields Eq. (14.28).

The Laplace transform method is very useful for the calculation of the response of an n-stage amplifier with identical stages. In analogy with Eq. (14.29), the gain function in Laplace notation is now

$$g_{vn}(s) = (g_{v0})^n \frac{s^n}{(s + 1/\tau_2)^n} \tag{14.31}$$

The Laplace transform $V_{0n}(s)$ of the output voltage $v_{0n}(t)$ is thus

$$V_{0n}(s) = \frac{g_{vn}(s)}{s} = (g_{v0})^n \frac{s^{n-1}}{(s + 1/\tau_2)^n} \tag{14.32}$$

Transforming back to the time domain, one obtains

$$v_{0n}(t) = \frac{d^{n-1}}{dt^{n-1}}\left[\frac{(t/\tau_2)^{n-1}\exp(-t/\tau_2)}{(n-1)!}\right] = 1 - \frac{nt}{\tau_2} + \cdots \qquad (14.33)$$

as can be found in any textbook in circuit theory. We thus see that the sag of an n-stage amplifier with identical stages is n times the sag of a single stage; in other words, *the sags of individual stages add linearly.*

14.1e. Miller Effect in Cascaded Amplifiers

In the discussion of the Miller effect in Chapter 4, we assumed that the load impedance was resistive and equal to R_{par}. In cascaded amplifiers this assumption is satisfied only for the last stage. The previous stage has the load resistance R_{par} shunted by a capacitance C_{par}. We thus have to replace R_{par} by $R_{par}/(1 + j\omega C_{par} R_{par})$ and hence obtain instead of Eq. (4.43) for the input current i_1

$$i_1 = v_1\left(1 + \frac{g_m R_{par}}{1 + j\omega C_{par} R_{par}}\right)j\omega C_{oi} \qquad (14.34)$$

so that the input admittance of the previous stage may be written as

$$Y_{in} = \left(1 + \frac{g_m R_{par}}{1 + j\omega C_{par} R_{par}}\right)j\omega C_{oi} \qquad (14.34a)$$

Therefore the Miller approximation

$$Y_{in} = (1 + g_m R_{par})j\omega C_{oi} \qquad (14.34b)$$

is not accurate at $f \simeq f_1$, where a large conductive part exists. As a consequence the results of Section 14.1a are only approximations, valid for situations in which the Miller effect is relatively small.

With more stages connected in cascade the effects become progressively more complicated when one goes back from the output circuit of a given amplifier stage to the output of the preceding stage. This problem is best handled with computer techniques. It is then found that the drop-off in response is somewhat faster than Eq. (14.2) would indicate.

14.1f. Feedback Methods for Improving the Frequency Response of Cascaded Amplifiers

We have already seen in Section 8.4b that the frequency response of an operational amplifier can be improved by feedback. We shall now go into greater detail. We shall assume that we have a three-stage operational amplifier in which the output of the last stage is completely resistive and the two interstage networks determine the frequency response. Let each interstage network have an upper cutoff frequency f_1 and let the midband gain of the amplifier without feedback be $-K_{v0}$; then the voltage gain is

$$-K_v = -\frac{K_{v0}}{(1 + jf/f_1)^2} \qquad (14.35)$$

Substituting into Eq. (8.46), for the voltage gain with resistive feedback we thus obtain

$$\frac{V_0}{V_s} = -\frac{K_v R_f}{R_s(1 + K_v) + R_f} = -\frac{K_{v0} R_f}{K_{v0} R_s + (1 + jf/f_1)^2(R_s + R_f)} \qquad (14.36)$$

It is now possible to define the upper cutoff frequency f'_1 of the feedback circuit by equating

$$\left| K_{v0} R_s + \left(1 + j\frac{f'_1}{f_1}\right)^2 (R_s + R_f) \right|^2 = 2(K_{v0} R_s + R_s + R_f)^2 \qquad (14.36a)$$

from which f'_1 can be determined. Since $K_{v0} R_s \gg R_s + R_f$, it is obvious that $f'_1 > f_1$ as stated in Section 8.4b.

We shall now let R_f go from ∞ (zero feedback) to zero. We then find that there is a particular value $R_f = R_{f0}$ for which the amplifier response is optimally flat. This is a very desirable situation in wideband amplifiers. For smaller values of R_f the frequency response shows a peak that becomes more and more pronounced when R_f is further decreased. This situation should be avoided in wideband amplifiers; in low-frequency amplifiers it may be perfectly tolerable, since the peak in frequency response lies at

$$f_3 \simeq f_1\left(K_{v0}\frac{R_s}{R_f}\right)^{1/2} \qquad (14.36b)$$

and this frequency may be so high that it is of no interest in the application.

It is beyond the scope of this book to give a detailed discussion of the frequency response of multistage cascaded amplifiers with feedback and of the ways of improving it.

14.2. TUNED MULTISTAGE AMPLIFIERS

In tuned multistage amplifiers we run into the same problem as in low-frequency amplifiers in that the bandwidth B_n of an n-stage tuned amplifier is much smaller than the bandwidth B_1 of a single stage. In analogy with the n-stage low-frequency amplifier case one would thus expect

$$B_n = B_1(2^{1/n} - 1)^{1/2} \qquad (14.37)$$

The numerical relationships developed for the n-stage low-frequency amplifier also apply.

To improve the situation, one tunes the interstage networks to different frequencies in the required passband $f_0 - B/2 < f < f_0 + B/2$ and gives each tuned circuit the appropriate tuned circuit impedance. By suitable choice of the individual tuning frequencies and the tuned circuit impedances, one can achieve an approximately flat response inside the passband and a sharp drop in response outside the passband. This method is called *stagger tuning*. Two types of response are of practical importance:

1. The relative frequency response is

$$\left(\frac{g_{vn}}{g_{v0n}}\right)^2 = \frac{1}{1 + [2(f - f_0)/B]^{2n}} \tag{14.38}$$

   This is called the Butterworth flat response.

2. The relative frequency response $|g_{vn}/g_{v0n}|^2$ fluctuates between $1/(1 + \epsilon)$ and unity for $|f - f_0| < B/2$ and drops rapidly outside that interval. This is called the *Chebyshev response.*

To understand these methods properly, one has to investigate the required pole location in the s-plane in each case.

14.2a. Pole Location for a Single-Tuned Circuit

According to Eq. (4.47), the gain function of an amplifier stage with tuned interstage network may be written as

$$g_v(f) = \frac{v_0}{v_1} = -\frac{g_m R_{par}}{1 + j\omega_0 C_{par} R_{par}(\omega/\omega_0 - \omega_0/\omega)} \tag{14.39}$$

Here g_m is the transconductance of the amplifying device, C_{par} and R_{par} are the parallel capacitance and resistance of the interstage network, respectively, and $\omega_0 = 2\pi f_0 = (LC_{par})^{-1/2}$, where f_0 is the tuning frequency and L the tuning inductance.

Going over to Laplace notation by putting $j\omega = s$ and substituting $\omega_0 C_{par} R_{par} = Q_0$, the so-called Q-factor of the interstage network, we obtain

$$g_v(s) = -\frac{g_m R_{par}}{1 + [(s/\omega_0) + (\omega_0/s)]Q_0} \tag{14.40}$$

To find the poles of the response function, we solve

$$1 + \left(\frac{s}{\omega_0} + \frac{\omega_0}{s}\right)Q_0 = 0 \quad \text{or} \quad s^2 + \frac{\omega_0 s}{Q_0} + \omega_0^2 = 0$$

or

$$s = -\frac{\omega_0}{2Q_0} \pm \left(\frac{\omega_0^2}{4Q_0^2} - \omega_0^2\right)^{1/2} \simeq -\frac{\omega_0}{2Q_0} \pm j\omega_0 \tag{14.41}$$

if $4Q_0^2 \gg 1$. Therefore the two poles of the tuned circuit are

$$s_1 = -\frac{1}{C_{par}R_{par}} + j\omega_0; \quad s = -\frac{1}{C_{par}R_{par}} - j\omega_0 \tag{14.41a}$$

Since we are mostly interested in the frequency response around $\omega \simeq \omega_0$, the second pole has relatively little effect on the response.

The result of this discussion is that the tuning frequency determines the imaginary part of s_1 and the time constant $C_{par}R_{par}$ determines the real negative part of s_1. If we know the pole location for the stagger-tuned amplifier, we can thus immediately find the required tuning frequency ω_0' and the tuned circuit impedance R_{par} of each interstage network.

14.2b. Stagger-Tuned Amplifiers with Butterworth Response

The Butterworth flat response of bandwidth B for an n-stage amplifier is defined by the response function

$$g(f)^2 = \frac{(g_{v01}g_{v02} \cdots g_{v0n})^2}{1 + [2(f - f_0)/B]^{2n}} \tag{14.42}$$

so that the half-power points of the gain function are $f = f_0 \pm B/2$ and $g_{v01}, \ldots, g_{v0n}$ are the voltage gains of the individual stages at the frequency f_0.

We shall now evaluate the location of the poles of the gain function $g(s)$. These poles lie in the negative half of the complex s-plane, clustered around $s_0 = +j\omega_0$ and $s_0 = -j\omega_0$; only the first cluster of poles is of interest. Putting $s = j\omega_0 + \Delta s$, where $\Delta s = 2\pi j(f - f_0)$, and substituting $p = 2\Delta s/(2\pi B)$, Eq. (14.42) may be written as

$$g(\Delta s)g(-\Delta s) = \frac{(g_{v01}g_{v02} \cdots g_{v0n})^2}{1 + (jp)^{2n}} \tag{14.42a}$$

The poles of $g(\Delta s)$ are thus the roots of

$$1 + (-1)^n p^{2n} = 0 \tag{14.43}$$

that lie in the negative half of the p-plane. Here we must distinguish between the cases of odd and even n:

1. n *odd* $(1, 3, 5, \ldots)$. Then Eq. (14.43) may be written as

$$p^{2n} = 1 = \exp(\pm m \cdot 2\pi j) \qquad (m = 0, 1, \ldots, n) \tag{14.43a}$$

which has the solution

$$p = \exp\left(\pm \frac{\pi m j}{n}\right) \qquad (m = 0, 1, \ldots, n). \tag{14.44}$$

There are actually only $2n$ solutions, since the two p-values for $m = 0$ coincide as well as the two p-values for $m = n$. Only the poles with a negative real part are actually the poles of the function $g(\Delta s)$; the others are the poles of the gain function $g(-\Delta s)$. We thus have

$$\Delta s = \pi B p = -\pi B \exp\left(\pi j \frac{k}{2n}\right) \qquad (k = 0, \pm 2, \ldots, \pm n - 1) \tag{14.44a}$$

2. $n = $ even $(2, 4, 6, \ldots)$. Then Eq. (14.43) may be written as

$$p^{2n} = -1 = \exp[\pm(2m + 1)\pi j] \qquad (m = 0, \ldots, n - 1) \tag{14.43b}$$

which has the solution

$$p = \exp\left(\pm \frac{2m+1}{2n}\pi j\right) \qquad (m = 0, 1, \ldots, n-1) \qquad (14.45)$$

Note that in this case there are again $2n$ roots. Half of them, those that lie in the negative half of the s-plane, are the poles of $g(\Delta s)$. Consequently

$$\Delta s = \pi Bp = -\pi B \exp\left(j\pi \frac{k}{2n}\right) \qquad [k = \pm 1, \ldots, \pm(n-1)]$$
$$(14.45a)$$

We shall now show that for an optimally flat n-plet the midband gain is given by

$$g_0 = \left(\frac{g_m}{2\pi BC_{\text{par}}}\right)^n \qquad (14.46)$$

that is, the midband gain is such as if each stage had a bandwidth B and was tuned at the frequency f_0.

To prove this, we observe that all poles lie on a semicircle of radius πB, so that $|\Delta s_i| = \pi B$. The gain function of a single-tuned circuit with a pole at $j\omega_0 + \Delta s_i$ is

$$g_{vi}(s) = -\frac{\frac{1}{2}g_m/C_{\text{par}}}{\Delta s - \Delta s_i} \qquad (14.47)$$

so that the gain at $\Delta s = 0$, or $\omega = \omega_0$, is

$$|g_{vi0}| = \frac{\frac{1}{2}g_m/C_{\text{par}}}{|\Delta s_i|} = \frac{g_m}{2\pi BC_{\text{par}}}$$

as had to be proved.

Figure 14.2(a)–(c) shows the pole location for an optimally flat pair,

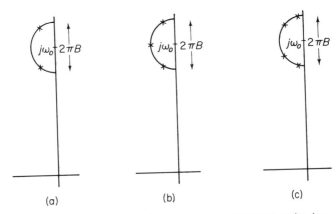

Fig. 14.2. (a) Pole location in a stagger pair. (b) Pole location in a stagger triplet. (c) Pole location in a stagger 4-plet.

an optimally flat triplet, and an optimally flat quadruplet, respectively. If one has a 10-stage amplifier, one usually does not make a stagger 10-plet, but one breaks the amplifier up into smaller units, say two stagger 5-plets or two stagger 4-plets plus a stagger pair, etc., with a relatively small loss in overall bandwidth.

14.2c. Chebyshev Response in *n*-Stage Tuned Amplifiers

Chebyshev response derives its name from the Chebyshev polynomials used in mathematical approximations. We shall first discuss what these polynomials are and then investigate how they can be used to improve the frequency response in *n*-stage tuned amplifiers.

Chebyshev polynomials $C_n(x)$ are defined as follows:

$$C_n(x) = \cos(n \cos^{-1} x) \qquad \text{for } |x| < 1;$$
$$C_n(x) = \cosh(n \cosh^{-1} x) \qquad \text{for } |x| > 1 \qquad (14.48)$$

is the *n*th Chebyshev polynomial ($n = 0, 1, 2, \ldots$). They have the property that $C_n(x)$ fluctuates between -1 and $+1$ for $|x| < 1$ and that $C_n(x)$ increases very rapidly with increasing $|x|$ for $|x| > 1$. Furthermore, the *n*th Chebyshev polynomial crosses the zero axis *n* times.

How can this be used in the frequency response of *n*-stage tuned amplifiers? To that end we define a gain function

$$g(j\omega) = \frac{g_0}{[1 + \epsilon C_n^2(x)]^{1/2}} \qquad (14.49)$$

where

$$x = Q_0 \left(\frac{\omega}{\omega_0} - \frac{\omega_0}{\omega} \right) \simeq 2 \frac{\Delta\omega}{2\pi B} \qquad (14.49a)$$

Q_0 is the Q-factor and $C_n(x)$ is the *n*th-order Chebyshev polynomial, $\omega_0 = 2\pi f_0$, where f_0 is the center frequency of the passband to be amplified and ϵ is the so-called *ripple parameter*. The gain function thus fluctuates between $g_0/(1 + \epsilon)^{1/2}$ and g_0 for $|x| < 1$ and decreases rapidly with increasing $|x|$ for $|x| > 1$. The response thus obtained is called the *equal ripple response* or *Chebyshev response*. The bandwidth is approximately equal to $f_0/Q_0 \simeq B$.

The Chebyshev polynomials satisfy the recursion formula

$$C_{n+1}(x) = 2xC_n(x) - C_{n-1}(x) \qquad (14.50)$$

The first four polynomials are

$$C_1(x) = x; \qquad C_2(x) = 2x^2 - 1;$$
$$C_3(x) = 4x^2 - 3x; \qquad C_4 = 8x^4 - 8x^2 + 1 \qquad (14.50a)$$

It is easily seen that the third and fourth Chebyshev functions satisfy the recursion formula.

If one now puts $s = j\omega$, one obtains the Laplace form of the gain function $g(s)$. By evaluating the poles of $g(s)$, one can determine the individual tuning frequencies and tuned circuit impedances. The poles are found to lie on an ellipse with its long axis along the $j\omega$-axis of the S-plane and its center at $\omega_0 = 2\pi f_0$, whereas its length is equal to $2\pi B$. It is beyond the scope of this book to go into greater details.

It should be noted that the case $n = 1$ corresponds to a single stage with a single-tuned circuit if $\epsilon = 1$. For the case $n > 1$, one usually chooses ϵ somewhat smaller.

PROBLEMS*

14.1. An amplifier has interstage networks with $C_{par} = 15 \, \text{pF}$ and $g_m = 10,000$ μmhos. Design a simple RC interstage network that has a gain of $-(10)^{1/2}$ per stage and determine its load resistance and upper cutoff frequency.

14.2. Connect 10 stages like those in Problem 14.1 in cascade and determine the gain and the overall bandwidth of the system.

14.3. The amplifier is now series-peaked with $m = \frac{1}{2}$. Determine the cutoff frequency of a single stage for the same gain per stage.

14.4. Connect 10 stages like those in Problem 14.3 in cascade and determine the overall bandwidth of the system.

*14.5. Design a series-peaked pair that has a total gain of 10 and has poles on a semicircle of radius $2\pi f_1'$ evenly distributed, i.e., at $-2\pi f_1' \exp(\pm j\pi/8)$ and at $-2\pi f_1' \exp(\pm j 3\pi/8)$, respectively. Determine the load resistors of each stage, the frequency response of the pair, and the upper cutoff frequency.

*14.6. If 5 pairs like those in Problem 14.5 are connected in cascade, what is the overall bandwidth?

14.7. If dual gate FETs are so designed that they have $g_m = 10 \times 10^{-3}$ mho and a gain-bandwidth product of 150 MHz, design a stagger-tuned pair tuned at the center frequency of 500 MHz that has a total gain of 3^2. Determine the tuning frequencies and the tuned circuit impedances.

14.8. Repeat Problem 14.7 for a stagger-tuned triplet tuned at the center frequency of 500 MHz and having a total gain of 3^3. Determine the tuning frequencies and the tuned circuit impedances.

14.9. Repeat Problem 14.7 for a stagger-tuned 4-plet tuned at the center frequency of 500 MHz and having a total gain of 3^4. Determine the tuning frequencies and the tuned circuit impedances.

*Problems designated by an asterisk are somewhat more difficult.

15

Negative
Feedback

In Section 12.1a we developed feedback theory and applied it to the discussion of oscillators. We saw that if an amplifier had a voltage gain μ and if part β of the output voltage was fed back to the input, the voltage gain *with* feedback was found to be [Eq. (12.4)]

$$\frac{\mu}{1 - \mu\beta} \tag{15.1}$$

We then applied this theory to the discussion of oscillators and showed that the condition for oscillation for the feedback amplifier was

$$\mu\beta = 1 \tag{15.2}$$

For real μ and real β it is easily seen that the voltage gain of the amplifier is increased by the feedback for $0 < \mu\beta < 1$; this is called *positive* feedback. However, for $\mu\beta < 0$ it can be seen that the voltage gain is decreased by the feedback; this is called *negative* feedback.

It is the aim of this chapter to discuss the merits of negative feedback in greater detail and to apply the negative feedback concepts to several circuits discussed earlier.

15.1. *MERITS OF NEGATIVE FEEDBACK*

We mentioned briefly in Section 12.1 that for $\mu\beta < 0$ and $|\mu\beta| \gg 1$ the voltage gain could be written as $-1/\beta$. Since this is independent of the properties of the amplifier, we concluded that very stable amplifiers could be built by application of the proper amount of negative feedback. We shall now discuss the merits of negative feedback in greater detail.

15.1a. Frequency Response

We shall now show that negative feedback decreases the lower cutoff frequency f_2 and increases the upper cutoff frequency f_1 of the amplifier. To demonstrate this, we must introduce the concept of *Bode plots*.

To do so, we shall consider an amplifier consisting of identical stages. We plot the logarithm of the absolute value of the voltage gain, $\ln|g_v|$, versus the logarithm of the frequency, $\ln f$ (Fig. 15.1). For the midband region,

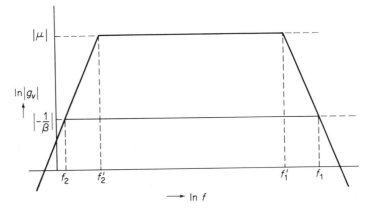

Fig. 15.1. Bode plots for amplifier without feedback and amplifier with feedback.

$|g_v|$ is independent of f and the plot is a horizontal line, which we shall call the *midband gain line*. For very low frequencies $\ln|g_v|$ increases linearly with $\ln f$, and the straight line thus obtained intersects the midband gain line at $f = f_2'$. For very high frequencies $\ln|g_v|$ decreases linearly with $\ln f$; the straight line thus obtained intersects the midband gain line at $f = f_1'$.

We can now represent the amplifier by the idealized gain versus frequency curve of Fig. 15.1. Such a plot is called a *Bode plot*. It holds for the case in which the amplifier stages are identical, i.e. when they have identical lower and upper cutoff frequencies. If this is not the case, the plots become somewhat more complicated.

If feedback is now applied, and the midband gain is lowered from $|\mu|$ to $|-1/\beta|$, the Bode plot of the feedback amplifier is obtained by lowering the midband gain line from $|\mu|$ to $|1/\beta|$. It is then seen immediately that f_1 is increased by the feedback.

Since μ, and perhaps also β, is complex at higher frequencies, care must be taken that the feedback circuit does not oscillate at some high frequency. Section 12.1a has mentioned the following rule for stability of feedback circuits: If $\mu\beta$ is plotted in the complex plane with the frequency as a parameter, one obtains a closed curve (Nyquist plot); the circuit is then stable if the curve does not encircle the point $(1, j0)$, whereas it is unstable if the curve encircles the point $(1, j0)$ once.

This is easily satisfied in a one-stage feedback amplifier with real positive $\beta = \beta_0$; since $\mu = \mu_0$ is negative at midband frequencies, the Nyquist plot lies in the negative half of a complex plane [Fig. 15.2(a)]. For a differential amplifier with two identical interstage networks of upper cutoff frequency f_1, lower cutoff frequency f_2, and real and positive $\beta = \beta_0$, this can also be made stable if the output terminal is chosen so that $\mu = \mu_0$ is negative for midband frequencies; the Nyquist plot for that case is shown in Fig. 15.2(b). For large $|\mu_0|\beta_0$ the frequency response shows a sharp peak at $f_3 \simeq f_1(|\mu_0|\beta_0)^{1/2}$ (compare Section 14.1f).

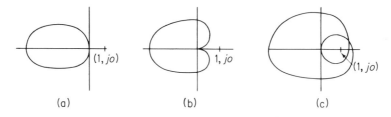

(a)   (b)   (c)

Fig. 15.2. Nyquist plots. (a) Amplifier with one interstage network. (b) Differential amplifier with two interstage networks and $\mu_0 < 0$. (c) Amplifier with three interstage networks.

However, for an amplifier with three identical interstage networks (upper cutoff frequencies f_1, lower cutoff frequencies f_2), real positive $\beta = \beta_0$, and negative midband gain $\mu = \mu_0$, the circuit is unstable unless $|\mu_0\beta_0|$ is relatively small because the point $(1, j0)$ will be encircled otherwise [Fig. 15.2(c)]. Large amounts of negative feedback cannot be applied in this case.

The situation can be remedied as follows. Suppose $\mu = \mu_0/(1 + jf/f_1)^3$ with μ_0 negative and $\beta = \beta_0(1 + jf/f_1)$ with β_0 positive; then $\mu\beta = \mu_0\beta_0/(1 + jf/f_1)^2$ with $\mu_0\beta_0$ negative. This gives a Nyquist plot as in Fig. 15.2(b), so that the circuit is perfectly stable. For large $\mu_0\beta_0$ the frequency response again shows a sharp peak at $f_3 = f_1(|\mu_0|\beta_0)^{1/2}$.

We can formulate this as follows. If we put $j\omega = s$, we see that the gain function $\mu(s)$ has three poles at $s = -2\pi f_1$ and the feedback function $\beta(s)$ has a zero at $s = -2\pi f_1$. We have thus compensated one pole of $\mu(s)$ by a corresponding zero of the feedback function $\beta(s)$.

In a three-stage feedback amplifier, with $\beta = \beta_0$ frequency independent, there is a frequency f_4 where the phase shifts is 180 degrees. If one increases the feedback factor β_0, one comes to the point where the feedback circuit oscillates at the frequency f_4. It is beyond the scope of this book to discuss the means that can be applied to avoid such a situation.

15.1b. Feedback and Harmonic Distortion

Suppose we have a preamplifier that can give a sufficiently large "clean" signal v_s, and we feed it into an amplifier that produces some disturbance like hum, noise or harmonic distortion. Then the effect of this disturbance can be considerably reduced by applying negative feedback, provided that the output voltage is kept constant in the process.

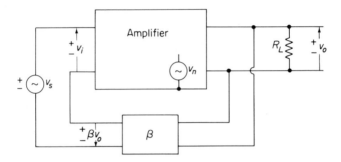

Fig. 15.3. Feedback amplifier with a disturbing signal described by an internal EMF v_n.

To demonstrate this, we consider Fig. 15.3. Let the disturbance be represented by an EMF v_n at the input of one of the stages of the amplifier. Let μ_n be the gain from that point to the output; without feedback the disturbance thus produces an output voltage $\mu_n v_n$. With feedback the disturbance produces an output voltage v_{0n} and an input voltage $v_{in} = \beta v_{0n}$, so that

$$v_{0n} = \mu_n v_n + \mu v_{in} = \mu_n v_n + \mu \beta v_{0n}$$

or

$$v_{0n} = \frac{\mu_n v_n}{1 - \mu \beta} \qquad (15.3)$$

Since the signal v_s gives an output voltage

$$v_0 = \frac{\mu v_s}{1 - \mu \beta} \qquad (15.4)$$

we have

$$\frac{v_{0n}}{v_0} = \frac{\mu_n v_n}{\mu v_s} \qquad (15.5)$$

which is independent of the feedback. We thus have the rule: *Feedback does not change the ratio v_{0n}/v_0 as long as v_s and v_n are the same with and without feedback.*

This result can be applied optimistically as follows. We now remember that we had a sufficiently large amount of clean signal v_s available. If, after applying feedback, the input signal v_s is now raised by the factor $(1 - \mu\beta)$, then the output signal *with* feedback is the same as the output signal *before* feedback was applied. The disturbance has, however, been reduced by the factor $1 - \mu\beta$. Therefore the ratio

$$\frac{\text{disturbing signal at output}}{\text{desired signal at output}} \tag{15.6}$$

has been reduced by the same factor.

Keeping v_0 constant in the process is particularly appropriate in the case of harmonic distortion. Since v_n increases with increasing v_0, feedback reduces v_n if v_s is kept constant. However, if v_0 is kept constant, v_n is also not altered. Since the ratio (15.6) is nothing but the harmonic distortion of the amplifier, we see that the harmonic distortion is reduced by the factor $(1 - \mu\beta)$.

This method works when a sufficiently large clean signal v_s is available at the input. This is not the case for a small-signal amplifier feeding from a signal source like an antenna. Such an amplifier produces noise, and this noise comes mainly from the input circuit and the first stage (see Chapter 13). Here feedback does not help, for v_s is fixed, and according to the previously established rule the noise/signal ratio is not altered by the feedback.

15.1c. Input and Output Impedance of a Feedback Amplifier

In Section 12.1a we calculated the input impedance Z_{in} "seen" by the input signal v_s, and we found [Eq. (12.6)]

$$Z_{\text{in}} = Z_i(1 - \mu\beta) \tag{15.7}$$

where Z_i is the input impedance of the amplifier itself. Negative feedback thus raises the input impedance which is "seen" by v_s.

The output impedance Z_{out} of a feedback amplifier is evaluated by representing the amplifier output by an EMF μv_i in series with the impedance Z_0. We further replace the load resistance R_L of Fig. 12.3 by an EMF v_0 and evaluate the current i flowing into the amplifier output (Fig. 15.4). We thus have $v_i = \beta v_0$; $\mu v_i = \mu\beta v_0$, and hence

$$i = \frac{v_0 - \mu v_i}{Z_0} = \frac{v_0 - v_0\mu\beta}{Z_0} = \frac{v_0}{Z_{\text{out}}} \tag{15.8}$$

so that

$$Z_{\text{out}} = \frac{Z_0}{1 - \mu\beta} \tag{15.9}$$

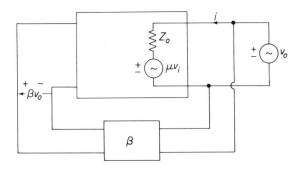

Fig. 15.4. Feedback circuit with arrangement for calculating the output impedance.

Negative feedback thus reduces the output impedance of the amplifier considerably. This principle is used in many regulated voltage supplies to make the output voltage practically independent of the output load.

15.1d. Other Forms of Feedback

In all the examples discussed so far we have considered feedback from the output voltage to the input voltage. There are three other forms of feedback possible:

1. From the output voltage to the input current.
2. From the output current to the input current.
3. From the output current to the input voltage.

If one goes through similar calculations one can show that the same beneficial effects of negative feedback can be obtained in these three cases as well. The input and output impedances must be investigated in greater detail for the individual forms of feedback.

15.2. APPLICATION OF THE FEEDBACK APPROACH TO OTHER CIRCUITS

We shall now show that the feedback approach can be applied to other circuits such as the source follower, FET amplifier with unbypassed source resistance, emitter follower, transistor amplifier with unbypassed emitter resistance and the operational amplifiers.

15.2a. Source and Emitter Followers

Figure 15.5a shows a source follower. If the FET has a transconductance g_m, and $R_{s1} + R_{s2} = R'_s$, then the voltage gain without feedback is $g_m R'_s$ and the feedback factor $\beta = -1$, since v_0 is fully fed back to the input in

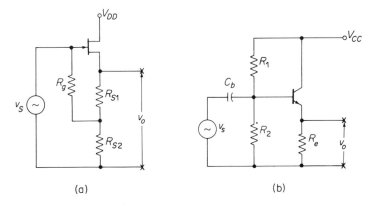

Fig. 15.5. (a) Source follower circuit. (b) Emitter follower circuit.

opposite phase. Hence the voltage gain of the source follower is

$$\frac{g_m R'_s}{1 + g_m R'_s} \qquad (15.10)$$

in agreement with Eq. (6.6). Since $1 - \mu\beta = 1 + g_m R'_s$, the distortion is reduced by a factor $(1 + g_m R_s)$. We must now investigate what this means.

Ignoring feedback, an input voltage v'_i would give an output voltage $g_m R'_s v'_i$, and the distortion would be

$$d'_2 = \frac{v'_i}{4(V_{GS0} - V_P)} \qquad (15.11)$$

With feedback we must keep the output voltage at $g_m R'_s v'_i$ by raising the input voltage to $v_i = v'_i(1 + g_m R'_s)$. Therefore the distortion with feedback is

$$d_2 = \frac{v'_i}{4(V_{GS0} - V_P)(1 + g_m R'_s)} = \frac{v_i}{4(V_{GS0} - V_P)(1 + g_m R'_s)^2} \qquad (15.12)$$

The total reduction factor is thus $(1 + g_m R'_s)^2$. This can be understood as follows. The voltage v_{gs} between gate and source is $v_i/(1 + g_m R'_s)$; this explains one factor $(1 + g_m R'_s)$. The feedback reduces the distortion due to v_{gs} by the other factor $(1 + g_m R'_s)$. Hence the total reduction is $(1 + g_m R'_s)^2$.

EXAMPLE 1: Suppose $d'_2 = 1\%$ for $v'_i = 100$ mV. If $g_m R'_s = 10$, find how much input signal can be tolerated in the source follower for 1% distortion.

ANSWER:

$$0.01 = \frac{v'_i}{4(V_{GS0} - V_P)}$$

$$4(V_{GS0} - V_P) = \frac{v'_i}{0.01} = 10 \text{ V}$$

Therefore

$$v_i = 0.01 \times 4(V_{GS0} - V_P)(1 + g_m R'_s)^2 = 0.01 \times 10 \times (11)^2$$
$$= 12.1 \text{ V amplitude}$$

We can apply the same considerations to the emitter follower (Fig. 15.5b). In analogy with Eq. (15.10) the voltage gain is

$$\frac{g_m R_e}{1 + g_m R_e} \tag{15.13}$$

This agrees with Eq. (8.28) if R_b is put equal to zero, $(\beta_F + 1)$ is replaced by $\beta_F = g_m r_\pi$, and numerator and denominator are then multiplied by g_m.

An input signal v'_i without feedback would give a distortion

$$d'_2 = \frac{v'_i}{4kT/e} \tag{15.14}$$

[compare Eq. (7.41)] and the output voltage ignoring feedback would be $g_m R_e v'_i$. *With* feedback we must keep the output voltage at $g_m R_e v'_i$ by raising the input voltage to $v_i = v'_i(1 + g_m R_e)$. Therefore the distortion with feedback is

$$d_2 = \frac{v'_i}{(4kT/e)(1 + g_m R_e)} = \frac{v_i}{(4kT/e)(1 + g_m R_e)^2} \tag{15.15}$$

EXAMPLE 2: Find how much input signal can be tolerated in the emitter follower for 1% distortion if $g_m = 40 \times 10^{-3}$ mho and $R_s = 2000\ \Omega$.

ANSWER:

$$d_2 = \frac{v_i}{(4kT/e)(1 + g_m R_s)^2} \quad \text{or} \quad v_i = \frac{4kT}{e} d_2 (1 + g_m R_s)^2$$
$$= 0.1 \times 0.01 \times (81)^2 = 6.6\ \text{V amplitude}$$

15.2b. FET Amplifier with Unbypassed Source Resistor

Figure 15.6a shows the circuit. Without feedback the ac drain voltage would be $-g_m R_d v_i$ and the ac source voltage $g_m R_s v_i$. Therefore $\mu = -g_m R_d$ and $\beta = -(-g_m R_d/g_m R_s) = R_d/R_s$. Hence the voltage gain with feedback is

$$\frac{\mu}{1 - \mu\beta} = \frac{-g_m R_d}{1 + g_m R_s} \tag{15.16}$$

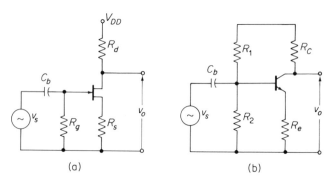

(a)   (b)

Fig. 15.6. (a) FET amplifier with unbypassed source resistor. (b) Transistor amplifier with unbypassed emitter resistor.

In analogy with Eq. (15.12) the distortion with feedback for an input signal v_i is

$$d_2 = \frac{v_i}{4(V_{GS0} - V_P)(1 + g_m R_s)^2} \tag{15.17}$$

Since R_s is not as large as in the source follower case the improvement in distortion is not as spectacular but it is very worthwhile. For example, if $g_m R_s = 2$, the improvement is by a factor of nine.

15.2c. Transistor Amplifier with Unbypassed Emitter Resistor

Fig. 15.6(b) shows this circuit. The only difference from the previous case is that R_d is replaced by R_c, R_s by R_e and $V_{GS0} - V_P$ by kT/e, whereas g_m is usually considerably larger. The gain with feedback is now [compare Eq. (15.16)]

$$\frac{-g_m R_c}{1 + g_m R_e} \tag{15.18}$$

which corresponds to Eq. (8.38) when one replaces $\beta_F + 1$ by β_F, substitutes $\beta_F = g_m r_\pi$ and divides denominator and numerator by r_π. In analogy with Eq. (15.15) the distortion is

$$d_2 = \frac{v_i}{(4kT/e)(1 + g_m R_e)^2} \tag{15.19}$$

15.2d. Operational Amplifier with Resistive Feedback

If we turn to Fig. 8.11(a) and assume that $R_f \gg R_s$, then $\mu = -K_v$ and $\beta = R_s/R_f$. Therefore the voltage gain with feedback is

$$K_{vf} = \frac{-K_v}{1 + (R_s/R_f)K_v} = \frac{-K_v R_f}{R_f + R_s K_v} \tag{15.20}$$

This agrees with Eq. (8.46) if $R_s \ll R_f$.

The harmonic distortion is reduced by the factor

$$1 - \mu\beta = 1 + \frac{R_s}{R_f}K_v = \frac{R_f + R_s K_v}{R_f} \tag{15.21}$$

This agrees with Eq. (8.57) if $R_s \ll R_f$.

PROBLEMS

15.1. If an operational amplifier has a gain $-K_v$ and an output impedance Z_0, show that the output impedance for resistive feedback is $Z_0/(1 + K_v R_s/R_f)$.

15.2. An FET amplifier without feedback is so operated that it gives 1% second harmonic distortion at $v_i' = 100$ mV.
   a) If the source resistor $R_s = 500\ \Omega$ is now unbypassed, and $g_m = 4 \times 10^{-3}$ mho, what input voltage v_i will give 1 % distortion?
   b) If $R_d = 4500\ \Omega$ what output voltage is obtained in this case?

15.3. A transistor amplifier without feedback gives 1% second harmonic distortion at $v_i' = 1.0$ mV.

    a) If the emitter resistor $R_e = 500\ \Omega$ is now unbypassed, and $g_m = 40 \times 10^{-3}$ mho, what input voltage v_i will give 1% distortion?

    b) If $R_c = 4500\ \Omega$, what output voltage is obtained in this case?

15.4. A transistor power amplifier has a transconductance of 10 mho at the quiescent operating point. The amplifier gives 15% distortion for full output power. What is the distortion for full output power if an unbypassed emitter resistor of 0.20 Ω is inserted?

15.5. A triode power amplifier has a transconductance of 4×10^{-3} mho at the quiescent operating point. The amplifier gives 5% distortion for full output power. The grid bias is supplied by a bypassed 400 Ω resistor in the cathode lead. What is the distortion at full output power if the cathode resistor bypass capacitor is removed?

Appendix A

Semiconductor Device Theory

A.1. BASIC LAWS OF SOLID-STATE ELECTRONICS

A.1a. Current Equations for Electrons and Holes

In expressions (1.7) and (1.9) for the current flowing in p-type and n-type semiconductors we introduce the current densities $J_p = I_p/A$ and $J_n = I_n/A$, where A is the cross-sectional area of the semiconductor bar in question and obtain

$$J_p = e\mu_p pE; \qquad J_n = e\mu_n nE \qquad \text{(A.1)}$$

where E is the field strength; p and n the carrier densities; μ_p and μ_n the mobilities for holes and electrons, respectively; and e the absolute value of the electron charge. The current densities thus introduced are called *drift currents*, i.e., currents produced by the drift of carriers under influence of an electric field.

Besides drift currents one can also have *diffusion currents*. In this case the current flow occurs because there is a gradient in the carrier density; carriers here flow from regions of high concentration to regions of low concentration by diffusion. If the gradients are in the X-direction,* the current

---

*If $p = p(x, y, z)$, then the gradient of p is a vector with components $\partial p/\partial x$, $\partial p/\partial y$, and $\partial p/\partial z$. If p depends only on x, $\partial p/\partial y = \partial p/\partial z = 0$, and the gradient of p points in the x-direction.

densities J_p and J_n for electrons and holes are found to be

$$J_p = -eD_p\frac{\partial p}{\partial x}; \qquad J_n = eD_n\frac{\partial n}{\partial x} \tag{A.2}$$

where D_p and D_n are the hole and electron *diffusion constants*, respectively; they are expressed in square centimeters per second. We use partial derivatives here because p and n may be functions both of the coordinate x and the time t.

Since the expressions for J_p and J_n differ only in sign and this difference comes about because the holes have a charge $+e$ and the electrons a charge $-e$, it is sufficient to prove the expression for J_p. To that end we turn to Fig. A.1. Let a semiconducting bar of cross-sectional area A be divided in small

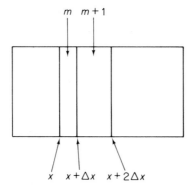

Fig. A.1. Diagram used to derive the diffusion law.

sections of length Δx. Then the number P_m of holes in the mth section Δx is $P_m = p(x)A\,\Delta x$ and the number in the $(m + 1)$th section is

$$P_{m+1} = p(x + \Delta x)A\,\Delta x = \left[p(x) + \frac{\partial p}{\partial x}\Delta x \right]A\,\Delta x$$

Since each hole has the same probability of moving from left to right as from right to left, the current $I_{m,\,m+1}$ flowing from section m to section $m + 1$ is proportional to eP_m and the current $I_{m+1,\,m}$ flowing from section $m + 1$ to section m is proportional to eP_{m+1}. The proportionality factor will be denoted by a; a is independent of x but may depend on Δx. For the net current flow I_p from left to right we thus have

$$I_p = ea(P_m - P_{m+1}) = -ea\frac{\partial p}{\partial x}A\,\Delta x^2 \tag{A.3}$$

However, I_p must be independent of the way in which the sample is divided up in slices. Therefore $a\,\Delta x^2$ must be a constant; calling this constant D_p, Eq. (A.3) may be written as

$$I_p = -eD_p\frac{\partial p}{\partial x}A \quad \text{or} \quad J_p = \frac{I_p}{A} = -eD_p\frac{\partial p}{\partial x}$$

which proves the equation for J_p.

If both an electric field E and a carrier density gradient are present, the current equations are found from a combination of Eqs. (A.1) and (A.2)

$$J_p = e\mu_p pE - eD_p \frac{\partial p}{\partial x} \qquad (A.4)$$

$$J_n = e\mu_n nE + eD_n \frac{\partial n}{\partial x} \qquad (A.5)$$

which say that in this case the currents are partly due to diffusion and partly due to drift.

Since carrier drift and carrier diffusion are both governed by collisions of carriers with the lattice, it is not surprising that the mobility μ and the diffusion constant D are related. We shall now show that for samples at the absolute temperature T

$$\frac{\mu_n}{D_n} = \frac{\mu_p}{D_p} = \frac{e}{kT} \qquad (A.6)$$

where k is Boltzmann's constant. This is called the *Einstein relation*.

We shall prove this for p-type material at equilibrium in which a gradient dN_a/dx in the acceptor concentration exists. Since all acceptors are ionized at room temperature, this sets up a hole density gradient dp/dx. This hole density gradient gives rise to a diffusion current, but since there is equilibrium, no net current can flow. Therefore an electric field must be set up in the sample that gives rise to a drift current that just balances the diffusion current everywhere. Thus from Eq. (A.4) we have

$$0 = \mu_p pE - D_p \frac{dp}{dx} \qquad (A.7)$$

However, if $\psi(x)$ is the potential at x, taken with respect to the reference point $x = 0$, we have $E = -d\psi/dx$, or

$$-\mu_p p \, d\psi = D_p \, dp \quad \text{or} \quad -\frac{\mu_p}{D_p} d\psi = \frac{dp}{p} \qquad (A.7a)$$

Integrating this equation under the conditions

$$p(x) = p(0) \quad \text{at } x = 0; \qquad \psi(x) = 0 \quad \text{at } x = 0$$

yields

$$-\frac{\mu_p}{D_p}\psi = \ln \frac{p}{p(0)} \quad \text{or} \quad p(x) = p(0)\exp\left[-\frac{\mu_p}{D_p}\psi(x)\right] \qquad (A.8)$$

However, the hole distribution $p(x)$ should be governed by a Boltzmann-type equation of the form $\exp(-E_a/kT)$, where $E_a = e\psi(x)$. That is,

$$p(x) = p(0)\exp\left[-\frac{e\psi(x)}{kT}\right] \qquad (A.8a)$$

Equating the exponents in both expressions yields Eq. (A.6).

We shall illustrate this discussion with examples.

EXAMPLE 1: In silicon $\mu_p = 480 \text{ cm}^2/\text{V-s}$ at $T = 300°\text{K}$ and $kT/e = 25.8 \times 10^{-3}$ V. Find D_p.

ANSWER: $D_p = (kT/e) \mu_p = 25.8 \times 10^{-3} \times 0.48 \times 10^3 = 12.4 \text{ cm}^2/\text{s}$.

EXAMPLE 2: In a short silicon $p^+ - n$ diode the hole density drops linearly from $10^{16}/\text{cm}^3$ to a very low value over a distance of 1 μm. Find J_p.

ANSWER:

$$\frac{dp}{dx} = -\frac{10^{16}}{10^{-4}} = -10^{20}/\text{cm}^4$$

However, according to Example 1, $D_p = 12.4 \text{ cm}^2/\text{s}$. Hence

$$J_p = -eD_p \frac{dp}{dx} = 1.6 \times 10^{-19} \times 12.4 \times 10^{20} = 19.8 \text{ A/cm}^2$$

indicating that large diffusion currents can flow if large density gradients exist.

A.1b. Continuity Equations

If in a semiconductor device an excess minority carrier concentration exists and a minority carrier current flows in the X-direction, then the minority carrier density can decrease in two ways:

1. By recombination with majority carriers.
2. More carriers are swept out of the volume ΔV than into it.

This is expressed mathematically by two equations, one for electrons and one for holes. They are called the *continuity equations* and give the rate of change of the carrier density in terms of processes 1 and 2. The equations are

1. For holes in n-type material

$$\frac{\partial p}{\partial t} = -\frac{p - p_n}{\tau_p} + \frac{1}{e}\frac{\partial J_n}{\partial x} \qquad (A.9)$$

2. For electrons in p-type material

$$\frac{\partial n}{\partial t} = -\frac{n - n_p}{\tau_n} + \frac{1}{e}\frac{\partial J_n}{\partial x} \qquad (A.10)$$

Here p and n are the minority carrier concentrations, p_n and n_p the equilibrium minority concentrations, τ_p and τ_n the lifetimes of the added minority carriers, and J_p and J_n the minority current densities. The first terms in Eqs. (A.9) and (A.10) represent process 1 and the second terms represent process 2.

Since the only difference between Eqs. (A.9) and (A.10) is a difference in the sign of the second term and since this difference comes about because the holes have a charge $+e$ and the electrons a charge $-e$, it is sufficient to prove Eq. (A.9).

The first term in Eq. (A.9) speaks for itself; it simply gives the rate of decay of the added carrier concentration $p - p_n$; if N_d is the donor density, τ_p is a constant as long as $p \ll N_d$. Therefore only the second term in Eq. (A.9) needs proof.

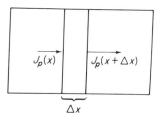

Fig. A.2. Diagram used to derive the continuity equation.

To do so, we turn to Fig. A.2, where a bar of semiconductor material of cross-sectional area A is divided up into sections Δx. We now look at the section between x and $x + \Delta x$. The current arriving at x is $J_p(x)A$, and the current leaving at $x + \Delta x$ is

$$J_p(x + \Delta x)A = \left[J_p(x) + \frac{\partial J_p}{\partial x} \Delta x \right]A$$

Therefore the rate of increase in the number of holes in the section Δx is

$$\frac{1}{e}[J_p(x) - J_p(x + \Delta x)]A = -\frac{1}{e}\frac{\partial J_p}{\partial x} A \Delta x = -\frac{1}{e}\frac{\partial J_p}{\partial x}\Delta V$$

where $\Delta V = A \Delta x$ is the volume of the section Δx. The rate of increase in the hole *density* is therefore $-(1/e)\,\partial J_p/\partial x$, which proves the second part of Eq. (A.9).

A.1c. Gauss's Law and Poisson's Equation

According to Gauss's law, a one-dimensional space-charge distribution $\rho(x, t)$ is accompanied by a field strength E such that

$$\frac{\partial E}{\partial x} = \frac{\rho(x, t)}{\epsilon \epsilon_0} \tag{A.11}$$

Since $E = -\partial \psi/\partial x$, where ψ is the potential, this equation may be written as

$$\frac{\partial^2 \psi}{\partial x^2} = -\frac{\rho(x, t)}{\epsilon \epsilon_0} \tag{A.12}$$

This is called *Poisson's equation.* Here ϵ is the relative dielectric constant of the material in question and $\epsilon_0 = 8.85 \times 10^{-14}$ F/cm, the electric conversion factor for MKS units. As soon as one knows $\rho(x, t)$, one can in principle evaluate $E(x)$ and $\psi(x)$.

In an n-type semiconductor material in which there are donors, electrons, and holes

$$\rho(x) = e(N_d + p - n) \tag{A.13}$$

and in a p-type semiconductor material in which there are acceptors, electrons, and holes

$$\rho(x) = -e(N_a + n - p) \tag{A.14}$$

Here N_d and N_a are the donor and acceptor densities and p and n the hole and electron densities, respectively.

A.1d. Dielectric Relaxation Effects

To illustrate what dielectric relaxation means, we shall consider a sample of semiconductor material of conductivity σ, dielectric constant ϵ, area A, and thickness d, with ohmic contacts made to the ends. The sample can be represented by a resistance R and a capacitance C in parallel (Fig. A.3). If now at

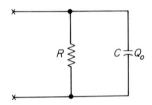

R $C \overset{}{=} Q_0$ Fig. A.3. Equivalent circuit of semiconductor slice, used to demonstrate dielectric relaxation effects.

$t = 0$ a charge Q_0 is applied to the electrodes, then the charge $Q(t)$ decays as

$$Q(t) = Q_0 \exp\left(-\frac{t}{RC}\right) = Q_0 \exp\left(-\frac{t}{\tau_d}\right) \tag{A.15}$$

The RC-time constant of the system is called the *dielectric relaxation time* τ_d of the system. We have, since

$$R = \frac{d}{\sigma A}; \quad C = \frac{\epsilon\epsilon_0 A}{d}; \quad \tau_d = RC = \frac{\epsilon\epsilon_0}{\sigma} \tag{A.15a}$$

that the dielectric relaxation time is a material constant independent of the dimensions of the material.

EXAMPLE 3: Find the dielectric relaxation time of $1\ \Omega$-cm silicon if silicon has $\epsilon = 12$.

ANSWER:

$$\tau_d = \frac{\epsilon\epsilon_0}{\sigma} = \frac{12 \times 8.85 \times 10^{-14}}{1} = 1.06 \times 10^{-12}\ \text{s}$$

This example shows that in semiconductor material normally used in semiconductor devices the dielectric relaxation time is very short; its effect shows up only at very high frequences.

Suppose, by some sort of mechanism, that a uniform excess electron density Δn is brought into n-type material; then it takes a time of the order of a few times the dielectric relaxation time τ_d before the excess electron density is distributed over the surface. Or, if by some other sort of mechanism, a uniform excess hole density Δp is brought into n-type material, then it takes a time of the order of a few times the dielectric relaxation time τ_d before the excess hole density is neutralized by the electrons; this gives rise to a positive surface charge. Afterwards the electrons and holes recombine in a time equal to a few times the minority carrier lifetime τ_p.

A.1e. Approximate Space-Charge Neutrality

In the space-charge regions of junction diodes and transistors a space charge exists which gives rise to a potential distribution in that region; this implies quite large electric fields. Outside such regions, the fields are much smaller, and only relatively small deviations from *complete* space-charge neutrality are needed to produce the necessary fields. It is then said that *approximate* space-charge neutrality exists in those regions.

First, consider an n-type sample in equilibrium that has a large gradient in the donor density $N_d(x)$. We saw that in such a case an electric field was developed to make the net current zero everywhere. According to Gauss's law, this implies that $N_d(x) - n(x)$ cannot be zero here. However, quite generally $[N_d(x) - n(x)] \ll N_d(x)$. That is what is meant by *approximate* space-charge neutrality.

Next consider a p^+-n junction diode that is so biased that holes are injected into the n-region. The hole density $p(x)$ is then approximately neutralized in a time equal to a few times the dielectric relaxation time of the n-region by electrons entering into the n-region through the ohmic contact. Approximate space-charge neutrality here means $(N_d - n + p) \ll p$, or $n \simeq N_d + p$, where N_d is the donor concentration and n the electron concentration.

A.2. JUNCTION DIODES

A.2a. Calculation of the Contact or Diffusion Potential V_{dif}

Let the space-charge region of a p-n junction extend from x_1 to x_2; then for an unbiased junction, $p(x_1) = N_a$ and $p(x_2) = p_n = n_i^2/N_d$. However, since a potential V_{dif} is developed across the space-charge region,

$$p_n = N_a \exp\left(-\frac{eV_{\text{dif}}}{kT}\right) = \frac{n_i^2}{N_d} \tag{A.16}$$

according to Eq. (A.8a). Consequently

$$\exp\left(\frac{eV_{\text{dif}}}{kT}\right) = \frac{N_a N_d}{n_i^2} \quad \text{or} \quad V_{\text{dif}} = \frac{kT}{e} \ln\left(\frac{N_a N_d}{n_i^2}\right) \tag{A.17}$$

This is called the *diffusion potential*.

EXAMPLE 1: A p^+-n junction has $N_a = 10^{19}/\text{cm}^3$ and $N_d = 10^{16}/\text{cm}^3$. What is the contact potential of the junction at $T = 300°\text{K}$ if $n_i = 10^{10}/\text{cm}^3$ at that temperature?

ANSWER:

$$V_{\text{dif}} = \frac{kT}{e} \ln\left(\frac{N_a N_d}{n_i^2}\right) = 25.8 \times 10^{-3} \ln 10^{12} = 0.89 \text{ V}$$

If now a voltage V_D is applied to the p-region and $V_D > 0$, the width of the space-charge region contracts. In this case, since a voltage $V_{\text{dif}} - V_D$ is now developed across the space-charge region, we have

$$p(x_2) = N_a \exp\left[-\frac{e(V_{\text{dif}} - V_D)}{kT}\right] = p_n \exp\left(\frac{eV_D}{kT}\right) \tag{A.18}$$

according to Eq. (A.8a). We need this expression for the evaluation of the diode current.

Finally, we shall evaluate $p(x)$ and $n(x)$ in the space-charge region of the junction diode. If $\psi(x)$ is the potential at point x in the space-charge region and $\psi(x_1)$ is taken as zero, we have, since $p(x_1) = N_a$,

$$p(x) = N_a \exp\left[-\frac{e\psi(x)}{kT}\right] \tag{A.19}$$

according to Eq. (A.8a). In the same way, since $n(x_2) = N_d$ and $\psi(x_2) = V_{\text{dif}} - V_D$,

$$n(x) = N_d \exp\left\{-\frac{e[V_{\text{dif}} - V_D - \psi(x)]}{kT}\right\} \tag{A.20}$$

Therefore $p(x)$ decreases rapidly with increasing x for $x > x_1$ and $n(x)$ decreases rapidly with decreasing x for $x < x_2$. These results are important for solving Poisson's equation in the space-charge region of a junction diode.

A.2b. Potential and Field Distribution in the Space-Charge Region

We shall evaluate the distribution of the electric potential for an abrupt junction, that is, a junction for which

$$N_a = \text{constant for } x < x_0; \qquad N_a = 0 \text{ for } x > x_0$$
$$N_d = 0 \text{ for } x < x_0; \qquad N_d = \text{constant for } x > x_0$$

We do so by solving Poisson's equation for the space-charge region:

$$\frac{d^2\psi}{dx^2} = -\frac{\rho(x)}{\epsilon\epsilon_0} \tag{A.21}$$

Figure A.4 shows the potential distribution for $V_D = 0$ (a) and $V_D \neq 0$ (b). Now, according to Eqs. (A.13) and (A.14),

$$\rho(x) = -e(N_a + n - p) \qquad \text{for } x_1 < x < x_0$$
$$= e(N_d + p - n) \qquad \text{for } x_0 < x < x_2 \tag{A.21a}$$

The space-charge distribution is shown in Fig. A.5(a). Since p and n are much smaller than either N_a or N_d for most of the space-charge region, these equa-

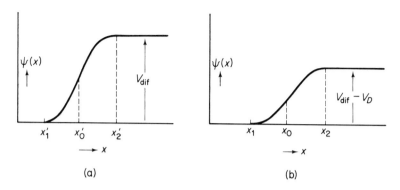

Fig. A.4. (a) Potential distribution in space-charge region of *p-n* diode for zero bias. (b) Potential distribution in space charge-region of *p-n* diode if bias V_D is applied.

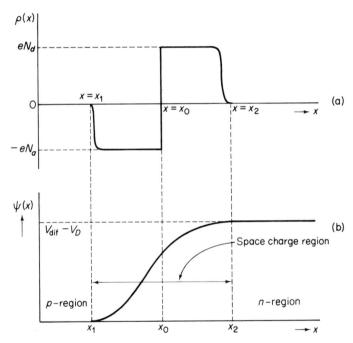

Fig. A.5. (a) Space-charge distribution in space-charge region. (b) Resulting potential distribution.

tions may be approximated as

$$\rho(x) = -eN_a \qquad \text{for } x_1 < x < x_0$$
$$\rho(x) = eN_d \qquad \text{for } x_0 < x < x_2 \qquad \text{(A.21b)}$$

Next we must settle the boundary conditions at x_1 and x_2. If we neglect the small electric fields associated with the current flow in the regions $x \leq x_1$ and $x \geq x_2$, we have the following: $d\psi/dx = 0$ for $x \leq x_1$ and $x \geq x_2$. Moreover, $\psi(x) = 0$ at $x = x_1$ and $\psi(x) = V_{\text{dif}} - V_D$ at $x = x_2$.

Poisson's equation must therefore be solved separately for the regions $x_1 < x < x_0$ and $x_0 < x < x_2$. Calling these solutions $\psi_1(x)$ for $x_1 < x < x_0$ and $\psi_2(x)$ for $x_0 < x < x_2$, respectively, we thus have

$$\frac{d^2\psi_1}{dx^2} = \frac{eN_a}{\epsilon\epsilon_0}; \qquad \frac{d^2\psi_2}{dx^2} = -\frac{eN_d}{\epsilon\epsilon_0} \tag{A.22}$$

$$\psi_1(x_1) = 0; \qquad \frac{d\psi_1}{dx} = 0 \qquad \text{at } x = x_1$$

$$\psi_2(x_2) = V_{\text{dif}} - V_D; \qquad \frac{d\psi_2}{dx} = 0 \qquad \text{at } x = x_2 \tag{A.22a}$$

Integrating once, bearing in mind that $d\psi_1/dx = 0$ at $x = x_1$ and that $d\psi_2/dx = 0$ at $x = x_2$, yields

$$\frac{d\psi_1}{dx} = \frac{eN_a}{\epsilon\epsilon_0}(x - x_1); \qquad \frac{d\psi_2}{dx} = \frac{eN_d}{\epsilon\epsilon_0}(x_2 - x) \tag{A.23}$$

Integrating once more, bearing in mind that $\psi_1(x_1) = 0$ and that $\psi_2(x_2) = V_{\text{dif}} - V_D$, yields

$$\psi_1(x) = \frac{eN_a}{2\epsilon\epsilon_0}(x - x_1)^2; \qquad \psi_2(x) = V_{\text{dif}} - V_D - \frac{eN_d}{2\epsilon\epsilon_0}(x_2 - x)^2 \tag{A.24}$$

We must now match these solutions at x_0. We must therefore require that at $x = x_0$ the fields as well as the potentials match. This yields

$$\left.\frac{d\psi_1}{dx}\right|_{x_0} = \left.\frac{d\psi_2}{dx}\right|_{x_0} \quad \text{and} \quad \psi_1(x_0) = \psi_2(x_0) \tag{A.25}$$

Applying this condition to the above solutions, from Eq. (A.23) we have

$$N_a(x_0 - x_1) = N_d(x_2 - x_0) \tag{A.26}$$

and from Eq. (A.24)

$$\frac{eN_a}{2\epsilon\epsilon_0}(x_0 - x_1)^2 = (V_{\text{dif}} - V_D) - \frac{eN_d}{2\epsilon\epsilon_0}(x_2 - x_0)^2 \tag{A.27}$$

Substitution of Eq. (A.26) into Eq. (A.27) yields immediately

$$d_p = x_0 - x_1 = \left[\frac{2\epsilon\epsilon_0(V_{\text{dif}} - V_D)}{eN_a(1 + N_a/N_d)}\right]^{1/2}$$

$$d_n = x_2 - x_0 = \left[\frac{2\epsilon\epsilon_0(V_{\text{dif}} - V_D)}{eN_d(1 + N_d/N_a)}\right]^{1/2} \tag{A.28}$$

and the thickness d of the space-charge region becomes

$$d = x_0 - x_1 + x_2 - x_0 = \left[\frac{2\epsilon\epsilon_0(V_{\text{dif}} - V_D)(N_a + N_d)}{eN_aN_d}\right]^{1/2} \tag{A.29}$$

which corresponds to Eq. (3.5). Moreover,

$$d_p = d\frac{N_d}{N_a + N_d}; \quad d_n = d\frac{N_a}{N_a + N_d} \qquad \text{(A.29a)}$$

which proves Eq. (3.5c). Figure A.5(b) shows the resulting potential distribution; it agrees with Fig. A.4(b).

We have already calculated the field distribution in the regions $x_1 < x < x_0$ and $x_0 < x < x_2$. Putting $E_1(x) = -d\psi_1/dx$ and $E_2(x) = -d\psi_2/dx$, we have

$$E(x) = -\frac{eN_a}{\epsilon\epsilon_0}(x - x_1) \qquad \text{for } x_1 < x < x_0;$$

$$\qquad\qquad\qquad\qquad\qquad\qquad\qquad\qquad \text{(A.30)}$$

$$E(x) = -\frac{eN_d}{\epsilon\epsilon_0}(x_2 - x) \qquad \text{for } x_0 < x < x_2$$

so that the field distribution is linear in x. The maximum value of $|E|$ occurs at $x = x_0$, where

$$E(x_0) = -\left[\frac{2eN_aN_d(V_{\text{dif}} - V_D)}{\epsilon\epsilon_0(N_a + N_d)}\right]^{1/2} = -2\frac{(V_{\text{dif}} - V_D)}{d} \qquad \text{(A.31)}$$

This is twice the value of the *average field*—$(V_{\text{dif}} - V_D)/d$—and comes about because $E(x)$ is linear in x. Figure A.6 shows this field distribution.

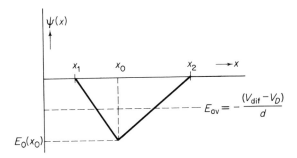

Fig. A.6. Field distribution in space-charge region.

If the junction has a cross-sectional area A, then there is a charge $+Q$ on the n-side of the junction ($x_0 < x < x_2$) and a charge $-Q$ on the p-side of the junction ($x_1 < x < x_0$), where

$$Q = eN_d(x_2 - x_0)A = \left[2e\epsilon\epsilon_0(V_{\text{dif}} - V_D)\frac{N_aN_d}{N_a + N_d}\right]^{1/2}A \qquad \text{(A.32)}$$

so that the small-signal capacitance of the space-charge region is

$$C = -\frac{dQ}{dV_D} = \left[\frac{\epsilon\epsilon_0}{2(V_{\text{dif}} - V_D)}\frac{N_aN_d}{N_a + N_e}\right]^{1/2}A = \frac{\epsilon\epsilon_0 A}{d} \qquad \text{(A.33)}$$

corresponding to Eq. (3.3).

EXAMPLE 2: A silicon p^+-n diode has $N_a = 10^{19}/cm^3$ and $N_d = 10^{16}/cm^3$. Find how the field at x_0 depends on the applied voltage V_D. Assume that $\epsilon = 12$ and take $\epsilon_0 = 8.85 \times 10^{-14}$ F/cm.

ANSWER: We saw in Section 3.1a that in this case $d = d_0(V_{dif} - V_D)^{1/2}$ and that for this example $d_0 = 0.37$ μm. Therefore

$$E_0 = -\frac{2}{d_0}(V_{dif} - V_D)^{1/2} = -5.4 \times 10^4(V_{dif} - V_D)^{1/2} \text{ V/cm}$$

EXAMPLE 3: Find the value of V_D for which $E_0 = -250,000$ V/cm.

ANSWER:

$$(V_{dif} - V_D)^{1/2} = \frac{2.5 \times 10^5}{5.4 \times 10^4} = 4.63; \qquad (V_{dif} - V_D) = 21.4 \text{ V}$$

Now

$$V_{dif} = \frac{kT}{e} \ln\left(\frac{N_aN_d}{n_i^2}\right) = 25.8 \times 10^{-3} \ln 10^{15} = 0.90 \text{ V}$$

since silicon has $n_i = 10^{10}/cm^3$ at $T = 300°K$. Therefore $0.9 - V_D = 21.4$; $V_D = -20.5$ V.

A.2c. Avalanche Breakdown

If the field strength in a back-biased silicon diode reaches a value of a few hundred-thousand volts per centimeter, avalanche multiplication sets in, in which the electrons or holes that make up the back current gain enough energy to create hole-electron pairs. The constituents of these pairs are both again accelerated and can create secondary hole-electron pairs and so forth. As a consequence an avalanche of hole-electron pairs is generated that can cause breakdown. We shall investigate this for a p^+-n diode; for the sake of simplicity we shall assume that the electron and the hole have equal ionizing power.

Let $-I_{rs}$ be the back current without multiplication; it is due to holes generated in the n-region and collected by the p-region. Let the hole, when traversing the space-charge region, produce p_{1h} hole-electron pairs and let each pair, in turn, produce p_2 secondary hole-electron pairs and so forth. Then the total current I is

$$I = -I_{rs}[1 + p_{1h}(1 + p_2 + p_2^2 + p_2^3 + \cdots)]$$
$$= -I_{rs}\left(1 + \frac{p_{1h}}{1 - p_2}\right) \qquad \text{(A.34)}$$

where p_{1h} and p_2 increase with increasing back bias $|V_D|$. There will therefore be a value of $|V_D|$ for which $p_2 = 1$. In that case I becomes infinite, a true breakdown occurs, and the current must be controlled by the external circuit. The value of $|V_D|$ at which this breakdown occurs is called the *breakdown voltage* V_b (see Section 3.1b).

Actually the electrons and the holes do not have equal ionizing power. Qualitatively this does not change anything, except that the breakdown will occur at a voltage somewhat different from the value expected from Eq. (A.34). When one of the components does not have any ionizing power whatsoever or is taken out of circulation before it can ionize, there is still avalanche multiplication, but it cannot result in a true breakdown.

A.3. CURRENT FLOW IN p-n JUNCTIONS

A.3a. Diffusion Equations

In *p-n* junctions and transistors in which one has carrier injection, it usually happens that for the injected carriers the diffusion term in the current density equations is much larger than the drift term. The latter can then be neglected and J_p and J_n become

$$J_p = -eD_p\frac{\partial p}{\partial x}$$

$$J_n = +eD_n\frac{\partial n}{\partial x}$$

(A.35)

Only in the case of very high injection can it happen that the drift terms become important.

If now Eq. (A.35) are substituted in the continuity Eqs. (A.9) and (A.10), one obtains

$$\frac{\partial p}{\partial t} = -\frac{p - p_n}{\tau_p} + D_p\frac{\partial^2 p}{\partial x^2}$$

$$\frac{\partial n}{\partial t} = -\frac{n - n_p}{\tau_n} + D_n\frac{\partial^2 n}{\partial x^2}$$

(A.36)

These equations are called the *time-dependent diffusion equations*. Under steady-state conditions, $\partial p/\partial t = \partial n/\partial t = 0$; Eqs. (A.36) then reduce to the *steady-state diffusion equations*

$$D_p\frac{d^2 p}{dx^2} = \frac{p - p_n}{\tau_p}; \qquad \frac{d^2 p}{dx^2} = \frac{p - p_n}{L_p^2}$$

$$D_n\frac{d^2 n}{dx^2} = \frac{n - n_p}{\tau_n}; \qquad \frac{d^2 n}{dx^2} = \frac{n - n_p}{L_n^2}$$

(A.37)

where $L_p^2 = D_p\tau_p$ and $L_n^2 = D_n\tau_n$. The parameters L_n and L_p are called the *diffusion lengths* for electrons and holes, respectively.

We shall apply these equations to the current flow in *p-n* junctions and transistors.

EXAMPLE 1: Holes in silicon have $D_p = 12.4 \text{ cm}^2/\text{sec}$. If $\tau_p = 10^{-6}$ s, find L_p.

ANSWER: $L_p = (D_p\tau_p)^{1/2} = (12.4 \times 10^{-6})^{1/2} = 35 \times 10^{-4}$ cm.

A.3b. Current Flow in Junction Diodes

First we shall consider the flow of holes in a *p-n* junction under influence of a bias voltage V_D applied to the *p*-region of the diode. We shall assume that the width w_n of the *n*-region is at least a few times the diffusion length L_p for the injected holes. We then have

$$\frac{d^2p}{dx^2} = \frac{p - p_n}{L_p^2} \tag{A.37a}$$

We take the origin of the coordinate system at the boundary between the space-charge region and the *n*-region [Fig. A.5(a)] and take as boundary conditions

$$p(0) = p_n \exp\left(\frac{eV_D}{kT}\right); \qquad p(\infty) = p_n \tag{A.37b}$$

The first condition corresponds to Eq. (A.18), and the second is equivalent to saying that the ohmic contact is sufficiently far away. This case is called the *long* diode case.

Since $\exp(x/L_p)$ and $\exp(-x/L_p)$ are solutions of Eq. (A.37a), the most general solution is

$$p(x) - p_n = A \exp\left(-\frac{x}{L_p}\right) + B \exp\left(\frac{x}{L_p}\right) \tag{A.38}$$

Applying the condition at $x = \infty$ yields $B = 0$; applying the condition at $x = 0$ yields $A = p(0) - p_n$. The full solution of Eq. (A.37a) satisfying the boundary conditions (A.37b) is therefore

$$p(x) - p_n = [p(0) - p_n] \exp\left(-\frac{x}{L_p}\right) \tag{A.39}$$

We now find the hole current density as

$$J_p(x) = -eD_p \frac{dp}{dx} = eD_p \frac{[p(0) - p_n]}{L_p} \exp\left(-\frac{x}{L_p}\right) \tag{A.40}$$

so that $J_p(x)$ has its maximum value at $x = 0$ and decreases exponentially with increasing x. The total current, however, is constant, so that the difference $J_p(0) - J_p(x)$ must be made up of electrons coming from the ohmic contact and recombining with holes on the way (Fig. A.7). If no electrons would be injected into the *p*-region, the total current would thus be, if A is the junction area,

$$I_D = J_p(0)A = \frac{eD_p[p(0) - p_n]A}{L_p} = \frac{eD_p p_n A}{L_p}\left[\exp\left(\frac{eV_D}{kT}\right) - 1\right] \tag{A.41}$$

in qualitative agreement with Eq. (2.9), so that

$$I_{rs} = I_{rsp} = \frac{eD_p p_n A}{L_p} = \frac{eD_p n_i^2 A}{L_p N_d} = \frac{en_i^2 A}{N_d}\left(\frac{D_p}{\tau_p}\right)^{1/2} \tag{A.41a}$$

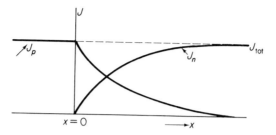

Fig. A.7. Current distribution in p^+-n diode.

In the same way we find that if all current is carried by electrons,

$$I_D = \frac{eD_n n_p A}{L_n}\left[\exp\left(\frac{eV_D}{kT}\right) - 1\right] \tag{A.42}$$

in qualitative agreement with Eq. (2.9), where

$$I_{rs} = I_{rsn} = \frac{eD_n n_p A}{L_n} = \frac{eD_n n_i^2 A}{L_n N_a} \tag{A.42a}$$

When both electrons and holes contribute to the current flow, the resulting current is

$$I_D = (I_{rsp} + I_{rsn})\left[\exp\left(\frac{eV_D}{kT}\right) - 1\right] \tag{A.43}$$

In p^+-n diodes $N_d \ll N_a$ and hence $I_{rsp} \gg I_{rsn}$, so that practically all current is carried by holes. In n^+-p diodes $N_a \ll N_d$ and hence $I_{rsn} \gg I_{rsp}$, so that practically all current is carried by electrons. The derivation in Section 2.1b has thus been fully verified.

To verify Eq. (3.12), we evaluate the stored hole charge Q_s with the help of Eq. (A.39). Obviously

$$Q_s = eA \int_0^\infty [p(x) - p_n] \, dx = eA[p(0) - p_n] \int_0^\infty \exp\left(-\frac{x}{L_p}\right) dx$$

$$= eA[p(0) - p_n]L_p = \frac{eD_p[p(0) - p_n]A}{L_p}\frac{L_p^2}{D_p} = I_D\tau \tag{A.44}$$

which proves Eq. (3.12).

EXAMPLE 2: A p^+-n diode has $D_p = 12.5$ cm²/s, $\tau_p = 1.25 \times 10^{-7}$ s, $N_d = 10^{16}$/cm³, $n_i = 10^{10}$/cm³ at $T = 300°$K, and $A = 10^{-4}$ cm². Find I_{rs} at $T = 300°$K.

ANSWER:

$$I_{rs} = \frac{en_i^2 A}{N_d}\left(\frac{D_p}{\tau_p}\right)^{1/2} = \frac{1.6 \times 10^{-19} \times 10^{20} \times 10^{-4}}{10^{16}}\left(\frac{12.5}{1.25 \times 10^{-7}}\right)^{1/2}$$

$$= 1.6 \times 10^{-15} \text{ A}$$

which is not too far from the value assumed in Section 2.1a.

As a preliminary to the calculation of the current flow in a *p-n-p* transistor, we shall now evaluate the current in a p^+-n diode with a very short n-region ($w_n \ll L_p$), so that practically no recombination occurs in the n-region. We must now specify what occurs at the ohmic contact to the n-region. We shall assume that it is a perfect sink for holes, so that $p(w_n) = p_n$ at all times.

If there is no recombination in the n-region, we have that

$$I_p(x) = -eD_pA\frac{dp}{dx} \tag{A.45}$$

is independent of x, so that $I_p(x) = I_D$, and dp/dx is a constant. Since $p(x) = p(0)$ at $x = 0$ and $p(x) = p_n$ at $x = w_n$,

$$\frac{dp}{dx} = -\frac{p(0) - p_n}{w_n}; \qquad I_D = \frac{eD_p[p(0) - p_n]A}{w_n} \tag{A.46}$$

This differs from Eq. (A.41) only in that L_p is replaced by the shorter length w_n. Therefore the length of the n-region only changes I_{rs} but does not alter the voltage dependence of I_D.

The diode has a much improved pulse response in comparison with the long diode ($w_n \gg L_p$), however, because the time constant involved in the current flow is now the diffusion time τ_d across the n-region rather than the lifetime τ_p of the carriers. As will be shown in Section A.5a, τ_d is given by

$$\tau_d = \frac{w_n^2}{2D_p} \tag{A.47}$$

EXAMPLE 3: Find τ_d for a silicon p^+-n diode having an n-region of 10^{-4} cm (1-μm) width. $D_p = 12.5$ cm^2/s.

ANSWER:

$$\tau_d = \frac{w_n^2}{2D_p} = \frac{10^{-8}}{25} = 4 \times 10^{-10}\text{s}$$

A.4. JUNCTION FET

We shall give here a more detailed derivation of the characteristic of a junction FET with n-type channel. To simplify the discussion, we shall use the geometry discussed in Chapter 5, in that we shall consider a thin slab of n-type material with two p^+-regions diffused in, one on each side, which are tied together externally to form the gate G (Fig. A.8). Two ohmic contacts are made to the side; one is the source S and the other the drain D.

We shall first evaluate the pinch-off voltage V_p of the channel by putting $V_{DS} = 0$. If V_{GS} is the voltage applied between gate and channel and $2a$ is the width of the n-region itself, then the thickness of the conducting part of the n-region will be $b = 2a - 2d$, where d is the width of the space-charge region

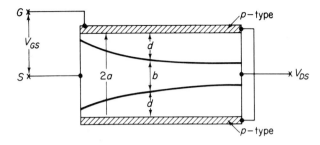

Fig. A.8. Cross section of JFET showing space-charge region and conducting channel.

of each p^+-n diode; since the diode is p^+-n, the space-charge region extends almost exclusively into the n-region.

As shown in Chapter 3, in this case we have

$$d = d_0(V_{\text{dif}} - V_{GS})^{1/2}; \qquad d_0 = \left(\frac{2\epsilon\epsilon_0}{eN_d}\right)^{1/2} \tag{A.48}$$

where V_{dif} is the diffusion potential of the junction. Consequently the thickness b of the conducting channel is

$$b = 2a - 2d = 2a - 2d_0(V_{\text{dif}} - V_{GS})^{1/2} \tag{A.49}$$

The pinch-off voltage V_P is now so defined that $b = 0$ at $V_{GS} = V_P$. This yields

$$(V_{\text{dif}} - V_P)^{1/2} = \frac{a}{d_0}; \qquad V_{\text{dif}} - V_P = \frac{a^2}{d_0^2} = \frac{a^2 e N_d}{2\epsilon\epsilon_0} \tag{A.50}$$

Therefore, for $V_{GS} > V_P(V_P$ is, of course, negative) the thickness of the conducting channel is

$$b = 2a\left(1 - \frac{d}{a}\right) = 2a\left[1 - \left(\frac{V_{\text{dif}} - V_{GS}}{V_{\text{dif}} - V_P}\right)^{1/2}\right] \tag{A.51}$$

since a/d_0 is given by Eq. (A.50). Consequently, since $\sigma = e\mu_n N_d$ is the conductivity of the channel, the conductance g per unit length of the channel is; if w is its width,

$$g = e\mu_n N_d bw = 2e\mu_n N_d aw\left[1 - \left(\frac{V_{\text{dif}} - V_{GS}}{V_{\text{dif}} - V_P}\right)^{1/2}\right] \tag{A.52}$$

We shall now turn on the voltage V_{DS}. As a consequence a voltage $V(x)$ is developed in the conducting channel; $V(x) = 0$ at the source S and $V(x) = V_{DS}$ at the drain D. The voltage between the gate and the channel at any point x is therefore $V_{GS} - V(x)$. We assume that $V_{GS} - V_{DS} < V_P$; the channel is then nowhere pinched off.

It is now usually assumed that Eqs. (A.51) and (A.52) remain valid, provided that it is taken into account that V_{GS} must be replaced by the actual voltage $V_{GS} - V(x)$ between the gate and the channel at x. If one does this, one finds that

$$b(x) = 2a\left[1 - \left(\frac{V_{\text{dif}} - V_{GS} + V(x)}{V_{\text{dif}} - V_P}\right)^{1/2}\right] \tag{A.51a}$$

$$g(x) = 2e\mu_n N_d aw\left[1 - \left(\frac{V_{\text{dif}} - V_{GS} + V(x)}{V_{\text{dif}} - V_P}\right)^{1/2}\right] \tag{A.52a}$$

This is called the *gradual channel approximation*.

When one makes a more accurate calculation, by solving the two-dimensional potential problem in the space-charge region between the channel and gate accurately, one comes to the conclusion that one may approximate Eq. (A.51a) as

$$b(x) = 2a\left[1 - \left(\frac{V_{\text{dif}} - V_{GS} + V(x)}{V_{\text{dif}} - V_P}\right)\right] = \frac{2a}{V_{\text{dif}} - V_P}[V_{GS} - V_P - V(x)] \tag{A.53}$$

so that the parameter b_0 introduced in Chapter 5 is given as

$$b_0 = \frac{2a}{V_{\text{dif}} - V_P} \tag{A.53a}$$

Saturation occurs if $b(x) = 0$ at $x = L$,

$$V_{\text{dif}} - V_{GS} + V_{DS} = V_{\text{dif}} - V_P \quad \text{or} \quad V_{DS} = V_{GS} - V_P \tag{A.53b}$$

We have now arrived at the basic starting point for the derivation of the quadratic characteristic of the JFET used in Chapter 5.

If one uses Eq. (A.52a), one obtains a somewhat different kind of characteristic. The resistance of a section Δx is $\Delta x/g(x)$ and hence, if $\Delta V(x)$ is the voltage developed across the section Δx, the current I_D flowing from drain to source is

$$I_D = \frac{g(x)}{\Delta x}\Delta V(x) \quad \text{or} \quad I_D \Delta x = g(x)\Delta V \tag{A.54}$$

Therefore, bearing in mind that I_D is constant throughout the channel, substituting for $g(x)$, and integrating over the length L of the channel,

$$\int_0^L I_D\, dx = I_D L = \int_0^{V_{DS}} 2e\mu_n N_d aw\left[1 - \left(\frac{V_{\text{dif}} - V_{GS} + V}{V_{\text{dif}} - V_P}\right)^{1/2}\right] dV$$

$$= 2e\mu_n N_d aw\left[V_{DS} - \frac{2}{3}\frac{(V_{\text{dif}} - V_{GS} + V_{DS})^{3/2}}{(V_{\text{dif}} - V_P)^{1/2}}\right.$$

$$\left. + \frac{2}{3}\frac{(V_{\text{dif}} - V_{GS})^{3/2}}{(V_{\text{dif}} - V_P)^{1/2}}\right]$$

so that

$$I_D = \frac{2e\mu_n N_d aw}{L}\left[V_{DS} - \frac{2}{3}\frac{(V_{\text{dif}} - V_{GS} + V_{DS})^{3/2}}{(V_{\text{dif}} - V_P)^{1/2}} + \frac{2}{3}\frac{(V_{\text{dif}} - V_{GS})^{3/2}}{(V_{\text{dif}} - V_P)^{1/2}}\right] \tag{A.55}$$

Pinch off sets in when $V_{DS} = V_{GS} - V_P$. Beyond pinch off, I_D stays practically constant, so that for $V_{DS} > V_{GS} - V_P$

$$I_D = \frac{2e\mu_n N_d a w}{L}\left[V_{GS} - V_P - \frac{2}{3}(V_{dif} - V_P) + \frac{2}{3}\frac{(V_{dif} - V_{GS})^{3/2}}{(V_{dif} - V_P)^{1/2}}\right] \quad (A.56)$$

In Chapter 5 these complicated expressions for I_D were replaced by simpler quadratic ones. It should be checked, of course, how good these approximations really are.

A.5. CURRENT FLOW IN TRANSISTORS

Since *p-n-p* transistors and *n-p-n* transistors differ only in the polarity and the directions of current flow, we shall discuss only one of them in detail. We have chosen the *p-n-p* transistor.

The emitter of a transistor is always much more heavily doped than the base. Therefore in a *p-n-p* transistor the emitter-base diode is a p^+-*n* diode, so that practically all current is carried by holes. Theoretically, there is also an electron current injected from the base into the emitter, but we shall see that the effect is often so small that it can be neglected.

In our calculations we shall put the *X*-axis of our coordinate system perpendicular to the emitter junction, assume the current flow to be one-dimensional (= current flow in parallel to the *X*-axis), and denote the width of the base region by *w*. The origin of the coordinate system is put at the boundary between the base region and the emitter-base space-charge region ($x = 0$); the boundary between the base region and the collector-base space-charge region is then at $x = w$ [Fig. A.9(a)]. To determine *w*, one should take the full width over which ionized donors exist and then subtract the parts of the space-charge regions of the two junctions that extend into the base. Therefore *w* depends somewhat on the bias conditions, especially on the collector-base voltage V_{CB}. We shall see that this is the cause of the collector conductance.

A.5a. Device Characteristic

We shall now assume that a voltage V_{EB} is applied to the emitter-base junction and that the voltage V_{CB} applied to the collector-base junction is first kept at zero. We then have as boundary conditions for the hole density

$$p(0) = N_a \exp\left[-\frac{e(V_{dif} - V_{EB})}{kT}\right] = p_n \exp\left(\frac{eV_{EB}}{kT}\right); \qquad p(w) = p_n \quad (A.57)$$

In good transistors there is very little recombination in the base region. This means that the current density in the base, at least in the case of one-dimensional current flow, is practically constant. Since the current in the base region flows by diffusion, it follows that dp/dx is practically independent of *x* and

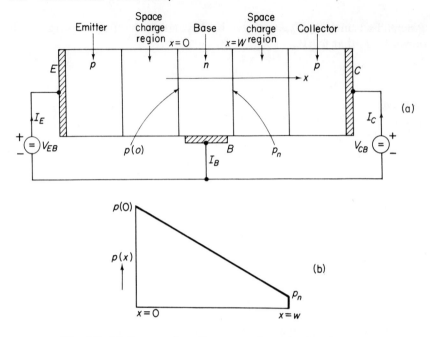

Fig. A.9. (a) Cross section of *p-n-p* transistor showing the space-charge regions. (b) Hole distribution in the base of a *p-n-p* transistor with zero voltage applied to the collector-base junction.

therefore equal to $-[p(0) - p(w)]/w$ [Fig. A.9(b)]. Consequently, if A is the emitter area, the emitter current I_E is

$$I_E = -eD_pA\frac{dp}{dx} = \frac{eD_pA}{w}[p(0) - p_n] = \frac{eD_pAp_n}{w}\left[\exp\left(\frac{eV_{EB}}{kT}\right) - 1\right]$$

$$= I_{ES}\left[\exp\left(\frac{eV_{EB}}{kT}\right) - 1\right] \tag{A.58}$$

where

$$I_{ES} = \frac{eD_pAp_n}{w} = \frac{eD_pAn_i^2}{wN_d} \tag{A.58a}$$

This current flows *into* the emitter contact.

To calculate the base current I_B, we first determine the excess hole charge Q_B stored in the base region. As is seen from Fig. A.9(b),

$$Q_B = e\cdot\tfrac{1}{2}[p(0) - p_n]Aw \tag{A.59}$$

since $\tfrac{1}{2}[p(0) - p_n]$ is the average excess hole charge in the base region and Aw is the base volume. This charge is compensated by an equal but opposite electron charge injected into the base through the ohmic contact to the base region. One can now assign a lifetime τ_p to these stored holes, and hence the recombination hole current disappearing in the base is Q_B/τ_p. The electrons that are eliminated by recombination must be replenished by a current flow through the ohmic contact; therefore the current flowing *out of* the base region is Q_B/τ_p, so that

$$-I_B = \frac{Q_B}{\tau_p} = \frac{1}{2}\frac{eAw[p(0) - p_n]}{\tau_p} = \frac{1}{2}\frac{eAwp_n}{\tau_p}\left[\exp\left(\frac{eV_{EB}}{kT}\right) - 1\right] \quad \text{(A.60)}$$

This may be written as

$$-I_B = \frac{w^2}{2D_p\tau_p}\frac{eD_pA[p(0) - p_n]}{w} = \frac{\tau_d}{\tau_p}I_E = (1 - \alpha_F)I_E \quad \text{(A.60a)}$$

where α_F is the forward current amplification factor

$$\tau_d = \frac{w^2}{2D_p}; \quad 1 - \alpha_F = \frac{\tau_d}{\tau_p} = \frac{1}{2}\frac{w^2}{L_p^2}; \quad \alpha_F = 1 - \frac{\tau_d}{\tau_p} \quad \text{(A.60b)}$$

and $L_p^2 = D_p\tau_p$. The collector current I_C flowing into the collector is therefore

$$I_C = -I_E - I_B = -\alpha_F I_E = -\left(1 - \frac{\tau_d}{\tau_p}\right)I_E \quad \text{(A.61)}$$

and the current amplification factor in the common emitter connection is

$$\beta_F = \frac{\alpha_F}{1 - \alpha_F} = \frac{1 - \tau_d/\tau_p}{\tau_d/\tau_p} = \frac{\tau_p}{\tau_d} - 1 \quad \text{(A.62)}$$

The meaning of the parameter τ_d can be illustrated as follows. According to Einstein, the mean square distance traveled by a particle in the X-direction by diffusion during the time t is

$$\overline{x^2} = 2Dt \quad \text{(A.60c)}$$

where D is the diffusion constant of the particles. Therefore τ_d is the average diffusion time of carriers through the base region.

EXAMPLE 1: A silicon p-n-p transistor has a base width of 10^{-4} cm. The diffusion constant for holes in silicon is 12.5 cm²/s and the lifetime of the holes in the base region is 10^{-7} s. Find τ_d and β_F.

ANSWER:

$$\tau_d = \frac{w^2}{2D_p} = \frac{10^{-8}}{2 \times 12.5} = 4 \times 10^{-10} \text{ s}$$

$$\beta_F = \frac{\tau_p}{\tau_d} - 1 = 249$$

EXAMPLE 2: Find I_{ES} for the transistor in Example 1 if $A = 10^{-5}$ cm², $N_d = 10^{16}/\text{cm}^3$ for the base region, and $n_i = 10^{10}/\text{cm}^3$ at $T = 300°$K.

ANSWER:

$$I_{ES} = \frac{eD_pAn_i^2}{wN_d} = \frac{1.6 \times 10^{-19} \times 12.5 \times 10^{-5} \times 10^{20}}{10^{-4} \times 10^{16}}$$

$$= 2 \times 10^{-15} \text{ A}$$

EXAMPLE 3: Find the turn-on voltage of the transistor in Example 2, i.e., the value of V_{EB} for which I_E is 1 mA.

ANSWER:

$$I_E = I_{ES} \exp\left(\frac{eV_{EB}}{kT}\right); \qquad V_{EB} = \frac{kT}{e} \ln\left(\frac{I_E}{I_{ES}}\right)$$
$$= 25.8 \times 10^{-3} \ln(5 \times 10^{11}) = 0.694 \text{ V}$$

Roughly speaking, silicon transistors thus have turn-on voltages of the order of 0.70 V.

The above theory does not fully apply to actual transistors, since they are usually not one-dimensional; generally, the collector junction is much larger than the emitter junction. This does not affect normal operation of the transistor too much but is quite significant for transistors operated in the reverse manner.

A.5b. Effects of the Electron Current and the Recombination Current in the Emitter Space-Charge Region

We shall first consider the effect of the electron current I_{En}, injected from the base into the emitter (Fig. A.10). We must now split the emitter current into

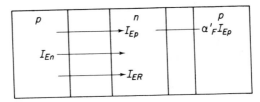

Fig. A.10. Current diagram, showing the effects of electron injection into the emitter and of recombination in the emitter-base space-charge region.

a part I_{Ep} due to holes, given by Eq. (A.58), and a part I_{En} due to electrons. According to Eq. (A.58),

$$I_{Ep} = \frac{eD_p A n_i^2}{wN_d}\left[\exp\left(\frac{eV_{EB}}{kT}\right) - 1\right] \qquad \text{(A.63)}$$

and according to diode theory (Section A.3b),

$$I_{En} = \frac{eD_n A n_i^2}{L_n N_a}\left[\exp\left(\frac{eV_{EB}}{kT}\right) - 1\right] \qquad \text{(A.64)}$$

We shall now introduce the collector efficiency $\alpha_F' = -I_C/I_{Ep} = 1 - \frac{1}{2}w^2/L_p^2$ for holes and define the current amplification factor α_F again by $\alpha_F = -I_C/I_E$. For $I_{En} \ll I_{Ep}$ we then have

$$\alpha_F = -\frac{I_C}{I_E} = \frac{\alpha_F' I_{Ep}}{I_{Ep} + I_{En}} = \frac{\alpha_F'}{1 + I_{En}/I_{Ep}} \simeq \alpha_F'\left(1 - \frac{I_{En}}{I_{Ep}}\right) \simeq \alpha_F' - \frac{I_{En}}{I_{Ep}}$$
$$\text{(A.65)}$$

since $1/(1 + x) \simeq 1 - x$ for small x; it is furthermore assumed that α'_F is sufficiently close to unity. We may thus write

$$1 - \alpha_F = 1 - \alpha'_F + \frac{I_{En}}{I_{Ep}} = \frac{1}{2} \frac{w^2}{L_p^2} + \frac{I_{En}}{I_{Ep}} \tag{A.65a}$$

The effect of the electron flow is therefore negligible if

$$\frac{1}{2} \frac{w^2}{L_p^2} \gg \frac{I_{En}}{I_{Ep}} \tag{A.66}$$

Substitution of Eqs. (A.63) and (A.64) into Eq. (A.66) yields

$$\frac{1}{2} \frac{w^2}{L_p^2} \gg \frac{D_n N_d w}{L_n D_p N_a} \quad \text{or} \quad \frac{1}{2} \frac{w}{L_p} \gg \frac{D_n L_p N_d}{D_p L_n N_a} \tag{A.66a}$$

This can be achieved by choice of the impurity ratio N_a/N_d, at least for currents that are not too large.

At large current densities, however, I_{En} increases strongly with respect to I_{Ep} for reasons that cannot be discussed here. Therefore α_F can be constant at lower currents but decreases at higher currents.

Another effect that can be quite significant at low currents is caused by recombination in the emitter space-charge region. Let this recombination current be denoted by I_{ER}; then $I_E = I_{Ep} + I_{ER}$. If, again, $I_C = -\alpha'_F I_{Ep}$, for $I_{ER} \ll I_{Ep}$, in analogy with Eq. (A.65) (see Fig. A.10), we have

$$\alpha_F = \alpha'_F - \frac{I_{ER}}{I_{Ep}} \tag{A.67}$$

It now turns out that I_{ER}/I_{Ep} is small in comparison with $1 - \alpha'_F$ at large currents but that it can appreciably increase at smaller currents. Also, the effect is much more pronounced at lower temperatures such as liquid nitrogen temperature. By proper design of the emitter-base junction the effect can be relatively small, even at emitter currents as low as 1 μA. For other transistors the effect is important, however, and causes a significant decrease in α_F when going to low currents.

Because of the two effects combined, α_F usually first increases with increasing I_E at low values of I_E, passes through a maximum at intermediate values of I_E, and decreases at high values of I_E. Since $\beta_F = \alpha_F/(1 - \alpha_F)$, the effect is much more pronounced in β_F.

The current dependence of α_F affects the ac operation of the transistor. We shall demonstrate this by calculating the variation ΔI_C in I_C due to a variation ΔI_E in I_E:

$$I_C = -\alpha_F I_E; \quad \Delta I_C = -\left(\alpha_F + I_E \frac{d\alpha_F}{dI_E} \right) \Delta I_E = -\alpha_f \Delta I_E \tag{A.68}$$

where

$$\alpha_f = \alpha_F + I_E \frac{d\alpha_F}{dI_E} \tag{A.69}$$

Therefore $\alpha_f > \alpha_F$ if $d\alpha_F/dI_E > 0$ and $\alpha_f < \alpha_F$ if $d\alpha_F/dI_E < 0$. The latter occurs at high currents. We have already established a similar relationship between the ac current amplification factor β_f and the dc current amplification factor β_F in Eq. (7.71b).

A.5c. Ebers-Moll Equations

We are now able to derive the Evers-Moll equations in greater detail. We shall do so for a *p-n-p* transistor. The current I_B is again assumed to flow *into* the base and the current I_C flows *into* the collector (Fig. A.11). For $V_{EB} \neq 0$,

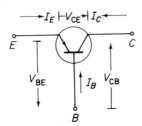

Fig. A.11. Voltages and currents in a *p-n-p* transistor.

$V_{CB} = 0$, we then have

$$I_E = I_{ES}\left[\exp\left(\frac{eV_{EB}}{kT}\right) - 1\right]; \qquad I_C = -\alpha_F I_{ES}\left[\exp\left(\frac{eV_{EB}}{kT}\right) - 1\right] \qquad (A.70)$$

In the same way, if $V_{CB} \neq 0$, $V_{EB} = 0$, we have

$$I_C = I_{CS}\left[\exp\left(\frac{eV_{CB}}{kT}\right) - 1\right]; \qquad I_E = -\alpha_R I_{CS}\left[\exp\left(\frac{eV_{CB}}{kT}\right) - 1\right] \qquad (A.71)$$

Here the reverse current amplification α_R is often much smaller than the forward current amplification factor α_F for two reasons:

1. The collector junction is often much larger than the emitter junction, so that for reverse operation a much larger current flows to the base than for normal operation, resulting in a reduction of α_R.
2. The collector region is often not very heavily doped, at least not in the vicinity of the base region; therefore I_{CS} contains a much larger contribution due to electron injection from the base than I_{ES}, again leading to a considerable reduction of α_R.

If both $V_{EB} \neq 0$ and $V_{CB} \neq 0$, by combination of Eqs. (A.70) and (A.71) we have

$$I_E = I_{ES}\left[\exp\left(\frac{eV_{EB}}{kT}\right) - 1\right] - \alpha_R I_{CS}\left[\exp\left(\frac{eV_{CB}}{kT}\right) - 1\right] \qquad (A.72)$$

$$I_C = -\alpha_F I_{ES}\left[\exp\left(\frac{eV_{EB}}{kT}\right) - 1\right] + I_{CS}\left[\exp\left(\frac{eV_{CB}}{kT}\right) - 1\right] \qquad (A.73)$$

$$I_B = -(1 - \alpha_F)I_{ES}\left[\exp\left(\frac{eV_{EB}}{kT}\right) - 1\right] - (1 - \alpha_R)I_{CS}\left[\exp\left(\frac{eV_{CB}}{kT}\right) - 1\right]$$

$$\text{(A.74)}$$

since $I_B + I_C + I_E = 0$ (Fig. A.11).

We shall now show that

$$\alpha_F I_{ES} = \alpha_R I_{CS} \tag{A.75}$$

This is called the *Onsager relation* for transistors.

The proof is simple. For $|eV_{EB}/kT| \ll 1$ and $|eV_{CB}/kT| \ll 1$, and bearing in mind that $-1 + \exp x = x$ for small x, Eqs. (A.72) and (A.73) may be written as

$$I_E = I_{ES}\frac{eV_{EB}}{kT} - \alpha_R I_{CS}\frac{eV_{CB}}{kT} \tag{A.72a}$$

$$I_C = -\alpha_F I_{ES}\frac{eV_{EB}}{kT} + I_{CS}\frac{eV_{CB}}{kT} \tag{A.73a}$$

However, these equations represent a passive network which shows reciprocity,* i.e.,

$$\frac{e\alpha_F I_{ES}}{kT} = \frac{e\alpha_R I_{CS}}{kT}$$

which proves Eq. (A.75).

We shall also show that there is a collector-saturated current I_{CB0} flowing in the collector lead for zero emitter current if the collector is back-biased so that $\exp(eV_{CB}/kT) \simeq 0$. In this case, Eqs. (A.72) and (A.73) become

$$I_E = I_{ES}\left[\exp\left(\frac{eV_{EB}}{kT}\right) - 1\right] + \alpha_R I_{CS} \tag{A.76}$$

$$I_C = -\alpha_F I_{ES}\left[\exp\left(\frac{eV_{EB}}{kT}\right) - 1\right] - I_{CS}$$

$$= -\alpha_F(I_E - \alpha_R I_{CS}) - I_{CS} = -\alpha_F I_E - I_{CB0} \tag{A.76a}$$

where

$$I_{CB0} = I_{CS}(1 - \alpha_F \alpha_R) \tag{A.76b}$$

This current is so small that it can be neglected for normal operation of the

*The reciprocity theorem states that if a four-terminal passive network, consisting of bilateral elements like resistors, capacitors, and inductors, is represented by the set of equations

$$I_1 = Y_{11}V_1 - Y_{12}V_2$$
$$I_2 = -Y_{21}V_1 + Y_{22}V_2$$

where V_1 and V_2 are the input and output voltages and I_1 and I_2 the input and output currents, respectively, then

$$Y_{12} = Y_{21}$$

The theorem can be directly applied to Eqs. (A.72a) and (A.73a) and yields Eq. (A.75).

transistor. It only becomes significant if the transistor is operated either at extremely small currents or at high temperatures. We were therefore justified in neglecting its effect in the previous chapters.

A.5d. Proof that $\mu = \mu'$ in Section 7.2a

According to Section 7.2a, the collector conductance $g_{ce} = 1/R_0 = g_m\mu$ and the feedback conductance $g_{cb} = 1/r_\mu = g_{be}\,\mu'$, where $\mu' = \mu$. We shall now prove the latter relationship for a *p-n-p* transistor. According to Eqs. (7.35c and d)

$$\mu = -\frac{kT}{e}\frac{1}{\alpha_F I_{ES}}\frac{\partial}{\partial V_{CE}}(\alpha_F I_{ES}); \qquad \mu' = \frac{kT}{e}\frac{1}{(1-\alpha_F)I_{ES}}\frac{\partial}{\partial V_{CE}}[(1-\alpha_F)I_{ES}]$$

(A.77)

However, according to Section A.5a,

$$I_{ES} = \frac{eD_pAn_i^2}{wN_d}; \qquad (1-\alpha_F) = \frac{1}{2}\frac{w^2}{L_p^2} = \frac{1}{2}\frac{w^2}{D_p\tau_p}$$

(A.77a)

so that

$$\alpha_F I_{ES} \simeq I_{ES} = \frac{eD_pAn_i^2}{wN_d}; \qquad (1-\alpha_F)I_{ES} = \frac{1}{2}\frac{weAn_i^2}{\tau_pN_d}$$

(A.77b)

where w increases with increasing V_{CB}.

We shall now observe that V_{BE} is kept constant in the partial derivative $\partial/\partial V_{CE}$. Since $V_{CE} = V_{CB} + V_{BE}$ (Fig. A.11), this means that $\partial/\partial V_{CE} = \partial/\partial V_{CB}$. Substituting into Eq. (A.77) and carrying out the differentiation of Eq. (A.77b) yields

$$\mu = -\frac{kT}{e}\frac{wN_d}{eD_pAn_i^2}\left(-\frac{eD_pAn_i^2}{w^2N_d}\right)\frac{\partial w}{\partial V_{CB}} = \frac{kT}{e}\frac{1}{w}\frac{\partial w}{\partial V_{CB}}$$

$$\mu' = \frac{kT}{e}\frac{2\tau_pN_d}{weAn_i^2}\frac{eAn_i^2}{2\tau_pN_d}\frac{\partial w}{\partial V_{CB}} = \frac{kT}{e}\frac{1}{w}\frac{\partial w}{\partial V_{CB}}$$

so that $\mu = \mu'$. Note that $\partial w/\partial V_{CB}$ is positive, so that μ and μ' are positive numbers.

A.6. HIGH-FREQUENCY BEHAVIOR OF THE n-p-n TRANSISTOR

We shall first derive an expression for the capacitance C_π of an *n-p-n* transistor. It may be written as

$$C_\pi = C_s + C_{je}$$

(A.78)

where C_s is the storage capacitance and C_{je} the capacitance of the emitter-base space-charge region. Now in analogy with Eq. (A.60) the excess electron charge stored in the base region is $-Q_B = I_B\tau_n$, where τ_n is the electron lifetime in the base region. Therefore the excess hole charge Q_B stored in the base region balances this, or

$$Q_B = I_B \tau_n \qquad \text{(A.79)}$$

The storage capacitance C_s must thus be defined as

$$C_s = \frac{dQ_B}{dV_{BE}} = \frac{dI_B}{dV_{BE}} \tau_n = \frac{\tau_n}{r_\pi} \qquad \text{(A.80)}$$

Therefore, from definition (8.95)

$$\frac{1}{f_\beta} = 2\pi C_\pi r_\pi = 2\pi(\tau_n + C_{je} r_\pi) \qquad \text{(A.81)}$$

and

$$\beta = \frac{\beta_F}{1 + jf/f_\beta} \qquad \text{(A.82)}$$

whereas

$$f_T = \beta_F f_\beta = \frac{\beta_F}{2\pi(\tau_n + C_{je} r_\pi)} \qquad \text{(A.83)}$$

For higher currents $\tau_n \gg C_{je} r_\pi$ and

$$f_\beta = \frac{1}{2\pi \tau_n}; \qquad f_T = \frac{\beta_F}{2\pi \tau_n} = \frac{1}{2\pi \tau_d} \qquad \text{(A.84)}$$

for according to Eq. (A.62), $\beta_F \simeq \tau_n/\tau_d$, where τ_d is the diffusion time through the base region. This is practically independent of current.

For low currents $\tau_n \ll C_{je} r_\pi$ and hence

$$f_\beta = \frac{1}{2\pi C_{je} r_\pi}; \qquad f_T = \frac{g_m}{2\pi C_{je}} \qquad \text{(A.85)}$$

Since g_m is proportional to I_B and C_{je} is practically independent of I_B, f_β and f_T are proportional to the current I_B at relatively low currents. This must be borne in mind when using transistors at very low currents (starved operation).

We shall now introduce a high-frequency current amplification factor α for the common base connection by the definition $\beta = \alpha/(1 - \alpha)$, so that

$$\alpha = \frac{\beta}{\beta + 1} = \frac{\beta_F/(1 + jf/f_\beta)}{1 + \beta_F/(1 + jf/f_\beta)} = \frac{\beta_F}{1 + \beta_F + jf/f_\beta} = \frac{\alpha_F}{1 + jf/f_\alpha} \qquad \text{(A.86)}$$

where

$$\alpha_F = \frac{\beta_F}{1 + \beta_F} \quad \text{and} \quad f_\alpha = f_\beta(1 + \beta_F) \simeq f_T \qquad \text{(A.86a)}$$

so that $|\alpha|$ decreases to $1/\sqrt{2}$ times the low-frequency value α_F at the frequency $f = f_\alpha$.

When one measures α, one finds that α can have a phase shift larger than 90 degrees, which can be represented as

$$\alpha = \frac{\alpha_F}{1 + jf/f_\alpha} \exp(-j\omega\tau) \qquad \text{(A.87)}$$

In our derivation of (A. 86) we assumed that the current flow was exclusively due to diffusion. In practice an electric field is present in the base region, partly from the distribution in acceptor concentration (built-in field) in the base and partly from small deviations in the space-charge neutrality condition. As a consequence, part of the current flow is due to drift. If the current flow through the base region were *exclusively* due to drift, the collection of current would result in a simple phase shift $\alpha = \alpha_F \exp(-j\omega\tau)$ and τ would be the drift time of the carriers. If the current flow through the base region were completely due to diffusion, Eq. (A.86) would be valid. If both diffusion and drift are present, it is therefore reasonable that Eq. (A.87) should hold, with τ appropriately chosen.

The fact that Eq. (A.87) rather than Eq. (A.86) is valid at high frequencies indicates that the equivalent circuit shown in Fig. 8.21a is not fully correct at those frequencies. It is beyond the scope of this book to go into details.

EXAMPLE: Find the current dependence of f_β for small base currents if C_{je} is 2 pF.

ANSWER: Since $r_\pi = 2.6 \times 10^4/I_B$, where I_B is in micro-amperes, we have

$$f_\beta = \frac{I_B}{2\pi \times 2 \times 10^{-12} \times 2.6 \times 10^4} \simeq 3 \times 10^6 I_B \text{ Hz} \qquad (I_B \text{ in } \mu a)$$

PROBLEMS

A.1. In silicon $\mu_n = 1480 \text{ cm}^2/\text{V-s}$ at $T = 300°\text{K}$ and $kT/e = 25.8 \times 10^{-3}$ V. Find D_n.

A.2. In a *p-n-p* silicon transistor with a junction area of 10^{-4} cm² the hole concentration on the emitter side of the base is kept at $10^{15}/\text{cm}^3$ and the hole concentration on the collector side is negligible. If the base has a width of 0.5 μm and the hole concentration decreases linearly in the base region, find the dc hole current passing through the base. $D_p = 12.5 \text{ cm}^2/\text{s}$.

A.3. An n^+-p junction has $N_d = 10^{20}/\text{cm}^3$ and $N_a = 10^{17}/\text{cm}^3$. Find the diffusion potential of the junction at $T = 300°\text{K}$ if $n_i = 10^{10}/\text{cm}^3$ at that temperature.

A.4. Design a *p-n* junction so that $d_p = 0.010d$ if $N_d = 10^{16}/\text{cm}^3$. Also find V_{dif} for that junction.

A.5. For the junction of Problem A.4, find (a) d_0, where $d = d_0(V_{\text{dif}} - V)^{1/2}$ is the width of the space-charge region; (b) E_0, where $|E| = E_0(V_{\text{dif}} - V)^{1/2}$ is the field in the space-charge region; and (c) the small-signal capacitance C_0 for $A = 10^{-4}$ cm², where

$$C = \frac{C_0}{(V_{\text{dif}} - V)^{1/2}}$$

$\epsilon_0 = 8.85 \times 10^{-14}$ F/cm; $\epsilon = 12$.

A.6. Find in Problem A.5 the applied voltage V when $E_0 = 150{,}000$ V/cm.

A.7. (a) In a silicon power rectifier of the p^+-n type the holes have a lifetime of 10^{-4} s. Find the diffusion length L_p for holes. (b) In a fast silicon pulse diode of the p^+-n type the holes have a lifetime of 10^{-9} s. Find the diffusion length L_p for holes. These two examples give some idea about the range of values of L_p expected for different p^+-n diodes. $D_p = 12.5$ cm^2/V-s.

A.8. (a) In silicon power rectifier of the p^+-n type the junction area is 1 cm^2. The n-region has a donor concentration of 10^{15}/cm^3, and hole lifetime in the n-region is 10^{-4} s. Find I_{rs} at $T = 300°$K. (b) What voltage must be applied to achieve a diode current of 10 A at $T = 300°$K? $n_i = 10^{10}$/cm^3, and $D_p = 12.5$ cm^2/s.

A.9. (a) Find a/d_0 in Eq. (A.50) for a silicon JFET so that $V_P = -4.0$ V. Assume that $V_{\text{dif}} = 0.90$ V. (b) Find the donor concentration in the channel if $a = 0.5$ μm. (c) Repeat if $a = 1.0$ μm.

A.10. Combining Eqs. (5.5a) and (A.53a), evaluate the value of K for the silicon JFET of Problem A.9(b) if $L = 10$ μm, $w = 100$ μm, and $\mu_n = 1480$ cm^2/V-s.

A.11. (a) Find I_d for the JFET of Problem A.10 at pinch off from Eq. (A.56) at $V_{GS} = 0$. (b) Find the transconductance at pinch off at $V_{GS} = 0$ V.

A.12. An n-p-n silicon transistor has planar geometry and a junction area of 10^{-4} cm^2. The emitter has $N_d = 10^{19}$/cm^3, the base has $N_a = 10^{16}$/cm^3, and the collector has $N_d = 10^{16}$/cm^3. The junctions are abrupt, and the width of the p-region is 2 μm. The actual width of the base region is less, since the emitter and the collector space-charge regions extend into the n-region. $\epsilon = 12$, $n_i = 1.0 \times 10^{10}$/cm^3 at $T = 300°$K, and $D_n = 38$ cm^2/s. (a) Calculate the diffusion potentials of the two junctions at $T = 300°$K. (b) Calculate the emitter bias V_{BE} needed to produce an emitter current of 1.0 mA, assuming a net base width w_B of 1 μm. (c) Calculate the part of the emitter space-charge region that extends into the base under the conditions of part (b). (d) Calculate the collector bias V_{CB} needed to produce a net base width w_B of 1 μ at a current of 1 mA. (e) Express the net base width w_B in terms of V_{CB} at $I_E = 1$ mA, neglecting the very small dependence of the width of the emitter space-charge region on V_{CB}. (f) Calculate $\mu = h_{re}$ at the V_{CB} value calculated in part (d).

A.13. A silicon p-n-p power transistor has a base width of 10 μm. If the lifetime of the holes in the base region is 2×10^{-6} s, find β_F and f_T. $D_p = 12.5$ cm^2/s.

A.14. A silicon p-n-p transistor has $N_d = 10^{16}$/cm^3, $N_a = 10^{19}$/cm^3, $\tau_n = \tau_p = 10^{-7}$ s, and a base width 2 μm. Compare $\frac{1}{2}(w^2/L_p^2)$ with I_{En}/I_{Ep}. $D_p = 12.5$ cm^2/s, and $D_n = 38$ cm^2/s.

A.15. Calculate the stored charge in the transistor of Problem A.13 at an emitter current of 10 A. Also calculate the hole storage capacitance of this transistor at that current.

Appendix B

Pulse Response of
Simple RC-Networks

The pulse response of many simple electronic circuits can be reduced to the response of the simple RC-circuits of Fig. B.1. In more complicated cases one should solve the transient problem by Laplace transform methods, but in simple circuits the solution is best obtained by solving the differential equation of the system directly.

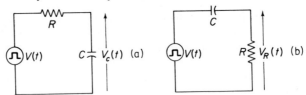

Fig. B.1. (a) Pulse response of RC circuit. (b) Pulse response of CR circuit.

We shall first discuss the transient response of the circuit shown in Fig. B.1(a). The differential equation of the circuit is

$$C\frac{dV_c}{dt} = \frac{V(t) - V_c(t)}{R} \quad \text{or} \quad \frac{dV_c}{dt} + \frac{V_c}{\tau} = \frac{V(t)}{\tau} \tag{B.1}$$

where V_c is the voltage developed across the capacitance C and $\tau = RC$ is the time constant of the network.

We shall now assume that the voltage $V(t)$ has been off for a long time and is turned on at $t = 0$. That is, $V(t) = 0$ for $t < 0$ and $V(t) = V_1$ for $t > 0$. We shall further assume that $V_c(t) = 0$ at $t = 0$. For $t > 0$, Eq. (B.1) may be written as

$$\frac{dV_c}{dt} + \frac{V_c}{\tau} = \frac{V_1}{\tau} \tag{B.2}$$

with the initial condition

$$V_c(t) = 0 \qquad \text{at } t = 0^+ \tag{B.2a}$$

We observe that $V_c = V_1$ is a solution of the inhomogeneous equation and that $A \exp(-t/\tau)$ is the full solution of the homogeneous equation, where A is an arbitrary constant. Therefore the full solution of Eq. (B.2) is

$$V_c(t) = V_1 + A \exp\left(-\frac{t}{\tau}\right) \tag{B.3}$$

However, we must now satisfy the condition (B.2a). This yields $V_1 + A = 0$ or $A = -V_1$. Hence the solution

$$V_c(t) = V_1\left[1 - \exp\left(-\frac{t}{\tau}\right)\right] \tag{B.4}$$

satisfies both the differential equation and the initial condition. This is called the *turn-on transient* of the circuit of Fig. B.1(a). We note that $V_c(\infty) = V_1$. This result could have been obtained directly from Eq. (B.2) by putting $dV_c/dt = 0$.

We shall next assume that the voltage $V(t)$ has been on for a long time and is turned off at $t = 0$. That is, $V(t) = V_1$ for $t < 0$ and $V(t) = 0$ for $t > 0$. We shall further assume that $V_c(t) = V_1$ for $t = 0$. For $t > 0$, Eq. (B.1) may then be written as

$$\frac{dV_c}{dt} + \frac{V_c}{\tau} = 0 \tag{B.5}$$

with the initial condition

$$V_c(t) = V_1 \qquad \text{at } t = 0^+ \tag{B.5a}$$

The full solution of Eq. (B.5) is

$$V_c(t) = A \exp\left(-\frac{t}{\tau}\right) \tag{B.6}$$

To satisfy condition (B.5a), we must now require that $V_1 = A$, so that

$$V_c(t) = V_1 \exp\left(-\frac{t}{\tau}\right) \tag{B.7}$$

satisfies both the differential equation and the initial condition. This is called the *turn-off transient* of the circuit of Fig. B.1(a).

It should be noted that the transient solutions for the stored charge q_s

in diodes and transistors are quite similar to the solution of the transient response of the voltage V_c in Fig. B.1(a). This is not surprising, for the differential equation of the one transient problem is transformed in the differential equation of the other problem by substituting q_s for V_c.

We shall now turn to the transient problem in Fig. B.1(b). For the *turn-on* transient we put $V(t) = 0$ for $t < 0$ and $V(t) = V_1$ for $t > 0$ with $V_c(t) = 0$ at $t = 0^+$; for the *turn-off* transient we put $V(t) = V_1$ for $t < 0$ and $V(t) = 0$ for $t > 0$ with $V_c(t) = V_1$ at $t = 0^+$. The solution is very easily obtained from the solution of Fig. B.1(a) by observing that the voltage $V_R(t)$ developed across R is given by

$$V_R(t) = V(t) - V_c(t) \tag{B.8}$$

From Eq. (B.4), for the *turn-on* transient for $t > 0$ we thus have

$$V_R(t) = V_1 \exp\left(-\frac{t}{\tau}\right) \tag{B.9}$$

and from Eq. (B.7) we have for the *turn-off* transient for $t > 0$ [since $V(t) = 0$ for $t = 0^+$]

$$V_R(t) = -V_1 \exp\left(-\frac{t}{\tau}\right) \tag{B.10}$$

A somewhat different problem occurs in the pulse response of interstage networks due to the coupling capacitor C between the stages. The equivalent circuit is shown in Fig. B.2; it differs from the one in Fig. B.1 in that it has two resistors.

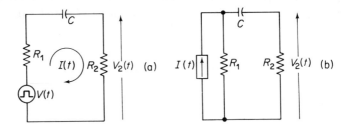

Fig. B.2. (a) Pulse response in *R-C-R*-circuit. (b) Alternate equivalent circuit of (a).

We shall first consider the circuit of Fig. B.2(a), with $V(t) = 0$ for $t < 0$ and $V(t) = V_1$ for $t > 0$ with $V_c(t) = 0$ at $t = 0^+$. From Eq. (B.4), for $V_c(t)$ we then have

$$V_c(t) = V_1\left[1 - \exp\left(-\frac{t}{\tau}\right)\right] \tag{B.11}$$

where $\tau = C(R_1 + R_2)$. The transient current $I(t)$ flowing through the circuit is therefore

$$I(t) = C\frac{dV_c}{dt} = \frac{C}{\tau}V_1 \exp\left(-\frac{t}{\tau}\right) \tag{B.12}$$

so that the voltage developed across R_2 is

$$V_2(t) = I(t)R_2 = \frac{R_2}{R_1 + R_2} V_1 \exp\left(-\frac{t}{\tau}\right) \tag{B.13}$$

since $CR_2/\tau = R_2/(R_1 + R_2)$, according to the definition of τ.

Usually, $V(t)$ comes from a step current generator $I_1(t)$ in parallel with the resistance R_1[Fig. B.2(b)]. Then

$$V(t) = I_1(t)R_1 \tag{B.14}$$

If now $I(t) = 0$ for $t < 0$ and $I(t) = I_1$ for $t > 0$, we have

$$V(t) = 0 \quad \text{for } t < 0; \qquad V(t) = I_1 R_1 \quad \text{for } t > 0 \tag{B.14a}$$

so that solution (B.13) may be written as

$$V_2(t) = I_1 \frac{R_1 R_2}{R_1 + R_2} \exp\left(-\frac{t}{\tau}\right) = I_1 R_{par} \exp\left(-\frac{t}{\tau}\right) \tag{B.15}$$

where $R_{par} = R_1 R_2/(R_1 + R_2)$ is the resistance of R_1 and R_2 connected in parallel. We used this result in Chapter 4.

We shall now return to Fig. B.1 but generalize the initial conditions. We put

$$V_c(t) = V_c(0^+) \quad \text{at } t = 0^+; \qquad V_c(t) = V_c(\infty) \quad \text{at } t = \infty \tag{B.16}$$

where $V_c(\infty)$ is obtained from Eq. (B.2) by putting $dV_c/dt = 0$. For both the turn-on and the turn-off transient, Eq. (B.2) becomes

$$\frac{dV_c}{dt} + \frac{V_c}{\tau} = \frac{V_c(\infty)}{\tau} \tag{B.17}$$

which has the general solution

$$V_c(t) = V_c(\infty) + A \exp\left(-\frac{t}{\tau}\right) \tag{B.18}$$

We must now satisfy the condition at $t = 0^+$. This yields

$$V_c(0^+) = V_c(\infty) + A \tag{B.18a}$$

so that

$$V_c(t) = V_c(0^+) \exp\left(-\frac{t}{\tau}\right) + V_c(\infty)\left[1 - \exp\left(-\frac{t}{\tau}\right)\right] \tag{B.19}$$

which may be written in the alternative form

$$V_c(t) = V_c(\infty) - [V_c(\infty) - V_c(0^+)] \exp\left(-\frac{t}{\tau}\right) \tag{B.19a}$$

so that $V_c(t)$ goes from $V_c(0^+)$ at $t = 0^+$ to $V_c(\infty)$ at $t = \infty$.

A similar case occurs in the transient response of diodes and transistors. In the case of transistors we have the following equation for the stored charge $q_b(t)$ in the base of an n-p-n transistor:

$$\frac{dq_b}{dt} + \frac{q_b}{\tau} = \frac{q_b(\infty)}{\tau} \tag{B.20}$$

with the initial condition $q_b(t) = q_b(0^+)$ at $t = 0^+$. In analogy with Eq. (B.19), the solution is

$$q_b(t) = q_b(0^+) \exp\left(-\frac{t}{\tau}\right) + q_b(\infty)\left[1 - \exp\left(-\frac{t}{\tau}\right)\right] \qquad \text{(B.21)}$$

It now turns out that for the turn-off transient $q_b(0^+)$ and $q_b(\infty)$ have opposite sign, with $q_b(0^+)$ positive and $q_b(\infty)$ negative. There is then a time t_0 after which $q_b(t)$ is negative. However, $q_b(t)$ is by its very nature a positive quantity. This therefore means that the transient stops at $t = t_0$. Hence

$$0 = q_b(\infty) + [q_b(0^+) - q_b(\infty)] \exp\left(-\frac{t_0}{\tau}\right)$$

or

$$\exp\left(\frac{t_0}{\tau}\right) = \frac{q_b(0^+) - q_b(\infty)}{-q_b(\infty)}$$

or

$$t_0 = \tau \ln n\left[\frac{q_b(0^+) - q_b(\infty)}{-q_b(\infty)}\right] \qquad \text{(B.22)}$$

In some cases it is inconvenient to refer the transient problem back to $V_c(t)$. It is often much better to refer back to another voltage. For example, in the case of Fig. B.2 it would be much more convenient to replace Eq. (B.19) by the corresponding equation for $V_2(t)$:

$$V_2(t) = V_2(0^+) \exp\left(-\frac{t}{\tau}\right) + V_2(\infty)\left[1 - \exp\left(-\frac{t}{\tau}\right)\right] \qquad \text{(B.23)}$$

An equation of this form was used in Chapter 11 to evaluate the wave form of monostable and astable multivibrators.

Solutions

1.2: (b) 3.1×10^{-6} mho/cm; **1.3:** 20×10^{-4} cm; **1.4:** 2.1×10^{-4} cm; **1.5:** 0.96×10^6 cm/s.

2.1: $V_D = 0.651$ V, $I_D = 0.150$ mA; **2.2:** (a) $0.77\,\Omega$, (b) -62 V, (c) 42.3%, (d) 8.2 A, (e) 9.15 h; (a) **2.3:** 344 W, (b) 3.14 KWH; **2.4:** $V_B = 5.7$ V; **2.5:** Breakpoints $V_i = 8$ V, 15 V, 30 V; $V_0 = 8$ V, 10 V, 20 V; **2.6:** $V_0 = 4.3$ V; **2.7:** $V_0 = 3.22$ V; **2.8:** (a) Clamping at 5 V only, $V_0(t) = -5 + 10 \cos \omega t$; **2.9:** (a) -20 V, -40 V, -20 V, (b) -40 V, -40 V, -20 V; **2.10:** 19.3 V, 19.3 V, 18.6 V; **2.11:** (a) 1670 μF, 830 μF, 830 μF, (b) 18.8 V, 18.8 V, 18.1 V; **2.12:** 40 kV, 10^4 pF; **2.13:** (a) $0.32\,\Delta V_0$, (b) 70 μF; **2.14:** (a) $0.32\,\Delta V_0$, (b) 17.5 μF; **2.15:** $R = 5000\,\Omega$, $C = 2.66\,\mu$F; **2.16:** (a) 9.5 V, (b) 3.2 mV, (c) 0.0266 μF; **2.17:** (a) $C_2 = 1.6\,\mu$F, (b) 0.25 MΩ; **2.18:** $C = 160$ pF, **2.19:** (a) 0.5 mA, (b) 0.32 mA.

3.1: $d = 1.15 \times 10^{-4}(0.80 - V_D)^{1/2}$ cm; $C = 9.2(0.80 - V_D)^{-1/2}$ pF; **3.3:** (a) $L = 0.262$ mH, (b) $V_D = -14.3$ V; **3.4:** (a) 1360 Ω, (b) 184 mW, (c) $0 \leq I_0 \leq 13.7$ mA; **3.5:** 0.683 V $\geq V_D \geq 0.563$ V; **3.6:** (a) 3000 Ω, 5 diodes, (b) 3.64 V, (c) 3.34 V, (d) 7.1 mV; **3.7:** $C_d = 3900$ pF; **3.8:** $C = 50$ pF; **3.9:** (a) 5300 Ω, (b) 6.8×10^{-9} s; **3.10:** (a) turn-on time 0.58×10^{-7} s, (b) turn-off time 2.15×10^{-7} s; **3.11:** 3.3×10^{-9} s.

4.1: 28 db; **4.2:** (b) 108 db; **4.3:** (b) 102 db, (c) 108 db; **4.4:** (a) $n = 10$, (b) 0.20 V; **4.5:** 2.0 V; **4.6:** 0.040 V; **4.7:** (a) 0.8 mV, (b) 1.6 mV; **4.8:** (a) $R_2 = 5000\,\Omega$, (b) $\Delta V_0 = \Delta V_1/(1 + \mu/5)$; **4.9:** (a) $V_0 = (V_1 + 10\,\mu)/(1 + \mu r)$, (b) 50.0 V $\geq V_0 \geq$

10.4 V, $\Delta V_0 = \Delta V_1/(1 + \mu r)$; **4.10:** $f'_1 = f_1(2^{1/n} - 1)^{1/2}$; $f'_2 = f_2/(2^{1/n} - 1)^{1/2}$;
4.11: (a) $R_L = 410 \, \Omega$, (b) $C_b = 8.7 \, \mu\text{F}$, (c) $|g_{v0}| = 16$; **4.12:** (a) $R_L = 530 \, \Omega$, (b)
$C_b = 1.6 \times 10^4 \, \text{pF}$, (c) $|g_{v0}| = 5.3$; **4.13:** $C_{\text{in}} = 27 \, \text{pF}$; **4.14:** (a) $R_c = 2000 \, \Omega$, (b)
$|g_{v0}| = 20$, (c) 1.4 μH.

5.2: (a) $\frac{1}{4}Kv_{gs0}^2$, (b) $0 \leq v_{gs0} \leq -V_T$, (c) no longer quadratic; **5.3:** $-v_1/V_P =$
$0.04p(1 - V_{GS0}/V_P)$ for $0 \leq V_{GS0}/V_P \leq 1$; **5.4:** (a) $R_s = 500 \, \Omega$, $R_d = 2000 \, \Omega$,
$|g_{v0}| = 8.0$; **5.5:** (a) yes, (b) $|g_{v0}| = 12.5$; **5.6:** (a) yes, (b) $|g_{v0}| = 10.4$; **5.7:** 1.6
$\times 10^4 \, \text{pF}$; **5.8:** (a) $10^5 \, \Omega$, (b) $C_b = 0.16 \, \mu\text{F}$; **5.9:** (a) $V_{GS0} = 6.67 \, \text{V}$, $R_d = 1870 \, \Omega$,
(b) 0.94%; **5.10:** $R_f = 2.1 \times 10^5 \, \Omega$; **5.11:** (a) $R_d = 1500 \, \Omega$, $|g_{v0}| = 6$, (b) $R_1 =$
135,000 Ω, $R_2 = 15,000 \, \Omega$, (c) $v_1 = 0.16 \, \text{V}$, (d) by a factor 2; **5.12:** (a) $v_1 = 4 \, \text{V}$,
(b) $R_d = 375 \, \Omega$, (c) 192 mW, (d) 32 V, (e) 1540 mW, 430 mW; **5.13:** (a) 1500 Ω,
(b) 56 V, (c) 770 mW, $\eta = 50\%$; **5.14:** (a) $Kv_1^2 \cos^2 \omega t$, (b) $\frac{1}{2}Kv_1^2$, $R_d = 1000 \, \Omega$;
5.15: 12.5 mA; **5.16:** (a) $R'_d = 2000 \, \Omega$, (b) 156 mW, (c) none, (d) 375 mW, (e)
42%, (f) 110 mW per device.

6.1: 6000 Ω; **6.2:** 0.72×10^{-3} V ampl; **6.4:** $|g_{v0}| = 40$; **6.5:** (b) 2.83×10^{-3} mho;
(c) $V_{D2} = V_{D3} = 4.83 \, \text{V}$; (d) $R_d = 7000 \, \Omega$ at $V_{D3} = 6.0 \, \text{V}$; **6.6:** (a) $\Delta V_2 =$
$-2.0 \, \text{V}$, (b) $\Delta V_2 = +2.0 \, \text{V}$, (c) $R_d = 3000 \, \Omega$, (d) $|g_{v0}| = 12$ and zero; **6.7:** (a)
$V_{GS1} = -6.0 \, \text{V}$, $v_i = 3.0 \, \text{V}$, $v_0 = -9.0 \, \text{V}$ (ampl), (b) $I_{D\min} = 0$, $I_{D\max} = -32.4 \, \text{mA}$,
(c) $V_0 = -24 \, \text{V}$, (d) $V_{GS1} = -32.4 \, \text{V}$, $I_D = -52.9 \, \text{mA}$; **6.8:** (a) $-5.0 \, \text{V}$, (b)
$V_{\text{on}} = -0.69 \, \text{V}$, $I_D = -2.5 \, \text{mA}$; **6.9:** $R_{\text{on}} = 106 \, \Omega$, choose $R_d = 5000 \, \Omega$, $V_0 =$
$-0.42 \, \text{V}$; **6.10:** (a) $R = 1500 \, \Omega$, yes, (b) $0.048\Delta V$.

7.2: $R = 5.3 \, \text{K}\Omega$; **7.3:** $V_{BE} = 0.760 \, \text{V}$; **7.4:** $V_{BE} = 0.718 \, \text{V}$; **7.5:** (a) $R_b =$
70,000 Ω, (b) $R_c = 4000 \, \Omega$, (c) $|g_{v0}| = 5.7$; **7.6:** (a) $r_\pi = 1030 \, \Omega$, $g_m = 97 \times 10^{-3}$
mho; (b) $R_0 = 51,500 \, \Omega$, $r_\mu = 2.06 \, \text{M}\Omega$; **7.7:** (a) $R_b = 330,000 \, \Omega$, (b) $R_c =$
2500 Ω, (c) $|g_{v0}| = 97$; **7.8:** (a) $R_c = 1333 \, \Omega$, (b) 19.4×10^{-3} mho $\leq g_m \leq 116.4 \times$
10^{-3} mho, $26 \leq |g_{v0}| \leq 156$; **7.9:** 12.6 mV ampl; **7.10:** $|g_{v0}| = 44$; **7.11:** (a) $R_e =$
2540 Ω, (b) $|g_{v0}| = 199$, (c) 5.0%, (d) 2.4%; **7.12:** $R = 5300 \, \Omega$; **7.13:** 250 Ω;
7.14: $R_0 = 20,000 \, \Omega$; **7.15:** $r_\mu = 2.6 \, \text{M}\Omega$, $R_0 = 26,000 \, \Omega$.

8.1: $R_d = 6700 \, \Omega$; **8.2:** (a) $R = 1600 \, \Omega$, $V_{BB} = 4.308 \, \text{V}$, (b) $V_e = 3.388 \, \text{V}$; **8.3:**
$R_{\text{in}} = 100,000 \, \Omega$, no; **8.4:** (a) $I_C = 1.0 \, \text{mA}$, (b) $R_c = 1400 \, \Omega$, (c) $|g_{v0}| = 27$, (d)
3.4 mA; **8.5:** (a) $I_{C2} = I_{C5} = 1.4 \, \text{mA}$, (b) $g_m = 54.4 \times 10^{-3}$ mho, (c) $|g_{v0}| = 740$,
(d) 6.0 mA; **8.6:** 0.70 V and 1.20 V; **8.7:** (a) $v_{\text{out}} = 0.1 \, \text{V}$, (b) $R_{\text{in}} = 10^9 \, \Omega$; **8.8:**
(a) $R_b = 4.6 \times 10^8 \, \Omega$, (b) $f_2 = 19 \, \text{Hz}$; **8.12:** $V_{CC} = 25 \, \text{V}$, $I_{C1} = 0.40 \, \text{A}$, $R'_L =$
62.5 Ω, $n = 3.55$, $P_{\text{ac}} = 5 \, \text{W}$; **8.13:** (a) 640 μF, (b) $R_2 = 475 \, \Omega$, $R_1 = 50 \, \Omega$;
8.14: (b) $V_{CE\min} = 2.17 \, \text{V}$, (c) $P_{\text{acmax}} = 46.7 \, \text{W}$; **8.15:** 18 mV; **8.16:** $\eta = 71\%$,
$P_{\text{ac}} = 17.0 \, \text{W}$, $P_{\text{supp}} = 23.9 \, \text{W}$; **8.17:** $f_\beta = 3.1 \times 10^{10} I_C \, \text{Hz}$ (I_C in A); **8.18:** (a)
$|g_{v0}| = g_m r_x(1 + \omega^2 C^2 r_x^2)^{1/2}/(\omega^2 C^2 r_x^2)$.

9.1: (a) $C_b = 1.6 \times 10^4 \, \text{pF}$, $R_c = 400 \, \Omega$, $C_c = 100 \, \mu\text{F}$, $R_2 = 50,000 \, \Omega$, $C_2 =$
4 μF, $V_A = 50 \, \text{V}$, $R_a = 25,000 \, \Omega$, $|g_{v0}| = 42 \, \text{db}$, (b) 3 stages, (c) 20 V ampl, (d)
15 mA, 2.25 W; **9.2:** (a) $R_c = 500 \, \Omega$, $C_c = 64 \, \mu\text{F}$, $R_a = 12,500 \, \Omega$; $|g_{v0}| = 25 \, \text{db}$,
(b) 5 stages, (c) 15 mA, 2.25 W; **9.3:** no; **9.5:** $R_L = 17,500 \, \Omega$, $\eta = 43.7\%$,
$P_{\text{ac}} = 2180 \, \text{W}$; **9.6:** $\eta = 48\%$, $P_{\text{ac}} = 2400 \, \text{W}$, $R_L = 4800 \, \Omega$; **9.7:** $R_L = 2500 \, \Omega$,
$P_{2\omega} = 1250 \, \text{W}$.

10.1: (a) $I_{D\text{max}} = 36$ mA, $R_{\text{on}} = 83\,\Omega$, (b) $V_{PO} = 3.0$ V, (c) 5.0×10^{-9} s, (d) 5.0×10^{-9} s; **10.2:** $C = 20$ pF; **10.3:** (a) 7.2×10^{-9} s, (b) 1.00×10^{-8} s, (c) 1.00×10^{-8} s, (d) 0.70×10^{-8} s, (e) total turn-on time 1.73×10^{-8} s, total turn-off time 1.70×10^{-8} s; **10.4:** (a) 1.3×10^{-9} s, (b) 1.00×10^{-8} s, (c) 3.14×10^{-8} s, (d) 4.80×10^{-8} s, (e) total turn-on time 1.13×10^{-8} s, total turn-off time 7.94×10^{-8} s; **10.5:** (a) 0.7×10^{-8} s, (b) 195 mA; **10.6:** $I_{B2} = 2.1$ mA, (b) 210 mA, (c) 0.48×10^{-9} s, (d) 100 mA, (e) 1.73×10^{-9} s.

11.1: Output changes from 1.7 V to 20.0 V; **11.2:** Output changes from $-2\frac{2}{3}$ V to -16 V; **11.3:** Output changes from 0.10 V to 2.90 V; **11.4:** (a) $C = 126$ pF, (b) 2.90 V, (c) -4.0 V; **11.5:** (a) 171 pF, (b) 4.23 V, (c) -6.00 V; **11.6:** (a) $C = 580$ pF, (b) -19.0 V, (c) $+19.0$ V; **11.7:** 6300 pF; **11.8:** 1960 pF.

12.6: (a) $1/(j\omega L) + j\omega C + 1/R - g_{d0}/(1 + v^2/v_1^2) = 0$, (b) 160 MHz, (c) 0.15 V; **12.7:** $R_2 = 42.9\,\Omega$.

13.1: Noise figures ≤ 13 can be measured; **13.2:** $500\,\Omega$; (a) $G_{\text{av}} = 3.3$; $F = 3.0$; **13.4:** $\beta_F = 75$.

14.1: $R_L = 316\,\Omega$, $f_1 = 33.6$ MHz; **14.2:** Gain 10^5, $f_{1n} = 9.0$ MHz; **14.3:** $f_1' = 47.5$ MHz; **14.4:** $f_{1n}' = 24.5$ MHz; **14.5:** $R_{L1} = 491\,\Omega$, $R_{L2} = 204\,\Omega$, $f_1' = 39.8$ MHz; **14.6:** $f_{1n}' = 31.4$ MHz; **14.7:** $f_{01} = 521.6$ MHz, $f_{02} = 478.4$ MHz; $R_{L1} = R_{L2} = 424\,\Omega$; **14.8:** $f_{01} = 500$ MHz, $f_{02} = 521.6$ MHz, $f_{03} = 478.4$ MHz, $R_{L1} = 300\,\Omega$, $R_{L2} = R_{L3} = 600\,\Omega$; **14.9:** $f_{01} = 509.6$ MHz, $f_{02} = 490.4$ MHz, $f_{03} = 523.1$ MHz, $f_{04} = 476.9$ MHz, $R_{L1} = R_{L2} = 325\,\Omega$, $R_{L3} = R_{L4} = 784\,\Omega$.

15.2: (a) $v_i = 0.90$ V; (b) $v_0 = 5.4$ V; **15.3:** (a) $v_i = 0.441$ V, (b) $v_0 = 3.78$ V; **15.4:** 5%; **15.5:** 1.9%.

A.1: $D_n = 38.2$ cm^2/s; **A.2:** 4.0 mA; **A.3:** 1.01 V; **A.4:** $N_a = 0.99 \times 10^{18}$/cm^3, $V_{\text{diff}} = 0.83$ V; **A.5:** (a) $d_0 = 0.367 \times 10^{-4}$ cm/volt$^{1/2}$, (b) 5.45×10^4 V$^{1/2}$/cm, (c) 2.89 pFV$^{1/2}$; **A.6:** -6.77 V; **A.7:** (a) 3.54×10^{-2} cm, (b) $L_p = 1.12 \times 10^{-4}$ cm; **A.8:** (a) $I_{rs} = 5.7 \times 10^{-12}$ A; (b) 0.725 V; **A.9:** (a) $a/d_0 = 2.22$; (b) $N_d = 2.5 \times 10^{16}$/cm^3, (c) $N_d = 0.63 \times 10^{16}$/cm^3; **A.10:** $K = 0.607$ mA/V^2; **A.11:** (a) 5.9 mA, (b) $g_m = 3.4 \times 10^{-3}$ mho; **A.12:** (a) $V_{\text{dif}} = 0.89$ V at emitter, $V_{\text{dif}} = 0.71$ V at collector, (b) $V_{BE} = 0.607$ V, (c) 0.19×10^{-4} cm, (d) 9.0 V, (e) $w_B = 1.81 - 0.27(0.7 + V_{CB})^{1/2}$, (f) $h_{re} = 1.08 \times 10^{-3}$; **A.13:** $\beta_F = 49$, $f_T = 4.0$ MHz; **A.14:** $\frac{1}{2}w^2/L_P^2 = 1.60 \times 10^{-2}$, $I_{En}/I_{EP} = 0.312 \times 10^{-3}$; **A.15:** $Q_B = 0.40\,\mu$C, $C_s = 15.5\,\mu$F.

Index

A

Acceptors, 4
ac filters (power supplies), 33
 LC, 33
 RC, 33
Amplification factor, 53, 54
 common base, 140
 common collector, 141
 common emitter, 142
 current, 53
 triode, 216
 voltage, 54
Amplifier, 52 *ff*
 direct coupled, 98
 FET cascaded, 130 *ff*
 ideal current, 53
 ideal linear, 53 *ff*
 ideal voltage, 54
 linear, 52 *ff*
 multistage, 298 *ff*
 power, 99 *ff*
 transistor, 146 *ff*
 tuned, 69, 307
Amplitude modulated signal, 35

Available power, 59
Available power gain, 61
Avalanche breakdown, 42, 88, 144, 334

B

Band, 3
 conduction, 3
 valence, 3
Band, structure of, 3
Bandwidth, 71
 effective, 289
Barkhausen condition, 271 *ff*
Base (of transistor), 10
Base current, 139
Base-emitter conductance, 149
Battery charger, 23
Beta cut-off frequency, 200
Bias circuits (FETs), 95 *ff*
Bias circuits (transistors), 155 *ff*
Bias line, 95
Binary arithmetic, 25
Boolean algebra, 25

Boolean algebra: (*Contd.*)
  addition, 25
  inversion, 25
  multiplication, 25
Bode plots, 314
Breakdown voltage, 334
Butterworth response, 308
Bypass capacitor, 95, 113, 157

C

Cascade connection, 54 *ff*, 289 *ff*
Cascode circuit, 72, 74, 121, 203
Charge control model:
  of diode, 47
  of transistor, 233
Chebyshev response, 308, 311
  polynomials, 311
Chopper:
  mechanical, 196
  transistor, 197
Clamp circuit, 26 *ff*
Clipper circuit, 26 *ff*
Collector, 10
  conductance, 150
  current, 138
  saturated current, 334
Common base connection, 141
Common collector connection, 141
Common emitter connection, 141
Common gate connection, 119
Common mode response, 123, 178
Constant current generator, 109, 165
Constant voltage generator, 166
Continuity equations, 326 *ff*
Controlled sources, 52 *ff*
  current source, 53
  voltage source, 53
COSMOS circuit, 231 *ff*
Current gain, 55
Cut-off frequency:
  in transistors, 201
  lower, 63
  upper, 67

D

Darlington circuit, 174 *ff*
dc restorer, 28
Decibel, 55

Delay time, 304, 305
Demodulation, 35
Depletion mode, 85
Dielectric relaxation, 328
  time, 328
Differential amplifier:
  FET, 122 *ff*
  transistor, 178 *ff*
Differential mode response, 123, 178
Diffusion, 323 *ff*
  constant, 324
  current, 323 *ff*
  equation, 335
  length, 335
  potential, 9, 20, 329
  time, 343
Diode, 9, 16 *ff*, 39 *ff*, 329 *ff*
  circuits, 15 *ff*
  detector, 35
  limiter, 23
  logic, 25-26
  small signal capacitance, 39
  small signal conductance, 43
  storage capacitance, 46
  turn-off transient, 46 *ff*
  turn-on transient, 46 *ff*
  voltmeter, 34-35
Diode connection (of transistors), 145
Distortion, 89, 153
Donors, 3
Double-base diode, 252 *ff*
Drain, 10-11, 78 *ff*
  conductance, 81, 85, 93
Drift, 323 *ff*
  current, 323 *ff*
Dual gate FET, 121

E

Ebers-Moll equations, 143, 346 *ff*
Einstein relation, 325
Electron, 3
Emitter, 10
  current, 138
  follower, 141, 171 *ff*, 207, 318 *ff*
Enhancement mode (of FET), 85
Equal ripple response, 311
Equivalent circuit:
  of diode, 45
  of FET, 88 *ff*, 118 *ff*
  of transistor, 148 *ff*, 199 *ff*

Equivalent circuit: (*Contd.*)
  of vacuum triode, 215 *ff*

F

Feedback, 182 *ff*, 272, 306, 313 *ff*
  conductance (of transistors), 150
  negative, 182, 272, 313 *ff*
  positive, 183, 273, 313 *ff*
FET (field effect transistor), 12 *ff*, 78 *ff*,
    338 *ff*
  junction gate, 12, 78 *ff*
  load, 115 *ff*, 124 *ff*
  n-channel, 78
  p-channel, 81
  MOS, 13, 78 *ff*
    n-channel, 83
    p-channel, 87
  switches, 128 *ff*
  tetrode circuit, 77, 121-122

G

Gain, 54 *ff*
  available, 54
  current, 55
  maximum, 62
  midband, 63, 70
  power, 60
  voltage, 54
Gain-bandwidth product, 63, 72, 131
Gate, 10-11, 78
Gauss's law, 327
Gradient, 323
Gradual channel approximation, 340

H

Half-power points, 71
Harmonic distortion, 89, 316 *ff*
  second, 89
Harmonic generators, 223-224
Heat sink, 159
Hole, 3
h-parameters, 160 *ff*

I

Ideal diode, 15
Ideal transformer, 60

Injection, 21
  of electrons, 21
  of holes, 21
Integrated resistor, 7
Interstage network, 62 *ff*, 131 *ff*, 204 *ff*
  tuned, 69 *ff*, 132-133, 206 *ff*
  untuned, 62 *ff*, 131 *ff*, 204 *ff*
Intrinsic material, 5
Inverter, 26, 126, 233

J

Junction diode, 9-10, 16 *ff*, 39 *ff*, 329 *ff*
Junction FET, 12 *ff*, 78 *ff*, 338 *ff*

L

Laplace transform, 304 *ff*
Life time, 6
Load line, 17, 42, 91, 101
Logic:
  cascaded, 130
  circuits, 25, 126 *ff*
  negative, 126
  positive, 25

M

Matching, 60
Midband gain, 63, 70
  line, 314
Miller effect, 69, 131, 202, 236 *ff*, 306
Mixer circuit, 124, 204
Mobility, 6-7
  electron, 7
  hole, 6
  surface, 84
Multivibrators
  astable, 266 *ff*
  bistable, 255 *ff*
  monostable, 262 *ff*

N

nand gate, 246
  complementary symmetry, 246 *ff*
Negative feedback, 182, 272, 313 *ff*
Neutralization, 75, 223

Noise, 286 *ff*
  figure, 293 *ff*
    of FET, 294
    of transistor, 295
  flicker, 287, 290
  resistance, 292
  shot, 286, 290
  thermal, 286, 290
  white, 290
n-type, 4
Nyquist condition, 271 *ff*
Nyquist plot, 273, 315
Nyquist's theorem, 290

O

Ohm's law, 6-7
Ohmic regime, 85
Ohmic region, 80
Onsager relation (for transistors), 347
Operational amplifier, 178 *ff*, 321
  with capacitive feedback, 186
  with resistive feedback, 181
Optimally flat response, 134, 135, 301 *ff*
or circuit, 26
Oscillator, 271 *ff*
  amplitude, 273 *ff*
  bridge type, 285
  Colpitts, 283
  Hartley, 283
  phase shift, 284
  tuned drain, 283
  tunnel diode, 284
  Wien bridge, 274
Output conductance:
  of FET, 81
  of transistor, 161
  vacuum triode, 215
Overshoot, 66, 134-135

P

Pentode, 218
Perveance, 213
Photoconductivity, 2
π - circuit (of transistor), 151, 200
  hybrid, 151, 201
Pinch-off mode, 85
Pinch-off region, 80, 85

Pinch-off voltage, 78
π - network, 52
Plate resistance, 216 *ff*
p-n-p-n switch, 251
Poisson's equation, 327
Poles, 302 *ff*, 308 *ff*
Positive feedback, 183
Power amplifiers:
  class A, 192, 221
  class B, 192
  class C, 223
  complementary symmetry, 194
  high-frequency, 221 *ff*
  with FETs, 99 *ff*
  with resistive output, 99 *ff*, 187 *ff*
  with transistors, 187 *ff*
  with transformer-coupled load, 189,
    218 *ff*
  with vacuum tubes, 218 *ff*
p-type, 4
Pulse response:
  of amplifiers, 65 *ff*
  of cascaded *RC* amplifiers, 303
  of diodes, 45 *ff*
  of FETs, 226 *ff*
  of transistors, 233 *ff*
Punch through, 144
Push-pull amplifiers, 75, 101

Q

Q-factor, 132

R

RC-coupled amplifiers, 62 *ff*, 94
Reciprocity theorem, 347
Recombination, 3
Recovery time, 242, 263, 265, 268
Rectified current, 89
Rectifier, 28 *ff*
  bridge, 29
  full wave, 29
  half wave, 28
  peak, 30 *ff*
  with capacitive output, 30 *ff*
Reverse voltage gain, 160
Reverse saturation current, 16, 21
Ripple parameter, 311
Rise time, 66, 304-305

S

Sag, 68, 305-306
Saturated transistor, 169, 170-171
Schottky's theorem, 290
Series peaking, 65, 133 *ff*
Sheet resistance, 8
Shunt peaking, 65, 133 *ff*
Silicon, 1
Silicon controlled rectifier, 251-252
Small-signal equivalent circuit:
  of diode, 45
  of FET, 88 *ff*, 118 *ff*
  of transistor, 148 *ff*, 199 *ff*
  of vacuum triode, 215 *ff*
Source (of FET), 10, 78
Source follower, 110 *ff*, 318 *ff*
  integrated, 116
Space-charge neutrality, 329
Space-charge region (of diode), 9, 20
Stability condition, 73
Stagger tuning, 72, 308 *ff*
Starved operation (of transistors), 349
Storage delay time, 238 *ff*
Substrate (in MOSFET), 83

T

Tapping, 132
Taylor expansion method, 92
Tetrode (FET), 77, 121 *f*
Tetrode (vacuum), 216 *ff*
Thermal generation, 4
Thermal noise, 286, 290
  half, 291 *ff*
Thermal resistance, 159
T-network, 52
Totem pole circuits, 248 *ff*
Transconductance, 53, 83, 93, 149, 215
Transresistance, 54

Transistors, 10-11, 138 *ff*
  n-p-n, 10, 142
  p-n-p, 10, 139
Triggering, 256, 259, 261 *ff*
Triode (vacuum), 213 *ff*
TTL gate, 245 *ff*
Tuned amplifiers, 69, 94, 307
Turn-off delay, 238 *ff*
Turn-off time, 231, 236 *ff*
Turn-on delay, 236, 240 *ff*
Turn-on time, 227 *ff*, 235 *ff*
Turn-on voltage:
  off diode, 17
  of FET, 78, 84
  of transistor, 140
Two-port network, 52

U

Unijunction transistor, 252 *ff*
Unit step function, 65, 67

V

Vacuum diode, 212-213
  saturated, 213
Vacuum triode, 213 *ff*
Voltage doubler, 32
Voltage gain, 54
  in FETs, 90 *ff*
  in transistors, 156 *ff*
Voltage quadrupler, 32
Volume control, 123, 204

Z

Zener diode, 42